W9-BWX-517

ZOOLOGICAL PHILOSOPHY

Translated by Hugh Elliot

With Introductory Essays by
David L. Hull and
Richard W. Burkhardt, Jr.

ZOOLOGICAL PHILOSOPHY

An Exposition with Regard to the Natural History of Animals

J. B. Lamarck

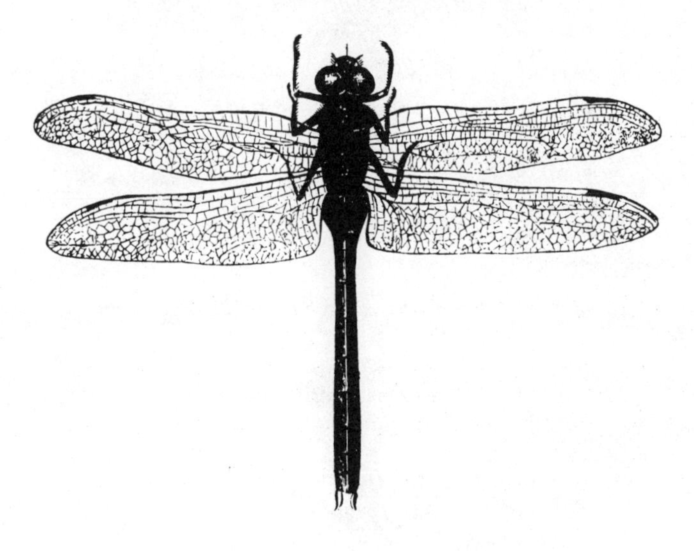

The University of Chicago Press

Chicago and London

The University of Chicago Press, Chicago 60637
The University of Chicago Press, Ltd., London

Printed in the United States of America

The selection from *Système des animaux sans vertèbres* is reprinted from *Annals of Science*, vol. 8, 1952, 231–54, by permission of Taylor & Francis, Ltd.

93 92 91 90 89 88 87 86 85 84 54321

Library of Congress Cataloging in Publication Data

Lamarck, Jean Baptiste Pierre Antoine de
Monet de, 1744–1829.
Zoological philosophy.

Translation of: Philosophie zoologique.
Includes index.
1. Zoology. 2. Physiology, Comparative.
3. Psychology, Comparative. I. Title.
QL45.L2313 1984 590 84–2570
ISBN 0–226–46809–7
ISBN 0–226–46810–0 (pbk.)

TABLE OF CONTENTS

PART III.

AN ENQUIRY INTO THE PHYSICAL CAUSES OF FEELING, INTO THE FORCE WHICH PRODUCES ACTIONS, AND LASTLY INTO THE ORIGIN OF THE ACTS OF INTELLIGENCE OBSERVED IN VARIOUS ANIMALS.

GLOSSARY

Giving the original French of some of the more important words used in the course of this translation.

English.	French.
affective,	*affectant.*
affinities,	(biol.) *rapports.*
	(chem.) *affinités.*
amphibian,	*amphibie.*
analytic,	*décomposant.*
arbitrary opinion,	*arbitraire.*
argument,	*raisonnement.*
artificial devices,	*parties de l'art.*
centre of communication,	*centre de rapport.*
classification,	*distribution, classification.*
complexity,	*composition.*
crude,	*brut.*
crura cerebri,	*pyramides du cerveau.*
digression,	*écart.*
ego,	*moi.*
environment,	*circonstances, milieux environnants, etc.*
erethism,	*éréthisme.*
exciting cause,	*cause excitatrice.*
factor,	*cause influente, cause essentielle, etc.*
faculty,	*faculté.*
feeling,	*sentiment.*
function,	(zool.) *faculté, fonction.*
fungus,	*champignon.*
ganglionic longitudinal cord,	*moelle longitudinale noueuse.*
hypocephalon,	*hypocéphale.*
inclination,	*penchant.*
inference,	*conséquence.*
inner feeling,	*sentiment intérieur.*
integral molecule,	*molécule intégrante.*
intellect,	*intelligence.*
kinship,	*parenté.*
laburnum,	*cytise.*
lacteals,	*vaisseaux chyleux.*
limb,	*membre.*
main medullary mass,	*masse médullaire principale.*
matrix,	*gangue, matrice.*

English.	French.
medulla oblongata,	*moelle allongée.*
mind,	*esprit.*
monas,	*monade.*
natural order,	*ordre naturel.*
need,	*besoin.*
nucleus (of sensations),	*foyer.*
orgasm,	*orgasme.*
peduncles of the cerebellum,	*jambes du cervelet.*
pelvis,	*bassin.*
procedure of nature,	*marche de la nature.*
propensity,	*penchant.*
proteus,	*protée.*
reproduction,	*génération.*
rudiment,	*ébauche.*
scale of nature,	*échelle de la nature.*
schematic classification,	*distribution systématique.*
sensibility,	*sensibilité.*
sensitive,	*sensible.*
serum,	*sanie.*
skill,	*industrie.*
soft radiarian,	*radiaire molasse.*
soul,	*âme.*
spinal cord,	*moelle épinière.*
spirits,	*esprits.*
spontaneous generation,	*génération directe* or *spontanée.*
stage,	*degré.*
synthetic,	*composant.*
radiarian,	*radiaire.*
root-collar,	*collet de la racine.*
unguiculate,	*onguiculé.*
ungulate,	*ongulé.*
vital knot,	*nœud vital.*
vital principle,	*arché-vitale.*
volatile,	*coercible.*
zoological philosophy,	*philosophie zoologique.*

THE ZOOLOGICAL PHILOSOPHY OF J. B. LAMARCK

RICHARD W. BURKHARDT, JR.

At the funeral services of the French zoologist Jean-Baptiste Lamarck (1744–1829), Lamarck's younger colleague Étienne Geoffroy Saint-Hilaire (1772–1844) evoked the tragic conditions of the last decade of Lamarck's life. Not only was Lamarck blind and impoverished, but he also had been the subject of envy and opposition, to which he submitted patiently but sorrowfully. From Geoffroy's funeral discourse derives the image of Lamarck being consoled by the reassuring words of his daughter that the future would avenge him (Geoffroy Saint-Hilaire 1838).

Today Lamarck is generally credited with having been the first major proponent of the idea of organic evolution. His most famous theoretical work, the *Philosophie zoologique* (1809), has been reissued in French and translated into other languages numerous times. It would appear, in short, that the words of his daughter have come true. But on closer scrutiny Lamarck does not fit well the role of the scientific martyr avenged by the progress of science. For one thing, the specific concepts and images with which his name is now associated bear only a slight resemblance to the particular concerns that informed his thinking. For another thing, his bold intellectual project of explaining the diversity of organic beings entailed not only ideas that appear strikingly prescient as we look back upon them, but also ideas that were coming to be regarded as indefensible in his own time (and have generally been regarded as indefensible ever since). Finally, when Lamarck's views on

Material in this introduction deriving from the author's book, *The Spirit of System: Lamarck and Evolutionary Biology*, is reprinted by permission of Harvard University Press. The author gratefully acknowledges Harvard University Press's permission to use the material here.

organic mutability are contrasted with the ideas of other naturalists of the beginning of the nineteenth century, it appears that the failure of his views to gain scientific respectability in his own day cannot be explained by stating that such views were virtually unthinkable at that time. Some of Lamarck's contemporaries, to be sure, were scandalized by his materialism and recoiled from it. But this does not mean that the idea of organic mutability was unthinkable at that time. Indeed, at least three of Lamarck's colleagues at the Muséum d'Histoire Naturelle in Paris—Faujas de Saint-Fond (1741–1819), Lacépède (1756–1825), and Geoffroy Saint-Hilaire—embraced it in one form or another. Lamarck's problem was that he was unable to relate his broad hypotheses to factual evidence in such a way as to cause his contemporaries to treat his hypotheses as profound insights rather than unfounded speculations. To say this is not to deny the great profundity of Lamarck's idea that the diversity of living beings was the product of natural causes operating over immense periods of time. It does suggest, however, that evaluating his work in its own historical context is not an easy matter.

The development of ideas over time frequently obscures the particular contexts in which these ideas were first formulated. The work of Charles Darwin and the debate between the "neo-Darwinians" and "neo-Lamarckians" of the late nineteenth century have had a profound influence upon the modern perception of Lamarck's place in history. Lamarck is remembered primarily as a pre-Darwinian evolutionist who proposed the wrong mechanism—the inheritance of acquired characters—to explain evolutionary change. But this view of Lamarck does not do justice to Lamarck's own conception of organic change, nor does it indicate how Lamarck's views on organic change related to the rest of his biological thinking or his scientific and philosophical work as a whole.

The general intellectual setting in which Lamarck's hope and passion to comprehend the world took shape was the milieu of the French Enlightenment. The particular locale of his daily work as a scientist was the naturalist's cabinet. Lamarck came to conceive of himself as playing a special intellectual role that mediated between the broad concerns of the philosopher and the detailed studies of the naturalist. As he strove to understand such subjects as the nature of life and the origin of the higher mental faculties, he insisted that these subjects were proper concerns for a zoologist. He believed, furthermore, that what he knew as a zoologist allowed him to go beyond the views of the philosophers who had interested themselves in such issues before him. When he distinguished his views from those of P.-J.-G. Cabanis (1757 – 1808), whose thoughts on the organic foundation of the mental faculties were very much like his own, he was quick to point out that the special strength of his own position was that it was based on insights gleaned from surveying the whole of the animal kingdom, not just from considering man or the animals most like man (Lamarck 1809, 1:414 – 20). As a self-styled naturalist-philosopher, Lamarck congratulated himself for having discovered "great truths, which the philosophers could not discover because they had not sufficiently observed nature, and which the zoologists have not perceived because they have occupied themselves too much with matters of detail" (Lamarck 1809, 2:295).

That Lamarck himself was able to deal with matters of detail is amply illustrated in his contributions to systematic botany, invertebrate zoology, and invertebrate paleontology. He did not wish to limit himself, however, to classificatory work. He sought to comprehend the broad processes by which nature brought about all her results. Not only in biology, but also in chemistry, geology, and meteorology, he endeavored to identify and explain the causes of what he called the "large facts" rather than the "small facts" of nature. He believed he could provide secure foundations for these various sciences where others before him had failed. While Lamarck's thoroughgoing naturalism stands out in contrast to the thought of many of his contemporaries, in his search for broad explanations of a vast range of different phenomena he proved all too ready to believe that the explanations for these phenomena were both near at hand and relatively simple. He supposed that he could account for a host of physical and chemical facts by the idea that the tendency of all compounds is to decompose. He thought he could explain the major features of the earth's surface as the result over immense periods of time of the natural effects of running water, the gradual displacement of the ocean basins, and the action of living organisms. He believed he could find a general pattern in the changes in the weather that correlated with the tidal action of the moon and sun on the atmosphere of the earth. His uniformitarian geology and his evolutionary biology were intellectual ventures comparable in scope and style to his physico-chemistry and his meteorology. Indeed his geology and biology employed some of the same concepts and explanatory models that are to be found in his physico-chemistry and meteorology. The reader coming to Lamarck's *Zoological Philosophy* expecting to find essentially a precursor to Darwin's *Origin of Species* may be quite surprised to find just how broad Lamarck's concerns in his book were. The *Zoological Philosophy,* as described in its subtitle, was "an exposition with regard to the natural history of animals," including "the diversity of their organization and the faculties which they derive from it ; the physical causes which maintain life within them and give rise to their various movements ; lastly, those which produce feeling and intelligence in some among them."

LAMARCK'S CAREER

Lamarck was born in 1744, the eleventh and last child of a family of noble lineage but modest circumstances. Though initially destined by his family for the priesthood, he found himself after his father's death with the chance to choose his own career and followed his older brothers into the military. Upon the conclusion of the Seven Years' War, his regiment was stationed in the south of France (1763–66). There he had the opportunity to make botanical excursions into the countryside, and his interest in natural history seems to date from this period. When an injury put an end to his military career in 1768, he went to Paris, where he first worked in a bank, then attended classes at the École de Médecine, and finally became a member of the circle of botanists and students at the Jardin du Roi (the King's Garden), the most important institution in France dedicated to the study of natural history in the late eighteenth century.

It was at the Jardin du Roi that Lamarck announced that he could provide a more efficient means of identifying the plants of France than any currently in existence. This promise was realized in his *Flore françoise* (1778 [1779]), a three-volume work which Buffon (1707–88), the head of the Jardin du Roi, arranged to have published at the government's expense. In 1779, thanks again to the support of Buffon and other influential parties, Lamarck was made a member of the Académie des Sciences in the botany section. Two years later, Buffon created the position of "correspondent" of the Jardin du Roi for Lamarck. The main purpose of this position seems to have been to provide Lamarck with an official title as he escorted Buffon's son on a tour of Europe. The position of "correspondent" was an unsalaried one, but it did at least provide Lamarck with an official connection to the Jardin du Roi. In 1789, the year after Buffon's death, Lamarck managed to secure from Flahault de la Billarderie, Buffon's successor at the Jardin du Roi, a new, modestly-salaried position, that of "Botanist of the King and Keeper of the Herbaria of the King's Garden."

When the Jardin du Roi was reorganized during the French Revolution and made into the Muséum National d'Histoire Naturelle (1793), all the officers of the old institution, Lamarck included, were named professors of the new one. The two professorships of botany at the Museum naturally went to the two distinguished scientists who had been the botanists of the Jardin du Roi, A.-L. de Jussieu (1748–1836) and R.-L. Desfontaines (1750–1833). As for Lamarck, he was cast in an unfamiliar role. He was made a zoologist, with the task of treating the organisms at the lower end of the animal scale. A list of the professors at the Museum in 1794 gives a capsule summary of Lamarck's situation at the time he embarked on his new career: "age 50, married for the second time, wife pregnant, six children, professor of the insects, worms, and microscopic animals."

There seems to have been some feeling on the part of the professors of the Museum that Lamarck had basically been awarded that area of natural history in which no one else at the institution was particularly interested. When Lamarck took up his professorship at the Museum, the lower animals (with the exception of the insects) had been the subject of relatively little systematic investigation. Lamarck's own knowledge of the lower animals was confined primarily to shells. The higher animals—the mammals, birds, reptiles, and fish—generally seemed to be more attractive subjects. As Lamarck himself later put it, "I thought then like the others that the less interesting part [of the animal kingdom] had fallen to me. It seemed to me in effect that there was much more profit and more interest to excite in describing the characters, manners, and habits of the lion than those of the earthworm" (Lamarck 1972:28). Lamarck soon came to argue, however, that his new province of study was as significant intellectually as any other part of nature's domain. Indeed, as it turned out, the investigation of the "animals without vertebrae" proved more enlightening to Lamarck than he had initially anticipated. In contemplating the natural processes revealed to him in his study of the invertebrates, he was led to explore fundamental questions about the causes of vital activity and to identify a broad, new science which he named "biology." One of the central features of his biology was a bold new explanation of the origin of living things. His first public exposition of this new theory was in the opening lecture he delivered to his students at the Museum in 1800.

It must not be supposed, however, that Lamarck's activities just before 1800 were limited to the study of invertebrate zoology. Between 1794 and 1797 he appears to have been primarily occupied with physico-chemical problems. In 1794 he published a two-volume work entitled *Recherches sur les causes des principaux faits physiques* [Studies on the causes of the principal physical facts]. He had actually written this work almost a decade and a half earlier, but when he presented it to the secretary of the Academie des Sciences in 1781, it was studiously ignored. To Lamarck's chagrin, the publication of 1794 met the same fate as did his next two publications, his *Refutation de la théorie pneumatique* [Refutation of the pneumatic theory] (1796) and his *Mémoires de physique et d'histoire naturelle* [Memoirs on physics and natural history] (1797). Lamarck's colleagues at the Institut de France (successor of the Académie des Sciences) did not disguise their boredom and annoyance when he began reading his series of memoirs to them in which he sought to explain why his physico-chemical views were superior to those of Lavoisier (1743–1794), Fourcroy (1755–1809), and other exponents of the new, pneumatic theory which was then revolutionizing the study of chemistry.

At the same time Lamarck was embarking upon his vain attempt to convince his contemporaries of the utility of his own system of physics, chemistry, and physiology, the young Georges Cuvier (1769–1832) had just come to Paris and was beginning his meteoric rise within the French scientific community. It bears noting that while Cuvier's work in comparative anatomy was effecting a major reform in the classification of the lower animals and providing a new rigor in the study of fossil and living vertebrates, Lamarck was fighting a hopeless, rear guard action against the impressive advances in modern chemistry. And while Cuvier was establishing a reputation for himself as a patient observer who never drew conclusions beyond those that the data permitted, Lamarck was gaining a reputation as a builder of groundless systems.

Lamarck's reputation as a "system-builder" was reinforced in the late 1790s by projects in addition to his physico-chemistry. In 1797 he told his colleagues at the Institut de France that the earth's atmosphere exhibited regular changes corresponding to the moon's elevation above, or declination below, the earth's equator. In 1799 he told the same body that the present features of the earth's surface were the result of common, everyday processes that had operated over immense periods of time. When in 1800 he told his students at the Museum that the origin and diversity of living things and the faculties they displayed could be explained in a relatively simple and wholly naturalistic fashion, he seemed to be advancing another "system" of the sort that his contemporaries had come to expect of him—and of which they did not approve. In sum, not only were they unimpressed by his particular ideas, they considered his whole approach to science to be retrogressive.

In the 1800s Lamarck proceeded to carry out studies on the systematics of living and fossil invertebrates. At the same time, he developed his thoughts on the causes of the origin, diversity, and faculties of living things. His concerns and activities did not stop there, however. In 1800 he began publishing a series of meteorological annuals which included his general observations on the causes of atmospheric events and his predictions—he acknowledged these were only "probabilities"—of the weather for the coming year (Lamarck 1800–1810). He continued the enterprise for eleven years, until Napolean's open ridicule forced him to stop. In 1802 he published his *Hydrogéologie*, a remarkable book (though less original than is commonly supposed) in which he set forth a uniformitarian view of geological change, maintaining that the age of the earth was "nearly incalculable," involving thousands or even millions of centuries. His chief zoological works were his *Système des animaux sans vertèbres* [System of invertebrate animals] (1801), his *Recherches sur l'organisation des corps vivans* [Studies on the organization of living bodies] (1802), his *Philosophie zoologique* [Zoological philosophy] (1809), and his *Histoire naturelle des animaux sans vertèbres* [Natural history of the invertebrate animals] (1815).

Lamarck's health began to fail in the first decade of the nineteenth century. In 1809, citing chronic illness, Lamarck turned down the chair of zoology at the Faculté des Sciences. Nine years later he lost his sight and had to turn the teaching of invertebrate zoology at the Museum over to his assistant, P.-A. Latreille. Lamarck's last major statement of his scientific philosophy, his *Système analytique des connaissances positives de l'homme* [Analytical system of man's positive knowledge] was published in 1820. The final volumes of his *Histoire naturelle des animaux sans vertèbres* appeared in 1822. He died in 1829.

Lamarck's Views on the Origin and Development of Organic Beings

Lamarck never used the word "evolution" to refer to the process of the origin and successive transformation and development of organic beings over time. Nor for that matter did he use the word "transformism." I will refer here, nonetheless, to Lamarck's "evolutionary" or "transformist" views, reminding the reader that while Lamarck did advance a theory of the origin and successive development of organic beings over time, his ideas and concerns should not be readily equated with the concerns of later scientists who called themselves "evolutionists" or "transformists" (or, for that matter, with the concerns of Charles Darwin [1809–82], who used the word "evolution" only sparingly in his own writings [Bowler 1975] and does not seem to have used the word "transformism" at all).

How Lamarck arrived at his theory of organic change is a topic that has been explored elsewhere (see Burkhardt 1972 and 1977). This writer believes that the primary concerns that brought Lamarck to his view of organic change were (1) his thinking on what constituted the essence of life in the simplest organisms (thinking which drew upon his view of the physical activity of the subtle fluids caloric [heat] and electricity), (2) his view of the "natural" way to arrange nature's productions, (3) his uniformitarian geology, and (4) his dissatisfaction with the idea that species had become extinct as the result of a global catastrophe.

The immediate stimulus for Lamarck's endorsement of the idea of the mutability of *species* seems to have been his work in invertebrate zoology and his refusal to believe in a wholesale destruction of species by extraordinary geological events. In addressing the question of how species change, he was able to call upon the influence of the environment on the individual, a theme that had been developed during the Enlightenment by *philosophes* and naturalists alike. Yet the broad view of organic change he advanced, as Hodge (1971) has pointed out, was not simply an extrapolation from his idea of species mutability. Lamarck's general scheme of organic development took its basic shape from his views on the nature of life and the natural way to arrange nature's productions.

Stated very briefly, Lamarck's theory of organic development (as he presented it in 1802 and afterward) involved (1) the spontaneous generation of the simplest forms of life at the respective bases of the plant and animal kingdoms, (2) the successive production from these simple forms of organisms of increasing complexity, and (3) the influence of particular "circumstances" that at least in some measure thwart the natural tendency toward increased complexity. This explanation of organic change was set in the context of a uniformitarian geology that provided the vast expanses of time necessary for imperceptibly small changes to produce over countless generations all the different forms of life on earth.

The Causes of Life

In 1794 Lamarck maintained that there were three principal causes of things that "man reasoning philosophically" would never be able to discover. These were (1) the cause that produced matter and all the essential qualities of matter; (2) the cause of "the existence of organic beings and that which constitutes the life and essence of these beings, since matter with all its qualities seems . . . to be in no way capable of producing a single being of this nature"; and (3) the cause of "the *activity* that is spread throughout the universe" (Lamarck 1794, 2:185). Two decades later, Lamarck would still regard the first and third topics as being beyond the range of what man could ever know positively through his study of nature. But his thoughts on the second topic—the essence of life and the "cause" of living things—appear to have changed significantly by 1797, and by 1800 he was endorsing the idea of spontaneous generation and maintaining that all life forms had been successively developed over time through natural causes.

In 1797 Lamarck announced that he did not consider the essence of life in a body to reside in any particular being, or soul, which vivified the body. Life, he insisted, could be known physically. He did not deny the existence of either the "immortal soul" of man or the "perishable soul" of beasts, but simply stated that since they could not be known physically, they were not the concern of the scientist. Life, as Lamarck defined it in 1797, was "nothing other than the movement in the parts of [organized] beings resulting from the execution of the functions of their essential organs, or the possibility of being in possession of this movement when it is suspended" (Lamarck 1797:255).

Yet while Lamarck in 1797 allowed that life could be defined physically, he portrayed nature on the one hand and "the power of life" on the other as two opposing forces—one breaking compounds down and the other building them up. He also maintained that life was always the product of life, that "everything comes from an egg" (Lamarck 1797:272). In 1800, in contrast, he was prepared to express the belief that life could be produced from non-life. The "sketches of animalization" at the "unknown end of the animal scale," he claimed, were organisms that were generated spontaneously or, as he put it, "directly." He regarded this direct generation, furthermore, as an ongoing or at least recurrent process which nature performed "with facility" under "favorable circumstances" (Lamarck 1801:41).

Lamarck had concluded by 1800 that the simplest animals were as dependent upon external influences for their vitality as plants were. He had decided that two particular physical agents, the "subtle fluids" of heat and electricity, constituted the stimuli upon which life—organic movement—in the simplest animals depended. And he had decided further that the very simplest organisms, the "mere sketches of animality," could be formed when conditions were favorable and destroyed when conditions were unfavorable. As he explained in 1802, through the physical agents of heat and electricity, nature was in fact able to imitate her own process of fertilization. The subtle fluids, heat and electricity, under the appropriate conditions, could do to certain gelatinous or mucilaginous materials just what the "*subtle vapor* of fertilizing materials" did to embryos, namely, organize them and make them ready for life (Lamarck 1802a:101).

Coming to believe in spontaneous generation was critical to the development of Lamarck's view of organic evolution. Others before him, to be sure, had believed in spontaneous generation without believing in evolution. But once Lamarck came to view the simplest forms of life—the forms in which the very essence of life was most clearly displayed—as being literally the *productions* of nature, he decided that nature's means were adequate to produce all other forms of life as well. As he put it in 1802, "once the difficult step [of admitting spontaneous generation] is made, no important obstacle stands in the way of our being able to recognize the origin and order of the different productions of nature" (Lamarck 1802a : 121 – 22 ; see also Lamarck 1815, 1 :180 – 81). His confidence in making this extraordinary claim stemmed from his belief that he had already identified "the cause which tends to make organization increasingly complex" and the way that the environment altered specific forms.

"The Cause Which Tends to Make Organization Increasingly Complex"

Though Lamarck is remembered today primarily for having endorsed the idea of the inheritance of acquired characters, his theory of organic change was a theory in which the inheritance of acquired characters played a secondary role. In 1809, in his *Philosophie zoologique,* he was quite explicit in portraying the diversity of animal forms as the result of two distinct processes. As he explained it:

> The state in which we now see all the animals is on the one hand the product of the increasing *composition* of organization, which tends to form a *regular gradation,* and on the other hand that of the influences of a multitude of very different circumstances that continually tend to destroy the regularity in the gradation of the increasing composition of organization (Lamarck 1809, 1 :221).

Likewise in 1815, in his last major exposition of his transformist views, he distinguished between "two very different causes" of organic diversity. The "first and predominant cause" represented the "natural" course of events or "plan of nature." The execution of this "plan," however, had been "thwarted here and there" by the constraining influence of particular "circumstances" (Lamarck 1815, 1 :133).

Lamarck's two-factor explanation of organic diversity reflected the guiding assumptions of his work as a naturalist classifying the different "productions of nature." Central to this work was the notion that within each of nature's kingdoms, the productions of nature could be arranged linearly. In his earliest work as a botanist, he set for himself the goal of arranging all the *species* of plants in a single series. Even though a number of naturalists, including the patriarch of the Museum, L.-J.-M. Daubenton (1716 – 1800), had dismissed the notion that a linear scale was the best representation of the "natural order," the idea of a general scale of being continued to dominate Lamarck's thinking as he began his zoological work. Lamarck undertook his study of the invertebrates with the conviction that if not the finer divisions then at least the major classes of the animal kingdom could be arranged in a scale of increasing (or decreasing) complexity of organization. Cuvier's pioneering anatomical research of the 1790s appeared to Lamarck to substantiate this conviction, and Lamarck was pleased to tell his students in 1800 that the invertebrates "show us even better than the [other animals] this surprising degradation in the composition of organisation" (Lamarck 1801:11).

The theory of organic mutability that Lamarck developed thus accomplished two things with respect to his commitment to a linear model of the natural order. On the one hand it maintained a general linear arrangement of the animal kingdom. On the other hand it explained why the gradation from the simplest animal to the most complex animal was not perfect.

The French naturalists of the late eighteenth century had decided that the principle of the "subordination of characters" was the key to ascertaining the natural relations among living things. In studying the organization of living things, it was evident that certain characters had to be given more prominence than others if one were to determine the "true affinities" among the different life forms. Cuvier used the principle of the subordination of characters in 1795 when he reformed the old Linnaean grouping of "worms" by dividing it into five different classes based on the circulatory and nervous systems of the organisms in question. Lamarck relied in no small measure upon Cuvier's anatomical researches but then went beyond Cuvier to provide a causal explanation of organic development. In his two-factor theory of organic change, Lamarck attributed the development of the most essential, internal characters of organisms—the characters upon which the zoologist formed the different families, orders, and classes of animals—to what he called the "power of life." The external characters of animals—the characters which the zoologist used to distinguish species or genera from one another—he ascribed to the influence of particular environmental "circumstances." He maintained in the *Philosophie zoologique* that contemporary misconceptions about the natural affinities of living things would vanish as soon as more was known about their internal organization and "especially when what belongs to the influence of places of habitation and to contracted habits is distinguished from that which results from the more or less advanced progress in the composition or perfecting of organization" (Lamarck 1809, 1:107).

The "Power of Life"

What was this "power of life" to which Lamarck attributed the tendency toward increased complexity of organization in the animal kingdom? Though the phrase "the power of life" sounds highly vitalistic, Lamarck, as indicated above, believed that life was "a very natural phenomenon, a physical fact." All his attempts to explain life as of 1800 were couched in mechanistic terms. But this is not to say that he attempted to *reduce* life to physics and chemistry. He always insisted that life was a phenomenon of *organized* beings and that there was a real hiatus between the organic and the inorganic worlds, a qualitative difference between organized beings and brute matter. Life, sensation, and thought, he maintained, were not attributes of matter as such, but were instead the result of the way matter was arranged in particular organized bodies. He was convinced, nonetheless, that the *production* of organization was a function of mechanical causes operating over time. The arrangement of the parts of living organisms could be explained in strictly mechanical terms.

We can sketch briefly Lamarck's explanation of how "the power of life" itself developed over time and how it simultaneously shaped organic forms. The explanation he offered relied upon the constructive action of fluids in motion. Maintaining that the excitatory cause and the "sole cause of conservation" of life in the simplest animals was the action of the subtle fluids, heat and electricity, Lamarck went on to claim that these subtle fluids were responsible not only for making life possible but also for diversifying life in its simpler stages. The constructive effects of these subtle fluids, he admitted, were not particularly apparent in the very first "sketches of animality" (the infusorians), or even in the polyps, which were still too tiny for the subtle fluids to shape. But once one reached the radiarians (the jellyfish), "where the body of each animal is much more ample and isolated," it was obvious, Lamarck said, that "these excitatory and expansive fluids, rushing without cease through the digestive organ of these animals," had modified the digestive organ "as well as the body itself." If, in the next higher class of animals, the echinoderms, the effects of the subtle fluids upon animal form were not as great as they had been in the jellyfish, this was hardly surprising, since the consistency of the body of an echinoderm was much greater than that of a jellyfish. But though the subtle fluids no longer had much effect on the shape of an organism of the consistency of an echinoderm, the ponderable fluids could have an effect, and it was through the motion of the ponderable, "essential," bodily fluids that the organization of animals continued to become increasingly complex (Lamarck 1815, 2:442–44; see also Lamarck 1809, 2:156–57). As Lamarck explained in 1802:

> the characteristic of the movement of fluids in the supple parts of the living bodies that contain them is to trace out routes and places for deposits and outlets; to create canals and the various organs, to vary these canals and organs according to the diversity of either the movements or nature of the fluids causing them; finally, to enlarge, elongate, divide, and gradually solidify these canals and organs. . . .
>
> The state of organization in each living body has been formed little by little by the increasing influence of the movement of fluids and by the changes continually undergone in the nature and state of these fluids through the usual succession of losses and renewals (Lamarck 1802a: 8–9).

In discussing what caused the movements and activities of living bodies, Lamarck explained that "the productive force of movement" was only external to the simplest forms of life. In the more complex forms it became internalized. But he denied that this meant that life was any less a physical phenomenon in the higher animals than it was in the lower ones. Though the development of animal organization did reach a point in the higher invertebrates where it allowed these animals to have a *sentiment intérieur* (internal feeling), and though this *sentiment intérieur* was "a power which, aroused by a felt need, causes the individual to act immediately" (Lamarck 1815, 1:17–18), Lamarck believed that the cause of this instinctive action was "uniquely mechanical" (Lamarck 1815, 3:239).

The Influence of the Environment, Habits, and the Inheritance of Acquired Characters

In his introductory discourse of 1800 Lamarck maintained that he could "pass in review . . . all the classes, all the orders, all the genera and the species of animals that exist, and show that the structure of the individuals and of their parts, that their organs, their faculties, &c. &c. are entirely the result of the circumstances to which the race of each species has found itself subjected by nature." He went on to say:

> I could prove that it is not at all the form either of the body or its parts that gives rise to habits, to the way of life of the animals, but that to the contrary it is the habits, the way of life and all the influential circumstances that have with time established the form of the bodies and the parts of animals (Lamarck 1801:14–15).

Though he repeated the above claims in 1802, by then he was attributing the general tendency toward increased complexity more to the natural effects of "organic movement" than to particular environmental circumstances. The latter were simply responsible for departures from the regular gradation in the animal scale. The influence of environmental conditions, in other words, accounted for the fact that species, unlike the general "masses" of organization, could not be arranged linearly, but instead formed "lateral ramifications" from the masses in such a way that the extremities of these ramifications were "truly isolated points."

In his explanations of how circumstances brought about organic change, Lamarck stressed the importance of the use and disuse of organs and took as a basic assumption the idea that is now most frequently associated with his name: the idea of the inheritance of acquired characters. But he neither originated this idea nor claimed special credit for it. Indeed, the reality of the inheritance of acquired characters was not an issue for him or for his contemporaries. As Lamarck himself stated in 1815:

> the law of nature by which new individuals receive all that has been acquired in organization during the lifetime of their parents is so true, so striking, so much attested by the facts, that there is no observer who has been unable to convince himself of its reality (Lamarck 1815, 1:200).

What Lamarck *did* claim credit for was having recognized that the law of use and disuse was of sufficient consequence to explain how organisms became diversified. He prized more, he said, "having been the first to recognize and determine [this]" than he prized the many classes, orders, genera, and species he had established in the course of his work as a systematist (Lamarck 1815, 1:191).

By offering examples of how changes in habits could bring about changes in structures, Lamarck left himself open to the unfortunate caricatures of his thinking that have continued to the present. Most influentional here, it would seem, was not the now-famous example of the giraffe (which Lamarck first mentioned only as an apparent afterthought in the index of his *Recherches sur l'organisation des corps vivans* of 1802), but rather the example of how the wading bird got long legs. Lamarck's words of 1800 are as follows:

> one may perceive that the bird of the shore, which does not at all like to swim, and which however needs to draw near to the water to find its prey, will be continually exposed to sinking in the mud. Wishing [*voulant*] to avoid immersing its body in the liquid, [it] acquires the habit of stretching and elongating its legs. The result of this for the generations of these birds that continue to live in this manner is that the individuals will find themselves elevated as on stilts, on long naked legs (Lamarck 1801:14).

In his notorious "eulogy" of Lamarck, Georges Cuvier (1832) greatly exaggerated the volitional character of Lamarck's explanation of organic development. So too did Charles Darwin when he wrote to J. D. Hooker in 1844: "Heaven forfend me from Lamarck[s] nonsense of a 'tendency to progression,' 'adaptations from the slow willing of animals,' etc.!" (Darwin 1903, 1:41). The importance of volition in Lamarck's explanation of change was likewise stressed in Alfred Russel Wallace's Linnean Society paper of 1858 (Darwin and Wallace 1958:277), and this misrepresentation of Lamarck's views continues to find its way into present-day historical treatments (e.g., Gruber 1974:307), despite numerous demonstrations of its inappropriateness (e.g. Russell 1916; Cannon 1957; Jordanova 1976; Burkhardt 1977 and 1981; and Richards 1982). Lamarck himself, as indicated above, bears some responsibility for this misinterpretation of his views. The fact remains, however, that he did not attribute a significant role in the evolutionary process of *wishing* or *willing* on the part of the individual organism. Indeed Lamarck regarded himself as one of the very first to assert that "wishing" or "willing" could not occur in any but the highest forms of animal life.

In 1809 Lamarck told his students that the standard definition of "animal" was not acceptable because it was "not true that animals are generally *sensate* beings [*êtres sensibles*], all endowed, without exception, with the power of producing *acts of volition,* and consequently with the faculty of moving voluntarily" (Lamarck 1972:210). It was only among the vertebrates, Lamarck maintained, and primarily among the birds and the mammals, that the faculty of willing came into being. But even then, he believed, most of the actions of the higher vertebrates (man included) took place without the will being involved at all. And what was largely true for the vertebrates—the "intelligent animals," as Lamarck called them—applied with full force to the invertebrates. Lamarck designated the higher invertebrates *les animaux sensibles,* the "sensate animals." These animals, he maintained, were endowed with a nervous system sufficiently complex to allow them the faculty of sensation, but insofar as they lacked a special organ of intelligence, they were incapable of thinking or any acts of will. The lower invertebrates were deprived of the faculty of sensation. These animals "without feeling" *(apathiques)* could neither think, nor will, nor—since they lacked any *sentiment intérieur*—could they even have instincts. The causes of their actions were entirely external to them. Had nature's work remained at this very imperfect level, Lamarck said, animals would have been merely "passive machines" (Lamarck 1809, 2:310).

What counted for Lamarck in the way animals responded to their environments was not the *desires* of the animals but rather their *habits.* All animals had habits. In the simplest invertebrates, Lamarck said, habits were the result of the subtle and expansive fluids of the environment acting mechanically upon the organisms in question. The subtle fluids traced different routes in the bodies they encountered, and, "once traced, . . . these routes [became] the immediate causes of a constant similarity in the actions and the nature of the movements of the individuals of each race" (Lamarck 1817a, 14:129). The physical characters acquired in one generation were passed on to the next, and along with the physical characters came the aptitudes for the same sorts of movements and habits displayed by the ancestors. In the higher invertebrates, where the nervous system became sufficiently complex to allow the development of the *sentiment intérieur* and instincts, habits were developed in response to felt needs. In certain forms, most notably the social insects and some spiders, the development of habits had been carried to such a degree that these forms exhibited activities comparable to those of animals of the highest intelligence. But Lamarck explicitly denied that intelligence could have anything to do with the activities of the invertebrates. The complex instincts displayed by these forms, he maintained, were the result of habits maintained over many generations. Instincts were caused by the physical organization an organism inherited from its ancestors, and this organization necessarily included the routes previously traced by nervous fluids as the ancestors acquired and developed their habits.

In Lamarck's explanation of organic change, habits were the cause not only of the instinctive behavior of animals but also of their visible structures. He insisted that habits preceded structures—not the other way around. His most daring discussion of the influence of habit upon form was in his suggestion of how the human species might have arisen from an ape-like form such as the orangutan. Couching his discussion in hypothetical terms, he described how a quadrumanous race that lived in trees might have come to walk on two feet if circumstances had demanded it. The need to dominate others and the need to see things would have caused the individuals of this race to stand and to walk upright. The need to communicate with other individuals of the same race would have led the members of the race to develop signs and eventually to articulate sounds. Over generations, they would even have developed vocal chords allowing them to speak. They would, in short, have gained the particular characteristic that René Descartes said distinguished humans from animals (Lamarck 1809, 1:355–57).

Lamarck and Finalism

Lamarck often used the phrase "nature's plan" in his discussion of how the development of complex forms over time was achieved. He was not, however, a finalist; that is, he did not suppose that the course of evolution represented the working out of a preestablished plan. When he took care to express himself precisely, he stated quite explicitly that nature was incapable of having any "plan" at all.

Lamarck did acknowledge that whatever happened in nature was consistent with the intentions of "the Supreme Author of all things." He did not presume to suggest, however, that man could know what those intentions were. Lamarck did maintain that the "power of life" necessarily brought about a gradual increase in the complexity of animal organization. But he did not claim that the process was a function of any predetermined goal. For one thing, as his studies progressed, he came to recognize that the process was more complicated than he had initially supposed. He found himself admitting more and more that the single series of increasing complexity he had used to "facilitate" his studies had to be distinguished from "the real or actual production of [the animals]" (1815, 1 :451). A single scale of increasing complexity, he decided, was simply untenable : the invertebrates were represented by at least *two* different series, with spontaneous generation taking place at the base of each of them. For another thing, while Lamarck represented man as the highest form of organization that nature had produced, he considered the human species as simply the most perfect form yet reached, with the endpoint, if there indeed was such a thing, remaining unknown and unknowable (Lamarck 1802b :90). Furthermore, Lamarck was prepared to recognize that the "perfection" of man's organization did not result in a being that was in every way perfect. Lamarck regarded man as a being that exhibited the worst possible qualities as well as the best, and it was in contemplating the former that Lamarck made the remarkable statement : "One would say that [man] is destined to exterminate himself after having rendered the globe uninhabitable" (1817b, 15 :271[fn.]).

Organization and Faculties

The occasion has already presented itself to stress Lamarck's concern with the correlation between the level of complexity of an animal and the *faculties* the animal possesses. In his very first lectures on invertebrate zoology, Lamarck identified the relation between organic complexity and the possession of particular faculties as one of the most significant topics that the study of invertebrate zoology could illuminate. In 1809, commenting on the different special faculties occurring in the animal kingdom, he maintained that each of these was the product of an organ or a system of organs, "so that every animal in which that organ or system of organs does not exist can in no way possess the [particular] faculty" (Lamarck 1809, 2 :166). As he had done earlier, he insisted that the laws of nature were no different for organic bodies than they were for inorganic bodies. The *results* observed in living bodies, however, were entirely different because of "the order and state of things" found there.

Lamarck's assertion that "every animal faculty whatsoever it be is an organic phenomenon" (Lamarck 1815, 1 :215) did not fail to attract the attention and arouse the anxieties of some prominent contemporaries. Especially controversial was his claim that even the highest of the mental faculties, the faculty of understanding, was dependent upon a special system of organs for its functioning. In England, for example, the Reverend William Kirby, a distinguished entomologist and observer of animal behavior, wrote : "We cannot help *feeling* that our thoughts are the attributes of an immaterial substance." Thought, Kirby said, was what enabled man to "take flights beyond the bounds of time and space, and enter into the Holy of Holies." "Who can believe," Kirby asked, "that such a faculty, so divine and godlike and spiritual, can be the mere result of organization?" The primary source of Lamarck's folly, Kirby felt, was evident :

> Lamarck's great error, and that of many others of his compatriots, is materialism ; he seems to have no faith in any thing but *body,* attributing every thing to a physical, and scarcely any thing to a metaphysical cause (Kirby 1835, 1 :xxvii).

The Response to Lamarck's Views

In concluding his *Philosophie zoologique,* Lamarck wrote : "men who strive in their works to push back the limits of human knowledge know well that it is not enough to discover and prove a useful truth that was previously unknown, but that it is necessary also to propagate it and get it recognized." Judged in his own terms, Lamarck was by and large unsuccessful. He was unsuccessful, it would seem, for a variety of reasons. He was unsuccessful because the weight of religious opinion in his day made his materialistic views unacceptable. He was also not helped by the fact that at the time he announced his evolutionary views, he had established a reputation for himself as a man who speculated without the proper facts to back him up. Furthermore, he proved unable to cultivate a circle of capable naturalists willing to champion his views.

Space does not permit a detailed examination here of the response to Lamarck's views. For more details on the subject, the reader may consult Landrieu (1909), Burkhardt (1977), and Corsi (1978). Here only two points will be made. The first is that with respect to the notion of species mutability,

Lamarck was not quite as alone in his thinking as is frequently supposed. The second is that there was not an abundance of evidence at Lamarck's disposal for demonstrating the reality of organic change.

It is often said that Lamarck's evolutionary theory was rejected in Lamarck's day because people at the beginning of the nineteenth century were unaccustomed to thinking in evolutionary terms. But Lamarck was by no means the only French naturalist who was attracted by the idea of organic mutability at the beginning of the nineteenth century. Indeed, of the zoologists at the Muséum d'Histoire Naturelle in the early 1800s—Lamarck, Lacépède, Geoffroy Saint-Hilaire, and Cuvier—only Cuvier was clearly unsympathetic with the idea. Faujas de Saint-Fond, the professor of geology, also appears to have entertained with some favor the notion of species change. Thus the striking thing about Lamarck appears not to have been that he was thinking about organic mutability at a time when no one else was, but rather that he was the only one at the time who chose to develop at length a broad hypothesis of organic change (Burkhardt 1977 :202 – 10).

As for the evidence that Lamarck was able to bring to bear in support of his hypothesis of evolution, there was very little that told strongly in his favor. To substantiate his claim that nature had begun with the simplest forms of life and successively produced all the rest, he could only point to his arrangement of the animal classes and maintain that this represented the true order of formation of the different living things. To substantiate his claim that changing circumstances could, over time, bring about changes in races and species, he had little evidence to offer besides evidence of variations *within* the limits of existing species. The fossil record was of no real use to him in demonstrating that organic change had taken place. He was able to point out the similarities between some modern forms and some fossils, but the fossil record was by no means complete enough in his day to provide him with any impressive sequences with which to substantiate his claims. Furthermore, since he believed that the order of formation of the different animals was adequately represented by the *existing* classes of animals, he had little reason to believe that fossil evidence had a decisive role to play in illuminating the development of life on earth. Unable to come up with compelling evidence for species change, he tried to shift the burden of proof to his opponents, saying that the apparent constancy of species simply reflected how long it took, by human standards, for the conditions of their existence to change. "The allegation that [the species] we see are constant, the circumstances in which they are observed being likewise, does in no way furnish the proof required" (Lamarck 1817, 10 :445). The burden of proof, however, was not on the exponents of species fixity but rather on Lamarck himself.

To say that Lamarck's explanation of the successive production of the different forms of life and the faculties attendant upon increased complexity were not generally accepted by his contemporaries is not to say that these views had no influence. If Cuvier chose not to dignify Lamarck's views with a systematic critique, he nonetheless perceived Lamarck's views as a threat. And if virtually no naturalists championed Lamarck's general account of transformism, certain individuals were attracted by particular aspects of his thinking (Corsi 1978), and Charles Lyell felt it appropriate to devote a significant portion of the second volume of his *Principles of Geology* (1830–33) to refuting Lamarck's views. Interestingly enough, Herbert Spencer, upon reading Lyell's objections to Lamarck's views, decided Lamarck was probably right. Spencer's autobiographical account indicates how akin he was in spirit to Lamarck :

> My inclination to accept [the conception of the natural genesis of organic forms] as true, in spite of Lyell's adverse criticisms, was, doubtless, chiefly due to its harmony with that general idea of the order of Nature towards which I had, throughout life, been growing. Supernaturalism, in whatever form, had never commended itself. From boyhood there was in me a need to see, in a more or less distinct way, how phenomena, no matter of what kind, are to be naturally explained. Hence, when my attention was drawn to the question whether organic forms have been specially created, or whether they have arisen by progressive modifications, physically caused and inherited, I adopted the last supposition ; inadequate as was the evidence, and great as were the difficulties in the way. Its congruity with the course of procedure throughout things at large, gave it an irresistible attraction ; and my belief in it never afterwards wavered, much as I was, in after years, ridiculed for entertaining it (Spencer 1904, 1 :201).

Conclusion

When he surveyed his own intellectual achievement, Lamarck felt that he had satisfactorily explained a host of important and interconnected issues, including how the simplest forms of life originated, how the organization of animals had become increasingly complex over time, how the higher animal faculties had emerged with the increasing perfection of organization, and how the influence of particular environmental circumstances had led to special habits and structures in animals. The overall view that he presented was both bold and profound.

Though we are perhaps better prepared than Lamarck's contemporaries were to appreciate the profundity of Lamarck's view of organic development, we are not in a position to confront Lamarck's works free of preconceptions. The questions we choose to ask about Lamarck today are inevitably a function of our present understanding of the nature of science and how science has developed, an understanding which is itself culturally and historically conditioned. Whatever questions one brings to the consideration of Lamarck's work, one will have to confront Lamarck's own writings. To his *Philosophie zoologique*, we have appended two short works, his *Discours d'ouverture* of 1800 as a first statement of his views on transformism, and Cuvier's infamous *Éloge* of Lamarck to indicate one important source of Lamarck's reputation, especially in the English-speaking world. These three pieces do not constitute all there is to Lamarck's life and work, but they do serve as a helpful introduction.

References

Bowler, P. J. 1975. The changing meaning of evolution. *J. Hist. Ideas.* 36:95–114.

Burkhardt, R. W., Jr. 1972. The inspiration of Lamarck's belief in evolution. *J. Hist. Biol.* 5:413–38.

———. 1977. *The spirit of system: Lamarck and evolutionary biology.* Cambridge: Harvard University Press.

———. 1981. Lamarck's understanding of animal behavior. In *Lamarck et son temps; Lamarck et notre temps,* 11–28. Paris: Vrin.

Cannon, H. G. 1957. What Lamarck really said. *Proc. Linn. Soc. Lond.* 168:70–85.

Corsi, P. 1978. The importance of French transformist ideas for the second volume of Lyell's *Principles of Geology. Brit. J. Hist. Sci.* 11:221–44.

Darwin, C. 1903. *More letters of Charles Darwin.* Ed. F. Darwin. 2 vols. London : John Murray.

Darwin, C. and Wallace, A.R. 1958. *Evolution by natural selection.* Cambridge, England.

Geoffroy Saint-Hilaire, E. 1838. *Fragments biographiques, précédés d'études sur la vie, les ouvrages et les doctrines de Buffon.* Paris : Pillet.

Gruber, H. E., and P. H. Barrett. 1974. *Darwin on man : A psychological study of scientific creativity, together with Darwin's early and unpublished notebooks.* New York : Dutton.

Hodge, M. J. S. 1971. Lamarck's science of living bodies. *Brit. J. Hist. Sci.* 5 :323 – 52.

Jordanova, L. J. 1976. *The natural philosophy of Lamarck in historical context.* Ph.D. diss., University of Cambridge.

Kirby, W. 1835. *On the power, wisdom and goodness of God, as manifested in the creation of animals, and in their history, habits and instincts.* 2 vols. London : William Pickering.

Lamarck, J.-B. 1778 (1779). *Flore françoise, ou description succincte de toutes les plantes qui croissent naturellement en France, disposée selon une nouvelle méthode d'analyse et à laquelle on a joint la citation de leurs vertus les moins équivoques en médecine, et de leur utilité dans les arts.* 3 vols. Paris : L'Imprimerie Royale.

———. 1794. *Recherches sur les causes des principaux faits physiques* . . . 2 vols. Paris : Maradan.

———. 1796. *Réfutation de la théorie pneumatique ou de la nouvelle doctrine des chimistes modernes* . . . Paris : Agasse.

———. 1797. *Mémoires de physique et d'histoire naturelle* . . . Paris : Agasse.

———. 1800 – 1810. *Annuaires météorologiques.* 11 vols. Paris : the author (1800 – 1805), Maillard (1806), Treuttel and Würtz (1807 – 1810).

———. 1801. *Système des animaux sans vertèbres* . . . Paris : Déterville.

———. 1802a. *Recherches sur l'organisation des corps vivans, et particulièrement sur son origine, sur la cause de ses dévelopemens et des progrès de sa composition, et sur celle qui, tendant continuellement à la détruire dans chaque individu, amène nécessairement sa mort* . . . Paris : Maillard.

———. 1802b. *Hydrogéologie ou recherches sur l'influence qu'ont les eaux sur la surface du globe terrestre ; sur les causes de l'existence du bassin des mers, de son déplacement et de son transport successif sur less différens points de la surface de ce globe ; enfin sur les changements que les corps vivans exercent sur la nature et l'état de cette surface.* Paris : Agasse.

———. 1809. *Philosophie zoologique, ou exposition des considérations relatives à l'histoire naturelle des animaux : à la diversité de leur organization et des facultés qu'ils en obtiennent ; aux causes physiques qui maintiennent en eux la vie et donnent lieu aux mouvemens qu'ils exécutent ; enfin, à celles qui produisent, les unes le sentiment, et les autres l'intelligence de ceux qui en sont doués.* Paris : Dentu.

———. 1815–22. *Histoire naturelle des animaux sans vertèbres ...* Paris : Verdière.

———. 1817a. Habitude. In *Nouveau dictionnaire d'histoire naturelle,* 14 :128–38. New ed. Paris : Déterville.

———. 1817b. Homme. In *Nouveau dictionnaire d'histoire naturelle,* 15 :270–76. New ed. Paris : Déterville.

———. 1820. *Système analytique des connaissances positives de l'homme restreintes à celles qui proviennent directement ou indirectement de l'observation.* Paris : Belin.

———. 1972. *Inédits de Lamarck, d'après les manuscrits conservés à la bibliothèque centrale du Muséum National d'Histoire Naturelle de Paris.* Eds. M. Vachon, G. Rousseau, and Y. Laissus. Paris : Masson.

Landrieu, M. 1909. *Lamarck, le fondateur du transformisme.* Paris : Société zoologique de France.

Richards, R. J. 1982. The emergence of evolutionary biology of behavior in the early nineteenth century. *Brit. J. Hist. Sci.* 15 :243–80.

Russell, E. S. 1916. *Form and function : A contribution to the history of animal morphology.* London : John Murray ; reprinted 1982, Chicago : University of Chicago Press.

Spencer, H. 1904. *An autobiography.* 2 vols. New York : Appleton.

LAMARCK AMONG THE ANGLOS

DAVID L. HULL

Scientists and their reputations are not always treated with total fairness during the course of science. Lamarck is one of those scientists who has proven to be perennially useful to other scientists in their ongoing pursuits. Because the name of Lamarck had such a disreputable sound when Darwin published his *Origin of Species,* Darwin tried to distance himself as much as possible from Lamarck, while his opponents tried to cast doubt on Darwin's theory by identifying it with Lamarck's foolish notions. Lamarck became an issue again in the dispute between the old guard Darwinians and the neo-Darwinians. G. J. Romanes (1848–94) fought to keep Darwinism "pluralistic" in the true spirit of Darwin, while August Weismann (1834–1914) argued that true Darwinism demanded the total exclusion of Lamarckian inheritance. The neo-Darwinians won. Since then, Lamarckism has periodically reared its head in Great Britain and the United States, but always as a suspect, unorthodox view. Any biologist who feels that his ideas are not getting sufficient attention can be guaranteed at least a snort of irate indignation if he terms them "Lamarckian."

Throughout the history of disputes over Lamarckism, what Lamarck really said has not been all that important. Scientists had Lamarck say what they needed him to say. One of the problems in dealing with Lamarck's views, as with those of any other scientist, is that time marches on. Each time the issue of Lamarckism has arisen, it has been in a decidedly new context. As Burkhardt explains in his contribution to this volume, Lamarck's acceptance of the inheritance of acquired characteristics was not in the least unusual for his time. Nor was the inheritance of acquired characteristics central to his transmutation theory. According to Lamarck, evolutionary change had a direction, but this direction was produced by a tendency toward increased complexity or perfection that Lamarck thought was inherent in all living creatures, not by the inheritance of acquired characteristics.

In order to set out the relevant differences between what Mayr (1982) calls "hard" and "soft" inheritance, two distinctions are necessary—between an organism and its environment and between the hereditary material and the rest of an organism. Not all biologists have made these distinctions, nor made them in the same way. For example, Lamarck says very little about the mechanisms that permit the inheritance of acquired characteristics. Darwin, to the contrary, at least sketches his theory of pangenesis (1868). According to Darwin, all cells in the body of an organism throw off minute gemmules characteristic of that cell. These gemmules somehow find their way to the gonads, where they are passed on in reproduction. For Darwin, gemmules are the hereditary material. Everything else belongs to the body of the organism. For Darwin, the inheritance of acquired characteristics meant the modification of gemmules through the modification of the cells that gave rise to them.

Weismann is famous for maintaining not only the continuity but also the purity of the germ plasm (1892, 1894). Germ cells produce other germ cells as well as somatic cells, while somatic cells never produce or affect the character of the cells in the germ line. However, as the controversy progressed, such easy distinctions rapidly broke down. Under certain conditions, somatic cells change into germ cells and produce new organisms, as in the case of vegetative reproduction. In such cases, changes in somatic cells might produce changes in the germ plasm. With the discovery of chromosomes, DNA, etc., the relevant distinctions had to be drawn and redrawn. For example, modifications in the body of a paramecium can be transmitted to later generations during fission independently of the organism's hereditary material. Is this a case of the inheritance of acquired characteristics?

In a sense, questions such as these are "only" terminological. What difference does it make what one terms these phenomena so long as the relevant distinctions are made? But language can be very seductive. The terms we use determine which phenomena we assimilate to each other. For example, because we have one term "cancer," we are led to search for a cure for cancer, when on a moment's reflection, it becomes obvious that "cancer" is numerous different diseases that are liable to have equally numerous cures. The distinction between "hard" and "soft" inheritance concerns the relationship between the character of the environmental or somatic causes of changes in the genetic material and the nature of those changes. Throughout the recent history of genetics, everyone has agreed that mutations are caused and that these causal chains begin outside the body. If all it takes for a mutation to be Lamarckian is that it arise from a change in the organism's environment and/or body, then all inheritance is Lamarckian. Soft inheritance entails much more than this.

One form of soft inheritance is the direct effect of the environment. According to this view, the environment can produce adaptive changes directly in the organism, e.g., mammals develop heavier coats as they migrate into colder climates. Migration in the opposite direction would reverse the process. According to the theory of use and disuse, changes induced by using or ceasing to use a structure can produce cumulative changes in these structures. People living at higher altitudes are liable to develop larger lungs, while the eyes of organisms raised in the dark are likely to atrophy. Throughout much of history, biologists thought that even mutilations could, in special circumstances, be inherited. Comparable observations hold for behavior. Certain behaviors must be elicited anew each time. Others are in some sense instinctual. In line with the preceding examples, habits through repeated occurrence might be converted into instincts (Richards 1982).

What all the preceding forms of soft inheritance have in common is that they assume some connection between the character of the causal chain producing the change and the nature of the change itself. A change in temperature produces a change in temperature tolerance. Changes in the use of an organ modify that organ. Advocates of hard inheritance maintain that no such correlations exist. An increase in temperature might produce mutations, but these mutations are just as likely to affect any structure or function. Whether or not the preceding forms of inheritance count as instances of the inheritance of acquired characteristics also depends on the distinction one draws between "germ line" and "soma." In the usual case, the assumption was that adaptive changes in the body of the organism were transmitted to the hereditary material and from there to later generations. The direct transmission of a phenotypic trait from one organism to another without touching the hereditary material was quite a different phenomenon.

I cannot pretend that the preceding distinctions are totally adequate to handle all the intricate differences of opinion that arose during the history of Lamarckism in the English-speaking world, but at least they are a good beginning. Nearly every type of hereditary phenomenon has been termed at one time or another Lamarckian. Some of this diversity can be explained in terms of the politics of science. At times Lamarckian was a pejorative term to be used to characterize the views of one's opponents. On rare occasions, it was used for just the opposite purpose. Some of this diversity also results from the radical changes that have taken place in our understanding of hereditary phenomena. "Germ plasm" meant one thing to a scientist who thought that it was parcelled out differentially in embryological development. It meant quite another thing to a scientist who thought that nearly all cells contain nearly all the same hereditary material. The situation was further complicated in the English-speaking world by the fact that few British or American scientists got their views of Lamarck by reading translations of Lamarck, let alone Lamarck's original works in French. In most instances, Lamarck's reputation was formed by descriptions of his views in secondary sources. In Darwin's day, the two chief sources of Larmarck's ideas in Great Britain and the United States were the second volume of Charles Lyell's *Principles of Geology* (1830–33) and an English translation of Cuvier's 1835 *Éloge* (1836).

Lyell and Cuvier on Lamarck

In his *Principles of Geology,* Lyell was concerned to set out a new theory of the formation of the earth's surface. More than this, he was urging on the scientific community his view of the proper way to do geology if not science at large. One element of Lyell's grand "uniformitarian" research program was the denial of any direction to terrestrial change, especially progressive change. Because the clearest example at the time of directional, progressive change was provided by the fossil record, Lyell was forced to confront the species problem. According to one view, God introduced species sequentially in time, improving from one stratum to the next on His basic designs. Lyell agreed that new species were introduced sequentially in time as other species became extinct, but he disagreed that such a sequence of events indicated any clearly defined direction or that the process was necessarily miraculous. Lyell argued that the causes of extinction were perfectly natural. The problem came when he turned to possible naturalistic explanations for the origins of new species. He dismissed spontaneous generation as a "fanciful notion left over from Aristotle" (Lyell 1830, 1:59). The only other plausible explanation was transmutation, and Lamarck's version of this theory was at the time the most cardfully worked out.

With the exception of an occasional reference to the "will" or the "desires" of an organism influencing its development, Lyell presented a reasonably accurate description of Lamarck's views. When all was said and done, however, Lyell was forced to conclude that the balance of evidence was opposed to Lamarck. Certainly acquired characteristics were sometimes inherited, but not to the extent necessary to transmute one species into another. And no evidence existed for Lamarck's hypothetical forces that supposedly drove organisms toward increased complexity and perfection. Although Lyell was committed to some naturalistic explanation for the origin of species, his rejection of all the naturalistic explanations proposed to date and his description of his own agnostic position in terms of "creation" allowed special creationists to adopt Lyell's geological theories without threatening their own theological beliefs any more than absolutely necessary.

Lyell's discussion of Lamarck in his *Principles* had a variety of effects; in a few cases, just the opposite of what he intended. It led Robert Chambers to conclude that possibly species did change through time. Chambers (1844) in turn brought the issue of the origin of species forcefully to the attention of Victorian society. In recognition of the low repute in which Lamarck was held at the time, Chambers felt obliged to make a few disparaging remarks about him, but to no avail. With the notable exception of Herbert Spencer and A. R. Wallace, scientists found Chambers's exposition even more unscientific than Lamarck's earlier efforts. T. H. Huxley, for example, began his review of the tenth edition of Chambers's *Vestiges of the Natural History of Creation* (1853) with Macbeth's observation, "Time was, that when the brains were out, the man would die" (Huxley 1854:425). Spencer's more "philosophical" advocacy (1852) went all but unnoticed. No one was roused either to attack Spencer's theory or to adopt it.

The most important effect of Lyell's exposition of Lamarck was, however, on Darwin (Gillispie 1959 ; Simpson 1961 ; Egerton 1976). So much attention has been paid by later commentators to matters of priority that the much more relevant issue of actual influence is all but lost in the polemics. Did Darwin come to adopt Lamarck's theory by reading it in Lyell's *Principles* ? Hardly. Substantial evidence indicates that Darwin was a special creationist when he read the second volume of Lyell's *Principles* and remained a special creationist for a good long time thereafter. What Lyell did do in the context of Lamarck's theory was to set out clearly and forcefully the species problem that Darwin eventually solved. In science, posing the right question in a tractable way is often half the battle. If nothing else, Lamarck is important in this connection. Lyell's rejection of Lamarck's theory did more to further the cause of evolutionism than did the advocacy of all previous evolutionists combined (Corsi 1978).

Cuvier's *Éloge* is quite another matter. Cuvier had the most profound disdain not only for the various theories that Lamarck propounded during his long career but even more so for the highly irresponsible way in which he presented these theories to the scientific community. Lamarck's views were not just mistaken, they were unscientific. Although Cuvier found some value in Lamarck's early and late work in classification and expressed pity for the poverty of his final years, a less sympathetic author for a eulogy would have been difficult to find. Because one of Lamarck's theories turned out to be in some sense right, present-day readers are likely to be much more interested in it than in any of his other theories. But the reception of Lamarck's theory of the transmutation of species cannot be separated from his theories in chemistry, meteorology, and mineralogy (Carozzi 1964 ; Burlingame 1981). The problem with Lamarck's theories was not that they were so far ahead of their time. In fact, they tended to be old-fashioned. Certainly respected naturalists in France before, at the time of, and after Lamarck advocated some form of transmutation. As Burkhardt has argued persuasively (1970, 1972, 1977), it was more Lamarck's style of theorizing than the content of his theories that prejudiced his contemporaries against him.

The greatest disservice that Cuvier did to Lamarck was to expand a facile misreading of the role of *besoins* in Lamarck's transmutation theory into an extremely effective caricature of that theory. Although Lyell (1830–33) made an occasional passing reference to the efficacy of an organism's will or desire on the modification of its structure, he did not exploit the propaganda value of parodies of this view to the extent that Cuvier did. If Lamarck had had the slightest interest in such matters, he might well have chosen his words with greater care. Although it is impossible to phrase one's views so carefully that they defy parody, there is no excuse for going out of one's way to invite easy ridicule. For example, in the first edition of the *Origin of Species,* Darwin observed that a race of bears which habitually caught insects by swimming with their mouths agape on the surface of water might be converted into a creature "as monstrous as a whale" (Darwin 1859 :184). Darwin's critics quickly showed the ease with which this example could be converted to their own use—bears evolving into whales, indeed! One of the few changes that Darwin made in the second edition of the *Origin* was to remove this example.

THAT WRETCHED BOOK

In present-day terminology, Lamarck presented a serious public relations problem for Darwin and the Darwinians. To some extent, Lamarck was Darwin's only genuine scientific precursor. Granted that other workers had suggested previously that species might change through time, but no one before Lamarck had set out a theory of transformism in anything like the detail that Lamarck did. It was equally true that Darwin's views on the evolution of species differed significantly from those of Lamarck. For example, Darwin introduced quite a different mechanism as the main cause of evolution—natural selection. The inheritance of acquired characteristics played an even more minor role in Darwin's theory than it had in Lamarck's, but more importantly, Darwin and Lamarck visualized phylogeny quite differently. For Darwin the vast majority of species that ever existed were extinct. Only the terminal twigs of the phylogenetic tree represented extant species. As Burkhardt (1977 :130) points out, Lamarck was driven to his theory of evolution by Cuvier arguing for extinction through global catastrophes. Lamarck argued to the contrary that few if any species had ever gone extinct. His trees of life were made up totally of extant forms. A species might go temporarily extinct, but the continued generation of simple forms at the base of Lamarck's trees guaranteed that no species would remain extinct for long. Darwin visualized phylogenetic trees as highly contingent results of chance variation and natural selection. Given different circumstances, very different trees would evolve. According to Lamarck, evolutionary development is part of the permanent make-up of the universe.

Time and time again, Darwin's critics attempted to tar him with the same brush as they used on Lamarck (Hull 1973). It was easy enough to cast doubt on Darwin's theory by equating it with Lamarck's views and then discrediting the latter. In this connection, Lyell did not help much by repeatedly referring to Darwin's theory as "Lamarck's theory as modified by Darwin." Darwin was able to take such abuse only so long. In exasperation he complained to Lyell:

> Lastly, you refer repeatedly to my view as a modification of Lamarck's doctrine of development and progression. If this is your deliberate opinion there is nothing to be said, but it does not seem so to me. Plato, Buffon, my grandfather before Lamarck, and others, propounded the *obvious* views that if species were not created separately they must have descended from other species, and I can see nothing else in common between the 'Origin' and Lamarck. I believe this way of putting the case is very injurious to its acceptance, as it implies necessary progression, and closely connects Wallace's and my views with what I consider, after two deliberate readings, as a wretched book, and one from which (I well remember my surprise) I gained nothing. But I know you rank it higher, which is curious, as it did not in the least shake your belief. But enough, and more than enough. Please remember you have brought it all down on yourself!! (Darwin 1899, 2:198–99)

Darwin's solution to the problem posed by Lamarck's reputation was to ignore him. He rarely refers to Lamarck by name in his published writings. In the *Origin,* Darwin mentions the adaptations in the sterile members of social insects as counting against the "well-known doctrine of Lamarck" (p. 242) and the distinction drawn by Lamarck between real affinities and analogies (p. 427). In the early pages of the *Origin,* Darwin also observes how "preposterous" it would be to account for the structure of mistletoe by the "effects of external conditions, or habits, or of the volition of the plant itself" (p. 3). Exactly who in the history of science ever thought that "volition" played a role in the evolution of plants Darwin does not say. In the historical sketch that he appended to the sixth edition of the

Origin (1872), Darwin begins his discussion of his precursors by refering to Lamarck as the "first man whose conclusions on the subject excited much attention" (p. 17). As the long passage quoted earlier indicates, Darwin was somewhat harder on Lamarck in his private correspondence.

Other Darwinians followed Darwin's lead in their treatment of Lamarck. For example, in a letter written in 1862 Huxley (1901, 1:215) chides Lyell for referring to evolutionary theory as a modification of Lamarck's views: "I should no more call his doctrine a modification of Lamarck's than I should call the Newtonian theory of the celestial motions a modification of the Ptolemaic system." In the final lecture to working men, Huxley (1863) repeats this objection, questioning the "capacity for forming a judgment" of anyone who viewed Darwin's theory as a "mere modification of Lamarck's," a harsh assessment of Lyell. But Huxley begins his discussion by remarking that "Lamarck was a great naturalist, and to a certain extent went the right way to work; he argued from what was undoubtedly a true cause of some of the phenomena of organic nature." He continued in much the same vein in a letter to G. J. Romanes in 1882:

> I am not likely to take a low view of Darwin's position in the history of science, but I am disposed to think that Buffon and Lamarck would run him hard in both genius and fertility. In breadth of view and in extent of knowledge these two men were giants, though we are apt to forget their services (Huxley 1901:2:42).

Finally, in the retrospective evaluation of the reception of the *Origin of Species* that he wrote for Darwin's *Life and Letters* (Darwin 1899, 1:543), Huxley was also respectful of Lamarck's place in the history of science, but even so found it necessary to remark that Lamarck's "one suggestion as to the cause of the gradual modification of species—effort excited by change of conditions—was, on the face of it, inapplicable to the whole vegetable world." Huxley might have been more hesitant. Given that Lamarck, early in his career, authored a flora of France (1778), it is unlikely that he had forgotten one of the two great kingdoms of living creatures. Possibly Huxley had misconstrued Lamarck's theory (Gillispie 1956; Hodge 1971a, 1971b; Mayr 1972; Burkhardt 1972, 1977; Richards 1982).

OF RAT TAILS, FORESKINS AND NUPTIAL PADS

The history of evolutionary biology after the publication of the *Origin* in Great Britain and the United States was marked by numerous disputes both between the Darwinians and their opponents and within the Darwinian camp itself. The relative importance of Lamarckian forms of inheritance was only one of the points at issue and initially not a very divisive one. From the beginning, Spencer (1852) was predisposed to attribute a greater effect to the inheritance of acquired characters than did Darwin, while Wallace (1870) went beyond even Darwin in the power he attributed to natural selection. Darwin's theory of pangenesis (1868) elicited very little excitement either among his fellow evolutionists or his opponents. Although Francis Galton (1871) ran a series of experiments which he thought refuted Darwin's claim that gemmules were circulated through the body, he did not totally reject the inheritance of acquired characters. Romanes (1892, 1893) found this "pluralistic" attitude toward Lamarckian inheritance to be in the true Darwinian tradition.

August Weismann's visit to the Manchester meetings of the British Association for the Advancement of Science in 1887 shattered the Darwinian compromise. For several years, Weismann had been working on a grand theory of heredity and evolution (1892, 1893). One aspect of this theory was the independence of what he termed the germ plasm from the somatoplasm. One consequence of the barrier that Weismann erected between the hereditary material and the rest of an organism's body was the noninheritance of acquired characters. Weismann's theory was attacked in Great Britain by Darwinians and non-Darwinians alike, especially by Spencer (1893a, 1893b) and Romanes (1892, 1893). One tactic that anti-Darwinians such as St. George Jackson Mivart (1871) had used with considerable success was to saddle Darwin with the view that natural selection was all sufficient. Hence, by showing that particular phenomena were inexplicable by selection alone, they were able to "refute" Darwin's theory. Romanes (1892) complained that in denying the inheritance of acquired characters, Weismann was playing into the hands of these critics.

In his native Germany, Weismann's theory had only limited success. As might be expected, Wallace (1889) adopted it enthusiastically, but in addition, for reasons that remain obscure, Weismann's ideas (1893, 1894) fired a group of young evolutionists in Great Britain to see exactly how far the power of selection could be pushed. Included among these "neo-Darwinians" were Thiselton-Dyer (1843–1928), E. R. Lankester (1847–1929), and E. B. Poulton (1858–1943). In reaction to these neo-Darwinians, numerous "neo-Lamarckians" tried to show that acquired characters could be inherited. With the exception of E. D. Cope (1840–97), A. Hyatt (1838–1902), and A. S. Packard (1839–1905) in the United States, the neo-Lamarckians did not form a group but worked largely in isolation from each other. Instead of viewing themselves as renegades, they tended to see themselves as conservatives championing a wider view of evolution against the overly restricted position of the neo-Darwinians (Bowler 1983).

One reason for the continuing confusion that plagued the dispute between the Darwinians, neo-Darwinians, and neo-Lamarckians was that none of the necessary distinctions were made with sufficient clarity and consistency. Weismann's early distinction between germ plasm and somatoplasm began to unravel as he attempted to accommodate numerous difficulties. Before he was done, Weismann seems to have abandoned his earlier hard line on soft inheritance (Churchill 1968, 1978; Ridley 1982). For their part, neo-Lamarckians were happy to call any sort of hereditary change in which the environment played any role whatsoever Lamarckian. Matters were only confused further by the rise of Mendelian genetics. For reasons that historians have set out at some length, early Mendelians found their principles of heredity incompatible with what they took to be the Darwinian theory of evolution. Sorting out the interrelations between all these schools of thought took scientist the better part of a decade.

Time and again, those scientists involved in the controversy over the inheritance of acquired characteristics maintained that it should be decided on experimental grounds, but as Galton discovered, the existence or nonexistence of such hereditary phenomena is not easy to document. Certainly the failure of Weismann to shorten the tails of mice by amputating them generation after generation proved nothing. Only if the absence of tails fulfilled some *need* would a Lamarckian expect such amputations to have any sort of a hereditary effect. After all, Jews had circumsized their male infants for thousands of years with no noticeable hereditary effect. The reputation of

Lamarckism was also not helped by recurrent claims of fraud, especially in the famous case of the midwife toad.

Most toads and frogs mate in the water. As an aid in holding onto their slippery mates during copulation, males develop rough patches of skin on their palms and fingers during the mating season. Toads that mate on dry land, such as the midwife toad, lack these nuptial pads. Paul Kammerer (1880–1926) set himself the task of seeing whether he could induce the formation of nuptial pads in the midwife toad by raising successive generations of the animal in water. Most eggs failed to develop, but Kammerer claimed to have reared several generations of toads until, finally, nuptial pads appeared in a few of the males (Kammerer 1924). Pluralist Darwinians and neo-Lamarckians were delighted. Neo-Darwinians and such Mendelians as William Bateson (1861–1926) were highly suspicious. The story of the ensuing fiasco is familiar. Eventually Kammerer's sole specimen was discovered to have been doctored with India ink, and soon thereafter Kammerer killed himself. As Koestler (1971) has argued persuasively, Kammerer's suicide was caused more by a failed romance than a failed experiment. As Gould (1972) also shows, even if Kammerer had produced nuptial pads in specimens of the midwife toad as he claimed, he would have still been a long way from demonstrating Lamarckian inheritance. Because the midwife toad evolved from ancestors that lived in water and possessed nuptial pads, the emergence of such pads under the extreme selective regimen to which Kammerer exposed his organisms could be explained within the Darwinian paradigm as an instance of atavism.

As if the Kammerer affair did not damage the reputation of Lamarckian inheritance enough, it was soon overshadowed by a controversy of even greater magnitude surrounding the work of a Russian agronomist, T. D. Lysenko (1898–1976). The Lysenko era in Russia was as tragic as the Kammerer affair was pathetic. Lysenko claimed to be able to produce extremely vigorous and productive strains of wheat by "vernalization," a method of treating seeds prior to planting. The need for increased grain production in Russia was acute. The methods being developed in the West for improving agricultural production gave every appearance of being slow and tedious. The only way in which Russian agriculture could hope to meet the needs of the Russian people was if Lysenko was right. As a result, the power of the state was put behind Lysenko and his research program. As it turns out, Lysenko was mistaken. The effects of Lysenkoism in Russia were tragic in two respects. First, the Russian people were hungrier than they need have been, and second, numerous scientists who opposed Lysenko's views were persecuted by the state. One moral that is commonly drawn from this sequence of events is that it is extremely dangerous to replace the system of checks and balances internal to science with an external system.

The Weismann Barrier

In the early decades of this century, Lamarckism was still a reputable position in science. When Packard wrote his *Lamarck: The Founder of Evolution* (1901), he wanted to set the record straight about what Lamarck really said. Too many people at the time, friend and foe alike, were attributing too many diverse views to the man. Packard also claimed for Lamarck the credit which he thought was due him for his theory of evolution. Neither he nor his opponents thought of the neo-Lamarckians as part of a lunatic fringe. During the next few decades, the situation in Great Britain and the United States changed dramatically. For example, when E. W. MacBride (1895) began publishing his Lamarckian views on embryology and evolution, he was contributing to a respectable position. By the end of his career, MacBride (1937) was viewed as an embarrassing eccentric (Burkhardt 1980). The inheritance of acquired characteristics was rejected during this period, not so much because its existence had been decisively refuted experimentally, but because it was incompatible with Mendelian genetics (Bowler 1983) and because the rapidly emerging synthetic theory of evolution explained apparent cases of Lamarckian inheritance without any recourse to Lamarckian mechanisms (Mayr and Provine 1980). The Darwinian orthodoxy was further supported by work in molecular biology. According to the Central Dogma of Watson and Crick, DNA codes for RNA, which codes for proteins. Biosynthetic pathways never go in the opposite direction as required by Lamarckian forms of inheritance.

By the time that H. G. Cannon (1957, 1958, 1959) published a flurry of papers and books on Lamarck, he was clearly an isolated voice railing against the Darwinian orthodoxy. In reading Cannon's defense of Lamarck, one gets the decided impression that his animosity is directed as much against his own contemporaries for ignoring his views as against the persecutors of his patron saint 150 years earlier. This same story has been repeated more recently in connection with the work of a young immunologist, E. J. Steele. According to Steele (1981), mutations can occur in the genetic material of somatic cells. In the case of the immunological system, at least, these mutated cells can come to predominate in an organism by a sort of natural selection. Finally, RNA retroviruses pick up some of these mutated genes, transport them to the gonads, and insert them in the genetic material of the germ cells.

Steele's proposal did not sound as outlandish as it would have ten years ago. In the past decade or so, molecular biologists have discovered that the genome is not the static, crystalline structure we once thought it was. Instead it is a veritable cauldron of activity, sections of DNA being snipped out and the remaining sections pieced back together, other genes jumping from place to place in the genome, even changes in RNA being read back into DNA (Schopf 1981, Lewin 1983). In the last instance, all that would be needed for Lamarckian inheritance would be changes in protein being read back either into RNA or else directly into the DNA itself. Although Steele suggested that his system might be more widely applicable, at least in the case of antibody formation, everyone had long known that the immunological system does some very strange things. If Lamarckian inheritance were to occur anywhere, the immunological system would be the right place. Finally, and most importantly, Gorczynski and Steele (1980) claimed that they had actually succeeded in transmitting immunological tolerance in mice from one generation to another. In the interim, several workers have succeeded in duplicating Gorczynski and Steele's results, though their proposed mechanism remains in doubt (Müllbacher, Ashman and Blandin 1983).

In reaction to the resistance that greeted his starling claims, Steele (1981 :7) once again tells the story of how Lamarck, the French philosopher and naturalist who "originated some 170 years ago" the "idea that environmentally induced characters can be inherited," was persecuted so unfairly by his contemporaries. Once again one is led to suspect some connection between Steele's own experience and his dismay at Cuvier's "character assassination" of Lamarck. Steele (1981 :11) also retells the Kammerer story as an example of the "social vindictiveness of the scientific community." However, character assassination and vindictiveness aside, Steele (1981 :11) thinks that "modern science's conservative position is the correct one."

Unlike Steele, early Lamarckians had not set out mechanisms to account for Lamarckian inheritance in sufficient detail.

Dawkins (1982:168) has pointed out in the meantime that even if the sequence of events postulated by Steele does occur, it is not an instance of Lamarckian inheritance. Certainly changes in the somatic cells are being transmitted to the germ cells, but these changes are in the *genetic material* of the somatic cells and their adaptive character is the result of a selection process, not environmental induction. Steele's proposed mechanism no more challenges the Darwinian orthodoxy than does vegetative reproduction.

If nothing else, the recent flap over Lamarckian inheritance shows that Lamarck is far from dead (Lewin 1981). The inheritance of acquired characters is one of those ideas that holds out eternal fascination. It seems so right. If only inheritance were Lamarckian, evolution would be rapid, orderly, and efficient. From Lamarck to the present, scientists have been so concerned with demonstrating the existence (or more usually nonexistence) of Lamarckian forms of inheritance that they have all but ignored an even more fundamental question. If inheriting acquired characters would be so adaptive, why has a mechanism not evolved to permit it to occur and, once evolved, rapidly become prevalent? One would think that in the six billion years or so that life has existed here on Earth, a Lamarckian form of inheritance would have cropped up at least once (Bateson 1963).

The God of the Galapagos

There is something profoundly disturbing about Darwinian evolution. It is cruel, wasteful, and opportunistic. It has no direction. No correlation exists between the mutations that an organism might need and those it might get. The only way that organisms can adapt to significant environmental change is through massive kill off. If a few organisms in a species already possess a mutation that they turn out to need, these organisms will contribute to the next generation and their species will survive. Otherwise, the species becomes extinct. As one might expect, extinction is the norm in Darwinian evolution. The God of the Galapagos is certainly not the Protestant God of waste not, want not.

Currently evolutionary theory is undergoing a period of fundamental upheaval. None of the basic postulates of evolutionary theory is remaining unexamined. If ever there were a period in which a Darwinian orthodoxy existed, it is not now. For example, certain evolutionists are arguing that evolution is not as gradual as Darwin thought it was (Eldredge and Gould 1972). After a period of intense disagreement, Darwinians now acknowledge that many of the changes that have accumulated at the molecular level may well be neutral with respect to selection (Stebbins and Ayala 1981). But the most profound puzzle about evolution currently exercising evolutionary biologists concerns the prevalence of sexual forms of reproduction. In the gene selectionist view of the evolutionary process, sexual reproduction has a 50 per cent cost at every locus at which a male and female differ. Several benefits have been suggested for sexual reproduction. The most common is increased genetic diversity, and that diversity in turn serves several functions. Since organisms cannot anticipate future changes in the environment, genetic diversity increases the likelihood that at least a few of the individuals in a species will turn out to have the mutations that the species needs. However, thus far, none of the explanations suggested for the prevalence of sexual reproduction totally satisfy even their proponents (Williams 1975 ; Maynard Smith 1978 ; Stanley 1979). As a result, Williams (1975 :103) has been driven to conclude that in higher animals "sexuality is a maladaptive feature, dating from a piscine or even protochordate ancestor, for which they lack the preadaptations for ridding themselves."

The situation with respect to Lamarckian inheritance is just the mirror image of that for sexual reproduction. Sexual reproduction would seem to be so maladaptive as to be rare, but it is common. Lamarckian inheritance would seem to be so adaptive as to be prevalent, but it is at best rare. One possible solution is the converse of Williams' counsel of despair. Evolution is extremely capricious. Just because some trait would be extremely adaptive were it to evolve, it does not follow that it will evolve. The absence of Lamarckian forms of inheritance is just one more unhappy accident of evolution. Another alternative is that Lamarckian inheritance would not be as advantageous as it first appears.

Several respects in which the inheritance of acquired characteristics would be maladative are readily apparent. As things now stand, organisms can adapt to relatively minor changes in their environment, but they cannot pass on these adaptations to their offspring. One function of the plasticity of organisms is to serve as a buffer between the genome and the environment.

Organisms change so that genomes do not have to. As a result, genomes can change only relatively slowly, only over the course of successive generations. As mentioned earlier, the solution to this problem is genetic diversity. If inheritance were Lamarckian, organisms could circumvent the need for genetic diversity by responding rapidly to environmental changes. The problem then becomes distinguishing short-term or haphazard changes in the environment from those that will turn out to be important. During recent geological periods, the earth has undergone cyclical changes in climate. Because these changes have occurred very slowly relative to the generation times of organisms, either system of heredity is likely to work. However, changes also take place during intermediate time spans. Lamarckian inheritance would be useful in adapting to these changes. However, the vast majority of environmental changes to which an organism is exposed are very short-term. For example, daily temperature extremes can vary as much as 50° F. throughout the temperate zones of the earth. Just when the genome of an organism has been modified to accommodate midday temperatures, night falls, and so does the ambient temperature. Thompson (1977 :125) has recently suggested yet another explanation for sex—to slow down selection, "to buffer populations against too rapid adaptation to shifting environments." The inheritance of acquired characters would be adaptive only if genes were prescient.

A second problem with the inheritance of acquired characters is that it would neutralize the effect of Mayr's founder principle (1963, 1982). Over the years Mayr has argued that speciation typically occurs when a small population becomes isolated from the body of its species. Because of the small numbers involved, there is no way that these isolated populations can possibly reflect the wide spectrum of genetic heterogeneity of the species at large. Because of the amount of inbreeding necessitated by isolation, such small populations become further differentiated from the species that gave rise to them. In the vast majority of cases, such isolates go extinct. Every once in a while, however, they serve as founder populations for new species. One thing that the inheritance of acquired characters would do is to decrease the genetic heterogeneity of a species. The only variations present in a species would be those that mirrored environmental variation. Hence, isolating small populations would be much less likely to result in speciation.

Finally, according to Williams (1966) and other gene selectionists, the chief goal of genes is to replicate themselves as perfectly as possible. A certain amount of change is an inevitable biochemical consequence of replication (Bernstein 1977), but if the gene selectionists are right, mutation rates should be very low. The existence of proofreading exonucleases to correct mistakes should come as no surprise, but the evolution of a genetic mechanism designed expressly to promote change would be highly problematic. The reason that cells in multicellular organisms can cooperate to the extent that they do is that they are genetically identical to each other—or nearly so. Any system by which mutated somatic cells would be preferentially converted to germ cells would threaten the very existence of multicellular organization.

Theorizing of the preceding sort is far from conclusive, but enough has been said to cast doubt on our intuition about how clearly adaptive the inheritance of acquired characters would be were it to evolve. One reason that Lamarckian inheritance is so rare or nonexistent may well be that it is strongly maladaptive. Lamarckian systems of inheritance may have been eliminated as fast as they evolved. I now turn to another rare phenomenon—sociality. Two sorts of sociality exist in the animal kingdom—cooperation among closely related kin, most of whom are neuters (eusociality), and among sexual organisms that are at most only distantly related (true sociality). The former is quite rare ; the latter even rarer. Why is sociality so rare (Richerson and Boyd 1984) ? The question should sound familiar. The existence of sociality is of interest for our present purposes because nearly everyone who addresses the topic remarks that sociocultural evolution, unlike biological evolution, is Lamarckian. If Lamarckian inheritance is so maladaptive in biological evolution, how is it possible for sociocultural evolution to be Lamarckian ?

The Lamarckian Character of Sociocultural Evolution

Prior to Darwin, several authors had suggested that languages evolve. Darwin (1859 :422) himself uses the evolution of languages to illustrate his own theory of biological evolution. The extension of Darwin's theory to include societies was inevitable. If species and languages evolve, why not societies ? In fact, Spencer was more interested in sociocultural evolution than in biological evolution (Smith 1982). Spencer, of course, thought that both biological and sociocultural evolution were Lamarckian. At present most biologists accept the fundamentally Darwinian character of biological inheritance. Today sociocultural evolution is thought to differ from biological evolution in three respects:

1. the character of the adaptations
2. the mode of transmission
3. the method of selection

All organisms exhibit both biological and physical characteristics. Frogs not only are alive but also have a determinate mass. However, many organisms also exhibit behavioral traits. They flee predators, dance to attract mates, and misinform their conspecifics. Of these organisms, some also exist in social groups which exhibit certain organizational characteristics. For example, some societies are patrilineal, some matrilineal. In one sense, a theory of sociocultural evolution concerns the origin, transmission, and proliferation of traits that characterize social relations and social groups. According to the Darwinians, strictly biological traits are always transmitted genetically. By this they do not mean that the environment is irrelevant, but only that the structure of highly complex biological traits is provided by the order of bases in the genetic material and not by the environment. Traditionally, scientists have recognized a continuum between inherited and learned behavior. All birds that can sing must have the "genes for singing." If they lack the genetic potential for song, they will not sing. At one extreme, certain species of bird will sing a quite complex song peculiar to their species even though they themselves have never heard it. At the other extreme, some species of bird must be taught to sing by their conspecifics. Within limits, they sing whatever song they first hear.

Sociobiologists have made themselves extremely unpopular among social scientists by their emphasis on the genetic component of behavior and their attempt whenever possible to attribute the structure exhibited by a behavioral or social trait to the genetic component. However, no one is claiming that the sociocultural traits that are transmitted genetically are in any sense Lamarckian. Male dominance in certain species may or may not be "programmed into the genes," but to the extent that it is, it is not Lamarckian. Claims of Lamarckian inheritance are made *only* for those characteristics that are *not* transmitted genetically. The only sociocultural traits for which Lamarckian claims are made are those transmitted by social learning.

The generalized ability to learn plane geometry is present in the genetic make-up of nearly all human beings. This ability is turned into actuality year after year in geometry classes around the world. The question then becomes, to what extent is social learning Lamarckian? Because genes are not involved in sociocultural transmission, it can be Lamarckian in only a metaphorical sense.

One goal of a research program commonly termed evolutionary epistemology is to develop a theory of sociocultural evolution modeled on the biological theory (Campbell 1974). In these theories, analogs must be found for genes, genetic transmission, the production of the phenome by the genome, etc. (Kary 1982). Although several terms have been coined for the analogs to genes, "memes" seems to be catching on (Dawkins 1976). Memes are transmitted in a variety of ways, including speech, the written word, recordings, and so forth, but the primary way seems to be by example. We seem to learn more from watching others behave than from simple individual learning (Richerson and Boyd 1984).

The interface between the genome and the phenome is more problematic. It is also crucial for the question at issue because the difference between Darwinian (hard) inheritance and Lamarckian (soft) inheritance depends on the relation between the genome and the phenome. At the biochemical level, this difference is captured in the distinction between transcription and translation. In transcription the information encoded in the DNA is transmitted largely intact. Whether DNA is producing DNA or RNA, the resulting correlation in structure is one-one. In the production of proteans (translation) considerable information is lost. In the past, those authors who have claimed that sociocultural evolution is obviously Lamarckian have, I think, run together the literal and metaphorical models. As Williams (1981:257) has remarked, cultural inheritance "may be considered Lamarckian only in the special sense that inherited features are not coded in the genes, rather than in the usual sense that phenotypic features are being transcribed onto the germ plan."

Perhaps a metaphor borrowed from Dawkins (1982) might help. The information necessary to make a particular sort of cake is contained in the recipe for that cake. If a cook follows a recipe carefully, variations in the resulting cake are usually minimal. However, given a particular cake, indefinitely many slight variations in the original recipe could have occurred without modifying the appearance of the cake. Cooks can pass on recipes. This is the analog to ordinary genetic inheritance. Cooks can also give one another cakes. This is not an analog to Lamarckian inheritance because the recipes remain unaffected. Perhaps organ transplants might be an example of such phenotypic transmission.

Now, in what sense is social learning Lamarckian? Quite obviously, each of us who understands Euclidean geometry has acquired this knowledge after we were born. We learned it from others. If this knowledge counted as a phenotypic character, then acquired characters can be transmitted. Unfortunately, memes (or ideas) are the analogs of genes, not characters. Social learning is an example of the inheritance of acquired memes and *not* an example of the inheritance of acquired characters.

Learning from experience is a better candidate for Lamarckian inheritance in sociocultural evolution. While baking a cake, a cook may make a mistake and use sour cream instead of sweet milk. If the cook likes the taste of the resulting cake and can discover what mistake caused the change in taste, he or she might alter the recipe accordingly. On this interpretation, applications of our ideas count as conceptual phenotypes. When we learn from experience, conflicts between our ideas and their applications cause us to change our memes. If such applications count as part of our conceptual phenotype, then sociocultural evolution is in this sense Lamarckian. It should be noted that in his most recent publication, Dawkins (1982) has argued that sociocultural products should be counted part of an organism's extended phenotype.

I have my doubts as to whether terming the ability to learn from experience and pass on that knowledge to others a form of Lamarckian inheritance is all that informative. It seems too open to easy misconstruction. Bears evolving into whales, indeed! However, this is not the characteristic of inquiry that most people have in mind when they term sociocultural evolution Lamarckian. Instead they seem to be pointing to its intentional character (Toulmin 1972; Cohen 1973; Losee 1977). The factors responsible for the generation of conceptual variants can also function in their selection. People in general and scientists in particular are problem solvers. They think up new ideas *in order to* solve problems. Sociocultural evolution is not a matter of chance variation and natural selection but of purposive variation and rational selection (Toulmin 1972; Rescher 1977; Steele 1981). For this reason, Richerson and Boyd (1984) call sociocultural evolution a matter of "directed variation."

Of course, genetic variation is not "chance" in an indeterministic sense. All mutations are caused by some physical process or other. In addition the structure of an organism's genome strongly constrains the mutations that are possible. The viability of the resulting organism adds further constraints. The only contingency that the term is designed to preclude is that an organism might tend to get those mutations it is going to need in the future. Genes are not clairvoyant. Occasional claims to the contrary, neither are people. To the extent that we understand natural processes, we can predict the future, but that is all. When evolutionary epistemologists like Campbell (1977) claim that sociocultural evolution is a matter of "blind" variation, they are concerned only to deny any human ability to foresee the future. Even the most talented scientist is not prescient, especially at the frontiers of knowledge. Although there is much more to how we learn from experience than simple trial and error, at the cutting edge of science the process of discovery approaches this extreme (Hull 1982).

It is certainly true that when all else fails, people are capable of acting rationally. They can understand the problems that confront them and go about solving them in a rational way. Although the failure rates in sociocultural evolution are probably of the same magnitude as the extinction rates in biological evolution, people would no doubt be much less successful if they did not try. To some extent at least, people succeed in solving problems because they strive to. However, terming sociocultural evolution "Lamarckian" on this account is returning once again to the most caricatured interpretation of Lamarck, as if giraffes have succeeded in increasing the length of their necks by striving to reach the leaves at the top of trees. Once again, Lamarck has proven himself useful to scientists in their ongoing efforts to get their views accepted and the views of their opponents rejected. I have no doubt that he will continue to fulfill this same function in the future. If nothing else, reading Lamarck's *Discours d'ouverture* (1800) [Introductory lecture for 1800] and *Philosophie zoologique* (1809) will allow the reader to see exactly how different Lamarck's views are from the wide variety of opinions which have been designated through the years as "Lamarckian."

A note of appreciation is owed to John Beatty, Ernst Mayr, and William Wimsatt for commenting on an earlier draft of this paper.

References

Bateson, G. 1963. The role of somatic change in evolution. *Evol.* 17:529–39.

Bernstein, H. 1977. Germ line recombination may be primarily a manifestation of DNA repair processes. *J. Theor. Biol.* 69:371–80.

Bowler, P. J. 1983. *The Eclipse of Darwinism.* Baltimore and London: The Johns Hopkins University Press.

Burkhardt, R. W., Jr. 1970. Lamarck, evolution, and the politics of science. *J. Hist. Biol.* 3:275–98.

———. 1972. The inspiration of Lamarck's belief in evolution. *J. Hist. Biol.* 5:413–38.

———. 1977. *The spirit of system: Lamarck and evolutionary biology.* Cambridge: Harvard Unviersity Press.

———. 1980. Lamarckism in Britain and the United States. In *The evolutionary synthesis,* ed. E. Mayr and W. B. Provine, 343–52. Cambridge: Harvard University Press.

Burlingame, L. J. 1981. Lamarck's chemistry: The chemical revolution rejected. In *The analytic spirit,* ed. H. Woolf, 64–81. Ithaca: Cornell University Press.

Campbell, D. 1974. Evolutionary epistemology. In *The philosophy of Karl R. Popper,* ed. P. A. Schillp, 1:413–63. LaSalle, IL: Open Court Publishing Company.

———. 1977. The natural selection model of conceptual evolution. *Phil. Sci.* 44:502–7.

Cannon, H. G. 1957. What Lamarck really said. *Proc. Linn. Soc. Lond.* 168:70–85.

———. 1958. *The evolution of living things.* Manchester: Manchester University Press.

———. 1959. *Lamarck and modern genetics.* Springfield: IL: Charles C. Thomas Publishers.

Carozzi, A. V. 1964. Lamarck's theory of the earth: Hydrogéoloque. *Isis.* 55:304–5.

Chambers, R. 1844. *Vestiges of the natural history of creation.* London: John Churchill Publishers.

Churchill, F. R. 1968. August Weismann and a break from tradition. *J. Hist. Biol.* 1:91–112.

———. 1978. The Weismann-Spencer controversy over the inheritance of acquired characters. *Proc. 15th Int. Congr. Hist. Sci.,* 451–68.

Cohen, L. J. 1973. Is the progress of science evolutionary? *Brit. J. Phil. Sci.* 24 :41 – 61.

Corsi, P. 1978. The importance of French transformist ideas for the second volume of Lyell's principles of geology. *Brit. J. Hist. Sci.* 11 :221 – 44.

Cuvier, G. 1835. Éloge de M. Lamarck. *Mémoires de l'Académie Royale des Sciences de l'Institut de France.* 13 :i– xxxi.

———. 1836. Biographical memoir of M. de Lamarck. *Edin. New Phil. J.* 20 :1 – 22.

Darwin, C. 1859. *On the origin of species.* London : Murray.

———. 1868. *The variation of animals and plants under domestication.* London : Murray.

———. 1891. *On the origin of species* 6th ed. London : Murray.

Darwin, F. 1899. *The life and letters of Charles Darwin.* New York : D. Appleton & Company.

Dawkins, R. 1976. *The selfish gene.* New York : Oxford University Press.

———. 1982. *The extended phenotype.* San Francisco : Freeman.

Egerton, F. N. 1976. Darwin's early reading of Lamarck. *Isis.* 67 :452 – 56.

Eldredge, N., and S. J. Gould. 1972. Punctuated equilibria : An alternative to phyletic gradualism. In *Models in paleobiology,* ed. T. J. M. Schopf, 82 – 115. San Francisco : Freeman.

Galton, F. 1871. Experiments in pangenesis, by breeding from rabbits of a pure variety, into whose circulation blood taken from other varieties had previously been largely transfused. *Proc. Roy. Soc. Lond.* 19 :393 – 410.

Gillispie, C. C. 1956. The formation of Lamarck's evolutionary theory. *Arch. int. hist. sci.* 9 :323 – 38.

———. 1959. Lamarck and Darwin in the history of science. In *Forerunners of Darwin,* ed. B. Glass, O. Temkin, and W. L. Straus, Jr., 265 – 91. Baltimore : Johns Hopkins Press.

Gorczynski, R. M. and E. J. Steele. 1980. Inheritance of acquired immunological tolerance to foreign histocompatibility antigens in mice. *Proc. Natl. Acad. Sci.* 77 :2871 – 75.

Gould, S. J. 1972. Zealous advocate. *Science.* 176 :623 – 25.

Hodge, M. J. S. 1971a. Lamarck's science of living bodies. *Brit. J. Hist. Sci.* 5 :323 – 52.

———. 1971b. Species in Lamarck. In *Colloque international "Lamarck,"* ed. J. Schiller, 31 – 46. Paris : Blanchard.

Hull, D. L. 1973. *Darwin and his critics : The reception of Darwin's theory of evolution by the scientific community.* Cambridge : Harvard University Press. Reprint, 1983. Chicago : University of Chicago Press.

———. 1982. The naked meme. In *Learning, development, and culture,* ed. H. C. Plotkin, 273 – 327. New York : John Wiley & Sons.

Huxley, L. 1901. *Life and letters of Thomas Henry Huxley.* New York : D. Appleton & Company.

Huxley, T. H. 1854. Review of Vestiges of the natural history of creation, 10th ed., 1853. *Brit. For. Medico-Chir. Rev.* 19 :425 – 39.

———. 1863. Six lectures to working men. *Nat. Hist. Rev.;* reprinted *On the origin of species or the causes of the phenomena of organic nature* (1968), Ann Arbor : University of Michigan Press.

Kammerer, P. 1924. *The inheritance of acquired characteristics.* New York : Boni & Liveright.

Kary, C. 1982. Can Darwinian inheritance be extended from biology to epistemology. In *PSA 1982,* ed. P. D. Asquith and T. Nickles, 356 – 69.

Koestler, A. 1971. *The case of the midwife toad.* New York : Random House.

Lamarck, J. – B. M. De. 1778. *Flore françoise,* Paris : Visse.

———. 1809. *Philosophie zoologique.* Paris : Dentu.

Lewin, R. 1981. Lamarck will not lie down. *Science.* 213 : 316 – 21.

———. 1983. How mammalian RNA returns to its genome. *Science.* 219 :1052 – 54.

Losee, J. 1977. Limitations of an evolutionary philosophy of science. *Stud. Hist. Sci.* 8 :349 – 52.

Lyell, C. 1830 – 33. *Principles of Geology.* London : Murray.

MacBride, E. W. 1895. Sedgwick's theory of the embryonic phase of ontogeny as an aid to phylogenetic theory. *Quart. J. Microsc. Sci.* 37 :325 – 42.

———. 1937. Mendel, Morgan, and genetics. *Nature.* 140 :348 – 50.

Maynard Smith, J. 1978. *The evolution of sex.* New York : Cambridge University Press.

Mayr, E. 1963. *Animal species and evolution.* Cambridge : Harvard University Press.

———. 1972. Lamarck revisited. *J. Hist. Biol.* 5 :55 – 94.

———. 1982. *The growth of biological thought.* Cambridge : Harvard University Press.

Mayr, E., and W. Provine, eds. 1980. *The evolutionary synthesis.* Cambridge : Harvard University Press.

Mivart, St. G. J. 1871. *On the genesis of species.* London : Macmillan.

Müllbacher, A., R. B. Ashman, and R. V. Blandin. 1983. Induction of T cell hyporesponsiveness to bebaru in mice, an abnormality in the immune responses of progeny of hyporesponsive males. *Austr. J. Exp. Biol. Med. Sci.*

Packard, A. S. 1901. *Lamarck : the founder of evolution.* London : Longmans.

Rescher, N. 1977. *Methodological pragmatism.* Oxford : Blackwell.

Richards, R. 1982. The emergence of evolutionary biology of behavior in the early nineteenth century. *Brit. J. Hist. Sci.* 15 :241 – 80.

Richerson, P. S., and R. Boyd. 1984. *Cultural transmission : Mechanisms and origins.* Forthcoming.

Ridley, M. 1982. Coadaptation and the inadequacy of natural selection. *Brit. J. Hist. Sci.* 15 :45 – 68.

Romanes, G. J. 1892. *Darwin, and after Darwin.* London : Longmans.

———. 1893. *An examination of Weismanism.* London : Longmans.

Schopf, T. J. M. 1981. Evidence from findings of molecular biology with regard to the rapidity of genomic change : Implications for species durations. In *Paleobotany, paleoecology and evolution,* ed. K. J. Niklas, 135 – 192. New York : Praeger.

Simpson, G. G. 1961. Lamarck, Darwin and Butler : Three approaches to evolution. *Am. Scholar.* 30 :238 – 49.

Smith, C. U. M. 1982. Evolution and the problems of mind. *J. Hist. Biol.* 15 :55 – 88.

Spencer, H. 1852. The development hypothesis. *The Saturday Analyst and Leader.* (March 20, 1852) :280 – 81 ; reprinted in *Essays, scientific, political and speculative* (1891), 1 :377 – 83.

———. 1893a. The inadequacy of natural selection. *Cont. Rev.* 63 :152 – 66 ; 439 – 56.

———. 1893b. A rejoinder to Professor Weismann. *Cont. Rev.* 64 :893 – 912.

Stanley, S. M. 1979. *Macroevolution, pattern and process.* San Francisco : Freeman.

Stebbins, G. L., and F. J. Ayala. 1981. Is a new evolutionary synthesis necessary? *Science.* 213 :967 – 71.

Steele, E. J. 1981. *Somatic selection and adaptive evolution.* 2d ed. Chicago : University of Chicago Press.

Thompson, V. 1977. Recombination and response to selection in Drosophila melanogaster. *Genetics.* 85 :125 – 40.

Toulmin, S. 1972. *Human understanding.* Princeton : Princeton University Press.

Wallace, A. R. 1870. *Contributions to the theory of natural selection.* London : Macmillan.

———. 1889. *Darwinism.* London : Macmillan.

Weismann, A. 1892. *Das Keimplasma : Eine Theorie der Vererbung.* Jena : Gustav Fischer.

———. 1893. *The germ-plasm : A theory of heredity.* Trans. W. N. Parker and H. Ronnfeldt. London : Walter Scott.

———. 1894. *The effect of external influences upon development.* London : H. Frowde.

Williams, G. C. 1966. *Adaptation and natural selection.* Princeton : Princeton University Press.

———. 1975. *Sex and evolution.* Princeton : Princeton University Press.

———. 1981. A defense of monolithic sociobiology and genetic mysticism. *Behav. Brain Sci.* 4 :257.

PREFACE.

EXPERIENCE in teaching has made me feel how useful a philosophical zoology would be at the present time. By this I mean a body of rules and principles, relative to the study of animals, and applicable even to the other divisions of the natural sciences; for our knowledge of zoological facts has made considerable progress during the last thirty years.

I have in consequence endeavoured to sketch such a philosophy for use in my lessons, and to help me in teaching my pupils; nor had I any other aim in view. But in order to fix the principles and establish rules for guidance in study, I found myself compelled to consider the organisation of the various known animals, to pay attention to the singular differences which it presents in those of each family, each order, and especially each class; to compare the faculties which these animals derive according to its degree of complexity in each race, and finally to investigate the most general phenomena presented in the principal cases. I was therefore led to embark upon successive inquiries of the greatest interest to science, and to examine the most difficult of zoological questions.

How, indeed, could I understand that singular degradation which is found in the organisation of animals as we pass along the series of them from the most perfect to the most imperfect, without en quiring as to the bearings of so positive and so remarkable a fact, founded upon the most convincing proofs? How could I avoid the conclusion that nature had successively produced the different bodies endowed with life, from the simplest worm upwards? For in ascending the animal scale, starting from the most imperfect animals, organisation gradually increases in complexity in an extremely remarkable manner.

I was greatly strengthened in this belief, moreover, when I recognised that in the simplest of all organisations there were no special organs whatever, and that the body had no special faculty but only those which are the property of all living things. As nature successively creates the different special organs, and thus builds up the animal organisation, special functions arise to a corresponding degree, and in the most perfect animals these are numerous and highly developed.

These reflections, which I was bound to take into consideration, led me further to enquire as to what life really consists of, and what are the conditions necessary for the production of this natural phenomenon and its power of dwelling in a body. I made the less resistance to the temptation to enter upon this research, in that I was then convinced that it was only in the simplest of all organisations that the solution of this apparently difficult problem was to be found. For it is only the simplest organisation that presents all the conditions necessary to the existence of life and nothing else beyond, which might mislead the enquirer.

The conditions necessary to the existence of life are all present in the lowest organisations, and they are here also reduced to their simplest expression. It became therefore of importance to know how this organisation, by some sort of change, had succeeded in giving rise to others less simple, and indeed to the gradually increasing complexity observed throughout the animal scale. By means of the two following principles, to which observation had led me, I believed I perceived the solution of the problem at issue.

Firstly, a number of known facts proves that the continued use of any organ leads to its development, strengthens it and even enlarges it, while permanent disuse of any organ is injurious to its development, causes it to deteriorate and ultimately disappear if the disuse continues for a long period through successive generations. Hence we may infer that when some change in the environment leads to a change of habit in some race of animals, the organs that are less used die away little by little, while those which are more used develop better, and acquire a vigour and size proportional to their use.

Secondly, when reflecting upon the power of the movement of the fluids in the very supple parts which contain them, I soon became convinced that, according as this movement is accelerated, the fluids modify the cellular tissue in which they move, open passages in them, form various canals, and finally create different organs, according to the state of the organisation in which they are placed.

Arguing from these two principles, I looked upon it as certain that, firstly, the movement of the fluids within animals—a movement which is progressively accelerated with the increasing complexity of the organisation—and, secondly, the influence of the environment, in so far as animals are exposed to it in spreading throughout all habitable places, were the two general causes which have brought the various animals to the state in which we now see them.

I have not merely confined myself in the present work to setting forth the conditions essential to the existence of life in the simplest organisations, and the causes which have given rise to the growing complexity of animal organisation from the most imperfect to the most perfect of animals; but, believing that there is some possibility of recognising the physical causes of feeling, which is possessed by so many animals, I have not hesitated to take up this question also.

I was indeed convinced that matter can never possess in itself the property of feeling; and I imagined that feeling itself is only a phenomenon resulting from the workings of an orderly system capable of producing it. I enquired therefore what the organic mechanism might be which could give rise to this wonderful phenomenon, and I believe I have discovered it.

On marshalling together the best observations on this subject, I recognised that for the production of feeling the nervous system must be highly complex, though not so highly as for the phenomena of intelligence.

Following out these observations, I have become convinced that the nervous system, when it is in the extremely imperfect condition characteristic of more or less primitive animals, is only adapted to the excitation of muscular movements, and that it cannot at this stage produce feeling. In this particular stage it consists merely of ganglia, from which issue threads. It does not present any ganglionic longitudinal cord, nor any spinal cord, the anterior extremity of which expands into a brain which contains the nucleus of sensations and gives origin to the nerves of the special senses, or at least to some of them. When the nervous system reaches this stage, the animals possessing it then have the faculty of feeling.

Finally, I endeavoured to determine the mechanism by which a sensation was achieved; and I have shown that nothing more than a perception can be produced in an individual which has no special organs, and moreover, that a sensation produces nothing more than a perception whenever it is not specially remarked.

I am in truth undecided as to whether sensation is achieved by a transmission of the nervous fluid starting from the point affected, or merely by a communication of movement in that fluid. The fact, however, that the duration of certain sensations is dependent upon that of the impressions which cause them, make me lean towards the latter opinion. My observations would not have thrown any satisfactory light upon the subjects treated, if I had not recognised and been able to prove that feeling and irritability are two very different organic phenomena. They have by no means a common origin, as has been supposed; the former of these phenomena constitutes a faculty peculiar to certain animals, and demanding a special system of organs, while the latter, which does not require any special system, is exclusively the property of all animal organisation.

So long therefore as these two phenomena continue to be confused as to their origin and results, it will be only too easy to make mistakes in proffering explanations of the causes of the general phenomena of animal organisation. It will be so especially in making experiments for the purpose of investigating the principle of feeling and of movement, and finally the seat of that principle in the animals which possess these faculties.

For instance, if we decapitate certain very young animals, or cut the spinal cord between the occiput and the first vertebra, or push in a probe, there occur various movements excited by the pumping of air into the lungs. These have been taken as proof of the revival of feeling by dint of artificial respiration; whereas these effects are due partly to the irritability not being extinct, for it is known that it continues to exist sometime after the death of the individual, and partly to certain muscular movements which can still be excited by the inhalation of air when the spinal cord has not been altogether destroyed by the introduction of a long probe right down its canal.

I recognised that the organic act which gives rise to the movement of the parts is altogether independent of that which produces feeling, although in both cases nervous influence is necessary. I notice that I can work several of my muscles without experiencing any sensation, and that I can receive a sensation without any movement resulting from it. But for these observations, I too might have taken the movements occurring in a young decapitated animal, or in one whose brain had been removed, as signs of feeling, and I should have fallen into error.

I think that if the individual is disabled by its nature or otherwise from giving an account of a sensation which it experiences, and that if it only indicates by cries the pain which it is made to undergo, we have no certain sign for inferring that it receives sensation except from knowing that the system of organs which gives it the faculty of feeling is not destroyed, but retains its integrity. Muscular movements excited from without cannot in themselves prove an act of feeling.

Having fixed my ideas on these interesting objects, I gave attention to the *inner feeling*, that is to say, that feeling of existence which is possessed only by animals which enjoy the faculty of feeling. I brought to bear on the problem such known facts as are relevant, in addition to my own observations, and I soon became convinced that this inner feeling constituted a power which it was essential to take into consideration.

Nothing in fact seems to me so important as the feeling which I have named, considered both in man and in the animals which possess a nervous system capable of producing it. It is a feeling which can be aroused by physical and moral needs, and which becomes the source whence movements and actions derive their means of execution. No one that I know had paid any attention to it; and this gap in our knowledge of one of the most powerful causes of the principal phenomena of animal organisation rendered all explanations inadequate to account for these phenomena. We have, however, a sort of clue to the existence of that inner power when we speak of the agitations which we ourselves are constantly experiencing; for the word *emotion*, which I did not create, is often enough pronounced in conversation to express the observed facts.

When I had considered that the inner feeling was susceptible of being aroused by different causes, and that it then constituted a power capable of exciting actions, I was so to speak struck by the multitude of known facts which attest the actual existence of that power; the difficulties which had long puzzled me with regard to the exciting cause of actions appeared to me entirely surmounted.

Admitting that I had been fortunate enough to alight upon a truth in attributing to the inner feeling of animals which have it the power which produces their movements, I had still only surmounted a part of the difficulties by which this research is hampered. For it is obvious that not all known animals do or can possess a nervous system; consequently, all animals do not possess the inner feeling of which I am speaking; and in the case of those which are destitute of it, the movements which they are seen to execute must have another origin.

I had reached this point when I reflected that without internal excitations plant life would not exist at all, nor be able to maintain itself in activity. I recognised the fact that the same consideration applied to a large number of animals; and as I had very frequently observed that nature varies her means when necessary in order to attain the same end, I had no further doubt about the matter.

I think therefore that the very imperfect animals which have no nervous system live only by the help of excitations which they receive from the exterior. That is to say, subtle and ever moving fluids contained in the environment incessantly penetrate these organised bodies and maintain life in them, so long as the state of these bodies permits of it. Now this thought is one which I have many times considered, which many facts appear to me to confirm, against which none of those that are known to me seem to conflict, and finally which appears to me obviously borne out by plant life. It was therefore for me a flood of light which disclosed to me the principal cause which maintains movements and the life of organised bodies, and to which animals owe all that animates them.

I combined this consideration with the two preceding ones, namely, that which concerns the result of the movement of fluids in the interior of animals and that which deals with the effects of a change that is maintained in the environment and habits of these beings. I could thus seize the thread which connects the numerous causes of the phenomena presented in animal organisation, and I soon perceived the importance of this power in nature which preserves in new individuals all the changes in organisation acquired by their ancestors as a result of their life and environment.

Now I remarked that the movements of animals are never directly communicated, but that they are always excited; hence I recognised that nature, although obliged at first to borrow from the environment the *excitatory power* for vital movements and the actions of imperfect animals, was able by a further elaboration of the animal organisation to convey that power right into the interior of these beings, and that finally she reached the point of placing that same power at the disposal of the individual.

Such are the principal conclusions which I have endeavoured to establish and develop in this work.

This *Zoological Philosophy* thus sets forth the results of my studies on animals, their characters both general and special, their organisation, the causes of their development and diversity, and the faculties which they thence derive. In its composition I have made use of the bulk of the material which I was collecting for a projected work on living bodies under the title of *Biology*. This work will now remain, so far as I am concerned, unwritten.

The facts which I name are very numerous and definite, and the inferences which I have drawn from them appeared to me sound and necessary; I am convinced therefore that it will be found difficult to replace them by any others.

The number of new theories expounded in the present work are likely to give the reader an unfavourable impression, if only from the fact that the commonly received beliefs do not readily give way to any new ones which tend to contradict them. Now, since the predominance of old ideas over new favour this prejudice, especially when there is some contributory personal interest, it follows that, whatever difficulties there may be in the discovery of new truths in nature, there are still greater difficulties in getting them recognised.

But these difficulties, arising from various causes, are on the whole more advantageous than otherwise to the general progress of knowledge. By means of a rigorous hostility to the admission of new ideas as truths, a multitude of more or less specious but unfounded ideas which appear, soon after fall into oblivion. Sometimes, on the other hand, excellent opinions and solid thoughts are for the same reasons discarded or neglected; but it is better that a truth once perceived should have a long struggle before obtaining the attention it deserves, than that all that is produced by the ardent imagination of man should be too readily received.

The more I meditate on this subject, and particularly on the numerous causes which may bring about a change in our opinions, the more am I convinced, that except for the physical and moral facts[1] that no one can question, all else is but opinion or argument; and we well know that arguments can always be met by others. Thus, although it is obvious that there are great differences in the probability and even the value of the opinions of different men, it seems to me that we should be wrong to blame those who refuse to adopt our own.

Should we recognise as well founded only those opinions that are most widely accepted? Experience shows clearly enough that persons with the most developed intellect and the highest wisdom constitute at all times an extremely small minority. The fact can scarcely be questioned. Authorities in the sphere of knowledge should weigh one another's worth and not count one another's numbers, although indeed a true estimation is very difficult.

Seeing how numerous and rigorous are the conditions required for forming a sound judgment, it is still uncertain whether the judgment of individuals who have been set up as authorities by public opinion is perfectly sound on the topics on which they pronounce.

There are then few positive truths on which mankind can firmly rely. They include the facts which he can observe, and not the inferences that he draws from them; they include the existence of nature, which presents him with these facts, as also the laws which regulate the movements and changes of its parts. Beyond that all is uncertain, although some conclusions, theories, opinions, etc., have much greater probability than others.

We cannot rely on any argument, inference or theory, since the authors of these intellectual acts can never be certain that they have taken into account the true data, nor that they have admitted these

[1] By *moral* facts I mean mathematical truths; that is to say, the results of calculations whether of quantities or forces, and the results of measurements; since it is through intelligence and not through the senses that these facts become known to us. Now these *moral* facts are just as much positive truths as are those relating to the existence of bodies that we can observe.

only. There is nothing that we can be positive about, except the existence of bodies which affect our senses, and of the real qualities which belong to them, and finally the physical and moral facts of which we are able to acquire a knowledge. The thoughts, arguments and explanations set forth in the present work should therefore be looked upon merely as opinions which I propose, with the intention of setting forth what appears to me to be true, and what may indeed actually be true.

However this may be, in giving myself up to the observations from which my theories have arisen, I have obtained the pleasure which their resemblance to truth has brought me, and I have obtained also the recompense for the fatigues entailed upon me by my studies and meditations. In publishing these observations, together with the conclusions that I have drawn from them, my purpose is to invite enlightened men who love the study of nature to follow them out, verify them, and draw from them on their side whatever conclusions they think justified.

This path appears to me the only one that can lead to a knowledge of truth or of what comes nearest it, and it is clear that such knowledge is more profitable to us than the error which might fill its place. I cannot doubt therefore that it is this path which we must follow.

It may be noticed that I have dwelt with special pleasure on the exposition of the second and especially of the third part of this work, and that I have been greatly interested in them. None the less, the principles bearing on natural history which I have studied in the first part should be looked upon as possibly the most useful to science, since they are in general most in harmony with the opinions hitherto received.

I might have considerably extended this work by developing under each heading all the interesting matter that it permits of; but I have preferred to confine myself to such exposition as is strictly necessary for the adequate comprehension of my observations. I have thus spared my readers' time without exposing them to the risk of failing to understand me.

I shall have attained my end if those who love natural science find in this work any views and principles that are useful to them; if the observations which I have set forth, and which are my own, are confirmed or approved by those who have had occasion to study the same objects; and if the ideas which they succeed in giving rise to, whatever they may be, advance our knowledge or set us on the way to reach unknown truths.

PRELIMINARY DISCOURSE.

To observe nature, to study her productions in their general and special relationships, and finally to endeavour to grasp the order which she everywhere introduces, as well as her progress, her laws, and the infinitely varied means which she uses to give effect to that order: these are in my opinion the methods of acquiring the only positive knowledge that is open to us,—the only knowledge moreover which can be really useful to us. It is at the same time a means to the most delightful pleasures, and eminently suitable to indemnify us for the inevitable pains of life.

And in the observation of nature what can be more interesting than the study of animals? There is the question of the affinities of their organisation with that of man, there is the question of the power possessed by their habits, modes of life, climates and places of habitation, to modify their organs, functions and characters. There is the examination of the different systems of organisation which are to be observed among them, and which guide us in the determination of the greater or lesser relationships that fix the place of each in the scheme of nature. There is finally the general classification that we make of these animals from considerations of the greater or lesser complexity of their organisation; and this classification may even lead us to a knowledge of the order followed by nature in bringing the various species into existence.

Assuredly however, it cannot be disputed that all these enquiries, and others also to which the study of animals necessarily leads, are of very great interest to anyone who loves nature and seeks the truth in all things.

It is a peculiar circumstance that the most important phenomena for us to consider have only been available since the time when attention was devoted to the study of the least perfect animals, and since the researches on the various complications in the organisation of these animals became the main object of study.

It is no less curious that the most important discoveries of the laws, methods and progress of nature have nearly always sprung from the examination of the smallest objects which she contains, and from apparently the most insignificant enquiries. This truth, already established by many remarkable facts, will receive in the course of this work a new accession of evidence, and should convince us more than ever that in the study of nature no object whatever can be disregarded.

The purpose of the study of animals is not merely to ascertain their different races, nor to determine all the distinctions among them by specifying their special characters. This study further aims at acquiring a knowledge of the functions which animals possess, the causes of the presence and maintenance of life in them, and of the remarkable progression which they exhibit in the complexity of their organisation, as well as in the number and development of their functions.

At bottom, the *physical* and *moral* are without doubt one and the same thing. It is by a study of the organisation of the different orders of known animals that this truth can be set in the strongest light. Now since these products from a common origin, at first hardly separated, become eventually divided into two entirely distinct orders, these two orders when examined at their greatest divergence have seemed to us and still seem to many persons to have nothing in common.

The influence of the physical on the moral has however already been recognised,[1] but it seems to me that sufficient attention has not yet been given to the influence of the moral on the physical. Now these two orders of things which have a common origin re-act upon one another, especially when they appear the most widely separated; and we are now in a position to prove that each affects the variations of the other.

It seems to me that we have gone the wrong way to work in the endeavour to show the common origin of the two orders of results which, in their highest divergence, constitute what is called the *physical* and the *moral*.

For the study of these two kinds of objects, apparently so distinct, has been initiated in man himself. Now his organisation, having reached the limit of complexity and perfection, exhibits the greatest complication in the causes of the phenomena of life, feeling and function. It is consequently the most difficult from which to infer the origin of so many phenomena.

After the organisation of man had been so well studied, as was the case, it was a mistake to examine that organisation for the purposes of

[1] See the interesting work of M. Cabanis entitled *Rapport du physique et du moral de l'homme*.

an enquiry into the causes of life, of physical and moral sensitiveness, and, in short, of the lofty functions which he possesses. It was first necessary to try to acquire knowledge of the organisation of the other animals. It was necessary to consider the differences which exist among them in this respect, as well as the relationships which are found between their special functions and the organisation with which they are endowed.

These different objects should have been compared with one another and with what is known of man. An examination should have been made of the progression which is disclosed in the complexity of organisation from the simplest animal up to man, where it is the most complex and perfect. The progression should also have been noted in the successive acquisition of the different special organs, and consequently of as many new functions as of new organs obtained. It might then have been perceived how *needs*, at first absent and afterwards gradually increasing in number, have brought about an inclination towards the actions appropriate to their satisfaction ; how actions becoming habitual and energetic have occasioned the development of the organs which execute them ; how the force which stimulates organic movements can in the most imperfect animals exist outside of them and yet animate them ; how that force has been subsequently transported and fixed in the animal itself ; and, finally, how it has become the source of sensibility, and last of all of acts of intelligence.

I may add that if this method had been followed, *feeling* would certainly not have been looked upon as the general and immediate cause of organic movements. It would never have been said that life is a consequence of movements executed by virtue of sensations received by various organs or otherwise ; nor that all vital movements are brought about by impressions received by sensitive parts (*Rapport du physique et du moral de l'homme*, pp. 38 to 39, and 85).

This cause would appear to be justified up to a certain point in the most perfect animals, but if it held good with regard to all bodies which enjoy life, they would all possess the faculty of feeling. Now it could hardly be shown that this is the case in plants ; it could hardly even be proved that it is the case in all known animals.

The supposition of such a general cause does not seem to me justified by the real methods of nature. When constituting life, she had no power to endow with that faculty the imperfect animals of the earlier classes of the animal kingdom.

With regard to living bodies, it is no longer possible to doubt that nature has done everything little by little and successively.

Hence, among the various subjects which I intend to discuss in the present work, I shall endeavour to make clear by the citation of recognised facts that nature, while ever increasing the complexity of animal organisation, has created in order the different special organs, as also the functions which the animals possess.

The belief has long been held that there exists a sort of scale or graduated chain among living bodies. Bonnet has developed this view; but he did not prove it by facts derived from their organisation; yet this was necessary especially with regard to animals. He was unable to prove it, since at the time when he lived the means did not exist.

In the study of all classes of animals there are many other things to be seen besides the animal complexity. Among the subjects of greatest importance in framing a rational philosophy are the effect of the environment in the creation of new needs; the effect of the needs in giving rise to actions, and of repeated actions in creating habits and inclinations; the results of increased or diminished use of any organ, and the means adopted by nature to maintain and to perfect all that has been acquired in organisation.

But this study of animals, especially of the least perfect animals, was long neglected; since no suspicion existed of the great interest which they exhibit. Moreover, what has been started in this respect is still so new that we may anticipate much more light from its further development.

When the study of natural history was actually begun, and naturalists inquired into both kingdoms, those who devoted their researches to the animal kingdom studied chiefly the vertebrate animals, that is to say *mammals*, *birds*, *reptiles* and, lastly, *fishes*. In these classes of animals the species are in general larger, and have their parts and functions better developed and more easily ascertainable than the species of invertebrate animals. Their study, therefore, seemed to present more of interest.

In fact the majority of invertebrate animals are extremely small, their functions are limited, and their organs much more remote from those of man than is the case of the more perfect animals. As a result, they have been to some extent despised by the vulgar, and down to our own time have only realised a very moderate amount of interest on the part of most naturalists.

We are beginning, however, to get over a prejudice so harmful to the progress of knowledge. During the few years that these singular animals have been closely examined, we have been compelled to recognise that the study of them is highly interesting to the naturalist and philosopher, because it sheds light, that could scarcely be otherwise obtained, on a number of problems in natural history and animal physics. It has been my duty in the Natural History Museum to attend to the exhibit of the animals which I called *invertebrate*, on account of the absence in them of a vertebral column. My researches on these numerous animals, the accumulated observations and facts, and finally the increased knowledge of comparative anatomy which I gained from them, soon inspired me with the highest interest in the subject.

The study of *invertebrate animals* must, in fact, be of special interest to the naturalist for four reasons :—(1) The number of the species of these animals in nature is much greater than that of vertebrate animals. (2) Since they are more numerous, they are necessarily more varied. (3) The variations in their organisation are much greater, more sharply defined and more remarkable. (4) The order observed by nature in the successive formation of the different organs of animals is much better expressed in the mutations which these organs undergo in invertebrate animals. Moreover, their study is more fertile in helping us to understand the origin of organisation, with its complexity and its developments, than could possibly be the case in more perfect animals such as vertebrates.

Convinced of these truths I felt that, in the instruction of my pupils, I should not plunge into detail straight away, but should above all show them the general principles which hold good of all animals. I tried to give them a view of the whole and of the essentials which appertained to it, with the intention of taking subsequent note of the main groups into which that whole appears to be divided for purposes of comparison and more intimate knowledge.

The real way, no doubt, of acquiring a thorough knowledge of an object, even in its smallest details, is to begin by inspecting it in its entirety. We should examine first its bulk, extent, and the various parts which compose it. We should enquire into its nature and origin, and its connection with other known objects. In short, we should enquire into the general principles involved, from all possible points of view. The subject is then divided into its chief parts for separate study and examination in all the bearings likely to be instructive. By further dividing and sub-dividing these parts, and inspecting each successively, we arrive at the smallest, where we do not neglect the least details. Once these researches finished, the effects have to be deduced from them, so that little by little the philosophy of science is established, modified and perfected.

It is by this method alone that human intelligence can gain knowledge (in any science) that is at once vast, solid and coherent. It is solely by this kind of analysis that science makes real progress, so that allied objects are never confused, but can be perfectly known.

Unfortunately this method is not sufficiently used in the study of natural history. The recognised necessity for close observation of special objects has produced a habit of not going beyond these objects with their smallest details. They have thus become for most naturalists the chief subjects of study. This would, however, not really be a drawback for natural science, were it not for the steady refusal to see in the observed objects anything besides their form, dimensions, external parts, colour, etc., but those who give themselves up to such a study are contemptuous of the higher ideals, such as the enquiry into the nature of the objects which occupy them, into the causes of the modifications or variations which these objects undergo, and into the relations of these same objects with each other and with all other known objects, etc., etc.

It is because the method which I have just named is insufficiently followed out that we find so much divergence in what is taught on this subject, both in works on natural history and elsewhere. Those who have gone in exclusively for the study of species find it very difficult to grasp the general affinities among objects; they do not in the least appreciate nature's true plan, and they perceive hardly any of her laws.

I am convinced that it is wrong to follow a method which so greatly limits ideas. I find myself on the other hand obliged to bring out a new edition of my *Système des animaux sans vertèbres*, since the rapid progress of comparative anatomy and the new discoveries of zoologists, together with my own observations, enable me to improve that work. I have accordingly collected into a special work, under the title of *Zoological Philosophy*, (1) the general principles at stake in the study of the animal kingdom; (2) the observed facts which require to be considered in that study; (3) the principles which regulate the most suitable *classification* of animals, and an arrangement of them in their natural order; (4) lastly, the most important of the results which flow naturally from the accumulated observations and facts, and which constitute the true foundation of the *philosophy* of science.

The *Zoological Philosophy* is nothing but a new edition, re-cast, corrected and much enlarged, of my work entitled *Recherches sur les corps vivants*. It is divided into three main divisions, and each of these divisions is broken up into separate chapters.

Thus, in the first division, which sets forth the essential observed facts and the general principles of the natural sciences, I shall begin by a discussion of what I call *artificial devices* used among the sciences in question. I shall deal with the importance of the consideration of affinities, and with the notion that should be conveyed, when we speak of *species* among living bodies. Afterwards, when I have treated of the general principles which concern animals, I shall adduce proof of the *degradation* of organisation which runs through the entire animal scale, placing the most perfect animals at the anterior extremity of that scale. On the other hand I shall show the influence of environment and habit on the organs of animals, as being the factors which favour or arrest their development. I shall conclude this division by a discussion of the *natural order* of animals, and by an account of their most suitable arrangement and classification.

In the second division I shall put forward my ideas as to the order and state of things which constitute the essence of animal life; and I shall indicate the conditions necessary for the existence of this wonderful natural phenomenon. Afterwards, I shall endeavour to ascertain the exciting cause of organic movements; of orgasm and of irritability; the properties of cellular tissue; the sole condition under which *spontaneous generation* can occur; the obvious effects of vital actions, etc.

Lastly, the third division will state my opinion as to the physical causes of feeling, of the power to act, and of the acts of the intelligence found in certain animals.

In this division I shall treat: 1st, the origin and formation of the nervous system; 2nd, the nervous fluid, which can only be known indirectly, but whose existence is attested by phenomena that it alone can produce; 3rd, physical sensibility and the mechanism of sensations; 4th, the reproductive power of animals; 5th, the origin of the will and the faculty of willing; 6th, ideas and the different kinds of them; 7th, lastly, certain peculiar acts of the understanding, such as attention, thoughts, imagination, memory, etc.

The reflections set forth in the 2nd and 3rd divisions doubtless comprise subjects that are very difficult to examine, and may even appear insoluble; but they are so full of interest that such attempts may possibly be profitable, either in the disclosure of unperceived truths or in pointing out the direction in which they may be sought.

PART I.

CONSIDERATIONS ON THE NATURAL HISTORY OF ANIMALS, THEIR CHARACTERS, AFFINITIES, ORGANISATION, CLASSIFICATION AND SPECIES.

CHAPTER I.

ON ARTIFICIAL DEVICES IN DEALING WITH THE PRODUCTIONS OF NATURE.

THROUGHOUT nature, wherever man strives to acquire knowledge he finds himself under the necessity of using special methods, 1st, to bring order among the infinitely numerous and varied objects which he has before him; 2nd, to distinguish, without danger of confusion, among this immense multitude of objects, either groups of those in which he is interested, or particular individuals among them; 3rd, to pass on to his fellows all that he has learnt, seen and thought on the subject. Now the methods which he uses for this purpose are what I call the *artificial devices* in natural science,—devices which we must beware of confusing with the laws and acts of nature herself.

It is not merely necessary to distinguish in natural science what belongs to artifice and what to nature. We have to distinguish as well two very different interests which incite us to the acquisition of knowledge.

The first is an interest which I call *economic*, because it derives its impetus from the economic and utilitarian needs of man in dealing with the productions of nature which he wants to turn to his own use. From this point of view he is only interested in what he thinks may be useful to him.

The other, very different from the first, is that *philosophic* interest through which we desire to know nature for her own sake, in order to grasp her procedure, her laws and operations, and to gain an idea of what she actually brings into existence. This, in short, is the kind of knowledge which constitutes the true naturalist. Those who approach the subject from this point of view are naturally few; they are interested impartially in all natural productions that they can observe.

To begin with, economic and utilitarian requirements resulted in the successive invention of the various *artificial devices* employed in natural science. When the interest of studying and knowing nature was first felt, these artificial devices continued to be of assistance in the prosecution of that study. These same artificial devices have therefore an indispensable utility, not only for helping us to a knowledge of special objects, but for facilitating study and the progress of natural science, and for enabling us to find our way about among the enormous quantity of different objects that we have to deal with.

Now the *philosophic interest* embodied by the sciences in question, although less widespread than that which relates to our economic requirements, compels us to separate what belongs to artifice from what is the sphere of nature. We have to confine within reasonable limits the consideration due to the first set of objects, and attach to the second all the importance that they deserve.

The artificial devices in natural science are as follows:

(1) Schematic classifications, both general and special.
(2) Classes.
(3) Orders.
(4) Families.
(5) Genera.
(6) The nomenclature of various groups of individual objects.

These six kinds of devices, commonly used in natural science, are purely artificial aids which we have to use in the arrangement and division of the various observed natural productions; to put us in the way of studying, comparing, recognising and citing them. Nature has made nothing of this kind: and instead of deceiving ourselves into confusing our works with hers, we should recognise that classes, orders, families, genera and nomenclatures are weapons of our own invention. We could not do without them, but we must use them with discretion and determine them in accordance with settled principles, in order to avoid arbitrary changes which destroy all the advantages they bestow.

It was no doubt indispensable to break up the productions of nature into groups, and to establish different kinds of divisions among them, such as classes, orders, families and genera. It was, moreover, necessary to fix what are called *species*, and to assign special names to these various sorts of objects. This is required on account of the limitations of our faculties; some such means are necessary for helping us to fix the knowledge which we gain from that prodigious multitude of natural bodies which we can observe in their infinite diversity.

But these groupings, of which several have been so happily drawn up by naturalists, are altogether artificial, as also are the divisions and sub-divisions which they present. Let me repeat that nothing of the kind is to be found in nature, notwithstanding the justification which they appear to derive from certain apparently isolated portions of the natural series with which we are acquainted. We may, therefore, rest assured that among her productions nature has not really formed either classes, orders, families, genera or constant species, but only individuals who succeed one another and resemble those from which they sprung. Now these individuals belong to infinitely diversified races, which blend together every variety of form and degree of organisation; and this is maintained by each without variation, so long as no cause of change acts upon them.

Let us proceed to a few brief observations with respect to each of the six artificial devices employed in natural science.

Schematic classifications.—By schematic classifications, general or special, I mean any series of animals or plants that is drawn up unconformably to nature, that is to say, which does not represent either her entire order or some portion of it. It is consequently not based on a consideration of ascertained affinities.

The belief is now thoroughly justified that an order established by nature exists among her productions in each kingdom of living bodies: this is the order on which each of these bodies was originally formed.

This same order is individual and essentially without divisions in each organic kingdom. It becomes known to us through the affinities, special and general, existing among the different objects of which these two kingdoms consist. The living bodies at the two extremities of that order have essentially the fewest affinities, and exhibit the greatest possible differences in their organisation and structure.

It is this same order, as we come to know it, that will have to replace those schematic or artificial classifications that we have been forced to create in order to arrange conveniently the different natural bodies that we have observed.

With regard to the various organised bodies recognised by observation, there was at first no other thought beyond convenience and ease of distinction between these objects; and it has taken the longer to seek out the actual order of nature in their classification, inasmuch as there was not even a suspicion of the existence of such an order.

Hence arose groupings of every kind, artificial systems and methods, based upon considerations of such an arbitrary character that they underwent almost as many changes in their principles and nature as there were authors to work upon them.

With regard to plants, the *sexual system* of Linnæus, ingenious as it is, presents a general schematic classification: and, with regard to insects, the *entomology* of Fabricius presents a special schematic classification. All the progress made in recent times by the philosophy of natural science has been necessary, in France at least, to carry the conviction that the natural method should be studied. Our classifications should conform to the exact order found in nature, for that order is the only one which remains stable, independent of arbitrary opinion, and worthy of the attention of the naturalist.

Among plants, the natural method is extremely difficult to establish, on account of the obscurity prevailing in the character of the internal organisation of these living bodies, and of the differences presented by plants of different families. Since the learned observations of M. Antoine-Laurent de Jussieu, however, a great step has been made in botany in the direction of the natural method ; many families have been constituted with direct reference to their affinities ; but the general position of all these families among themselves, and consequently of the whole order, remains to be determined. The fact is that we have found the beginning of that order ; but the middle, and especially the end, are still at the mercy of arbitrary opinion.

The case is different with regard to animals ; their organisation is much more pronounced, and presents different systems that are easier to grasp. The work has, therefore, in their case made greater progress ; as a result, the actual order of nature in the animal kingdom is now sketched out in its main outlines in a stable and satisfactory manner. It is only the boundaries of classes and their orders, of families and genera, that are still abandoned to arbitrary opinion.

If schematic classifications are still found among animals, these classifications are only minor, since they deal with objects belonging to one class. Thus, the hitherto received classifications of fishes and birds are still schematic classifications.

With regard to living bodies, the farther one descends from the general to the particular the less constant become the characters serving to determine affinities, and the more difficult to recognise is the actual order of nature.

Classes.—The name *class* is given to the highest kind of general divisions that are established in a kingdom. The further divisions of these receive other names : we shall speak of them directly. The more complete is our knowledge of the affinities between the objects composing a kingdom, the better and more natural are the classes established as the primary divisions of that kingdom, so long as attention has been paid to recognised affinities in forming them. Nevertheless, the boundaries of these classes, even the best of them, are clearly artificial ; they will therefore continue to undergo arbitrary variations at the hands of authors so long as naturalists will not agree and submit themselves to certain general principles on the subject.

Thus, even though the order of nature in a kingdom should be thoroughly known, the classes which we are obliged to establish in it will always be fundamentally artificial divisions.

It is true, especially in the animal kingdom, that several of these divisions appear to be really marked out by nature herself; and it is certainly difficult to believe that mammals, birds, etc., are not sharply isolated classes formed by nature. This is none the less a pure illusion, and a consequence of the limitation of our knowledge of existing or past animals. The further we extend our observations the more proofs do we acquire that the boundaries of the classes, even apparently most isolated, are not unlikely to be effaced by our new discoveries. Already the *Ornithorhyncus* and the *Echidna* seem to indicate the existence of animals intermediate between birds and mammals. How greatly natural science would profit if the vast region of Australia and many others were better known to us!

If classes are the first kind of division that can be established in a kingdom, it follows that the divisions which can be established among the objects of one class cannot themselves be classes; for it is obviously inappropriate to set up class within class; that, however, is just what has been done: Brisson, in his *Ornithologie*, has divided the class of birds into various special classes.

Just as nature is everywhere governed by laws, so too artifice should be subjected to rules. If there are none, or if they are not followed, its products will be vacillating and its purpose fail.

Some modern naturalists have introduced the custom of dividing a class into several sub-classes, while others again have carried out the idea even with genera; so that they make up not only sub-classes but sub-genera as well. We shall soon reach not only sub-classes but sub-orders, sub-families, sub-genera and sub-species. Now this is a thoughtless misuse of artifice, for it destroys the hierarchy and simplicity of the divisions, which had been set up by Linnæus and generally adopted.

The diversity of the objects belonging to a class either of animals or plants is sometimes so great as to necessitate the formation of many divisions and sub-divisions among the objects of that class; but it is to the interest of science that artificial devices should always have the greatest possible simplicity. Now that interest allows, no doubt, of any divisions and sub-divisions that may be necessary; but it is opposed to each division having a special denomination. A stop must be put to the abuses of nomenclature; otherwise the nomenclature would become more difficult to understand than the objects themselves.

Orders.—The name order should be given to the main divisions of the first rank into which a class is broken up. If these divisions leave scope for the formation of others by further sub-division, these sub-divisions are no longer orders ; and it would be very inappropriate to give them the name.

The class of molluscs, for example, are easily divided into two large main groups, one having a head, eyes, etc., and reproducing by copulation, while the other has no head, eyes, etc., and carry out no copulation to reproduce themselves. *Cephalic* and *acephalic* molluscs should be regarded as the two orders of that class ; meanwhile, each of these orders can be broken up into several remarkable groups. Now this fact is no sufficient reason for giving the name order or even sub-order to each of the groups concerned. These groups, therefore, into which orders are divided should be regarded as sections or as large families, themselves susceptible of still further subdivisions.

Let us maintain in our artifical devices the great simplicity and beautiful hierarchy established by Linnæus. If we are under the necessity to make many sub-divisions of orders, that is to say, of the principle divisions of a class, by all means let us make as many as may be necessary, but do not let us assign to them any special denomination.

The orders into which a class is divided should be determined by the presence of important characters extending throughout the objects comprised in each order ; but no special name should be assigned to them that is applicable to the objects themselves.

The same thing applies with regard to the sections that we have to form among the orders of one class.

Families.—The name family is given to recognised parts of the order of nature in either of the two kingdoms of living bodies. These parts of the natural order are, on the one hand, smaller than classes and even than orders, but, on the other hand, they are larger than genera. But however natural families may be and however well constituted their genera are according to their true affinities, the boundaries of these families are always artificial. The more indeed that the productions of nature are studied, and new ones observed, the greater the continual variations in the boundaries of families that are made by naturalists. Some divide one family into several new ones, others combine several families into one, while others again make additions to a family already known, increase it, and thus thrust back the boundaries which had been assigned to it.

If all the races (so-called *species*) belonging to a kingdom of living bodies were thoroughly known, as well as their true affinities, so that the sorting out of these races and their allocation in various groups were in conformity with their natural affinities, the classes, orders, sections and genera would be families of different sizes, for all these divisions would be larger or smaller parts of the natural order.

On such an assumption, nothing doubtless would be more difficult than to assign the boundaries between these different divisions; arbitrary opinion would produce incessant variation, and there would be no agreement except where gaps in the series made clear demarcations.

Fortunately for the practicability of the artifice which we have to introduce into our classifications, there are many races of animals and plants that are still unknown to us, and will probably remain so, since insuperable obstacles are placed in our way by the places where they live and other circumstances. The gaps thence arising in the series, whether of animals or plants, will leave us for a long time still, and perhaps for ever, the means of setting up the majority of the divisions.

Custom and indeed necessity require that a special name should be given to each family and to each genus so as to be applicable to the objects it contains. It follows that alterations in the boundaries, extent and determination of families will always cause a change in their nomenclature.

Genera.—The name of genus is given to combinations of races or so-called species that have been united on account of their affinities, and constitute a number of small series marked out by characters arbitrarily selected for the purpose.

When a genus is well made, all the races or species comprised in it resemble one another in their most essential and numerous characters. They differ only among themselves in characters less important, but sufficient to distinguish them.

Well made genera are thus really small families, that is to say, real parts of the actual order of nature.

Now we have seen that the series to which we give the name of family are liable to vary as to their boundaries and extent, according to the opinions of authors who arbitrarily change their guiding principles. In the same way the boundaries of genera are exposed to infinite variation because different authors change at will the characters employed to determine them. Now a special name has to be assigned to each genus; and every change in the constitution of a genus involves nearly always a change of name. It is difficult therefore to exaggerate the injury done to natural science by perpetual alterations of genera, which multiply synonymy, overburden nomenclature, and make the study of these sciences difficult and disagreeable.

When will naturalists agree to abide by general principles for uniform guidance in the constitution of genera, etc., etc.? The natural affinities, which they recognise among the objects which they have brought together, mislead them nearly all into the belief that their genera, families, orders and classes actually exist in nature. They do not notice that the good series which they succeed in forming by study of affinities do in truth exist in nature, for they are large or small parts of her order, but that the lines of demarcation which they are obliged to set up at intervals do not by any means so exist.

Consequently, genera, families, sections of various kinds, orders and even classes are in truth artificial devices, however natural may be the series which constitute these different groups. No doubt they are necessary and have an obvious and indispensable utility; but if the advantages, which these devices bring, are not to be cancelled by constant misuse, the constitution of every group must be in accordance with principles and rules that naturalists once for all have agreed to follow.

Nomenclature.—We come now to the sixth of the artificial devices which have to be employed in natural science. By nomenclature is meant the system of names assigned either to special objects, such as a race or a species, or to groups of these objects, such as a genus, family or class.

Now nomenclature is confined to the names given to species, genera, families and classes. It has therefore to be distinguished from that other artificial device called *technology*, which refers solely to the denominations applied to the parts of natural bodies.

"All the discoveries and observations of naturalists would necessarily have fallen into oblivion and been lost to society, if the objects observed and determined had not each received a name to serve as a recognition mark when speaking of them or quoting them." (*Dict. de Botanique*, art. "Nomenclature.")

It is quite clear that nomenclature in natural history is an artificial device, and is a means that we have to resort to for fixing our ideas in the sphere of natural observed productions, and to enable us to pass on either these ideas or our observations on the objects concerned.

No doubt this artificial device should like the others be controlled by settled rules that are generally adhered to; but I am bound to remark that its universal misuse, of which complaints are so justly made, arises principally from extrinsic causes which daily increase also in the other artificial devices already named.

In fact, lack of settled rules as to the formation of genera, families and even classes, exposes these artificial devices to all the vagaries of arbitrary judgment; nomenclature undergoes a continuous succession of changes. It never can be fixed so long as this lack of rules continues; and synonymy, already immense, will continue to grow and become more and more incapable of repairing a confusion which annihilates all the advantages of science.

This would never have happened if it had been recognised that all the lines of demarcation in the series of objects composing a kingdom of living bodies are really artificial, except those which result from gaps to be filled. But this was not perceived: there was not even a suspicion of it. Almost to the present day naturalists have had no further object in view than that of setting up distinctions. Here is evidence of what I mean:

"In fact, in order to procure and keep for ourselves the services of all natural bodies within our reach, that we can subordinate to our needs, it was felt that an exact and precise determination of the characters of each body was necessary, and consequently that the details of organisation, structure, form, proportion, etc., etc., should be sought out and determined, so that they could for all time be recognised and distinguished from one another. This is what naturalists are now doing up to a certain point.

"This part of the work of naturalists has made the most advance. Immense efforts have rightly been made for about a century and a half to perfect it, because it assists us to a knowledge of what has been newly observed, and serves as reminder of what was previously known. Moreover, it fixes our knowledge with regard to objects whose properties are or will hereafter become of use to us.

"But naturalists attach too much weight to forming lines of demarcation in the general series both of animals and plants; they devote themselves almost exclusively to this kind of work, without considering it under its true aspect or coming to any agreement as to the framing of settled rules in this great enterprise for fixing the principles of determination. Hence the intrusion of many abuses; for each one arbitrarily changes the principles for the formation of classes, orders and genera: and numerous different groupings are incessantly being set before the public. Genera undergo continual variation without limit, and the names given to nature's productions are constantly being changed as a result of this thoughtless proceeding.

"As a result, synonomy in natural history is now terribly widespread. Science every day becomes more obscure; she is surrounded by almost insurmountable difficulties; and the finest effort of man to set up the means of recognising and distinguishing the works of nature is changed into an immense maze, into which most men naturally hesitate to plunge." (*Discours d'ouvert. du cours de* 1806, pp. 5 and 6.)

Here we have a picture of the results of omitting to distinguish what really belongs to artifice from what is in nature, and of not having endeavoured to discover rules for the less arbitrary determination of the divisions which have to be established.

CHAPTER II.

IMPORTANCE OF THE CONSIDERATION OF AFFINITIES.

AMONG living bodies the name affinity has been given to features of analogy or resemblance between two objects, that are compared in their totality, but with special stress on the most essential parts. The closer and more extensive the resemblance, the greater the affinities. They indicate a sort of kinship between the living bodies which exhibit them; and oblige us in our classification to place these bodies in a proximity proportional to their affinities.

How great has been the progress of natural science since serious attention began to be given to affinities, and especially since their true underlying principles have been determined!

Before this change, our botanical classifications were entirely at the mercy of arbitrary opinion, and of artificial systems of any author. In the animal kingdom the invertebrate animals comprising the larger part of all known animals were classified into the most heterogeneous groups, some under the name of insects, some under the name of worms; where the animals included are from the point of view of affinity widely different from one another.

Happily this state of affairs is now changed; and, henceforth, if the study of natural history is continued, its progress is assured.

The principle of natural affinities removes all arbitrariness from our attempts at a methodical classification of organised bodies. We have here the law of nature which should guide us to the natural method. Naturalists are forced to agree as to the rank which they assign, firstly to the main groups of their classification, and afterwards to the individuals of which these groups are composed; finally, they are obliged to follow the actual order observed by nature in giving birth to her productions.

Thus, everything that concerns the affinities of the various animals should be the chief object of our researches, before making any division or classification among them.

The question of affinities does not apply only to species; for we have also to fix the general affinities of all the orders into which groups are united or divided from the comparative point of view.

Affinities, although possessing very different values according to the importance of the parts exhibiting them, can none the less be extended to the conformation of the external parts. If the affinities are so great that not only the essential parts, but also the external parts present no determinable difference, then the objects in question are only individuals of the same species. If on the other hand, notwithstanding a large degree of affinity, the external parts exhibit appreciable differences, though less than the essential resemblances; then the objects in question are different species of the same genus.

The important study of affinities is not limited to a comparison of classes, families, or even of species; it includes also a consideration of the parts of which individuals are composed. By comparison together of corresponding parts we obtain a firm basis for recognising either the identity of individuals of the same race or the difference between individuals of distinct races.

It has, in truth, been noted that the proportions and relations of the parts of all individuals composing a species or a race always remain the same, and so appear to be preserved forever. From this it has been rightly inferred that, by examining detached parts of an individual, one could decide to what species, old or new, these parts belong.

This power is very favourable to the progress of knowledge at the present time. But the conclusions drawn from it can only hold good for a limited period; since the races themselves undergo changes in their parts, in proportion to any considerable change in the circumstances which affect them. As a matter of fact, since these changes only take place with an extreme slowness, which makes them always imperceptible, the proportions and relations of the parts always appear the same to the observer, who does not really see them change. Hence, when he comes across any species which have undergone these changes, he imagines that the differences which he perceives have always existed.

It is none the less quite true that by a comparison of corresponding parts in different individuals, their affinities, nearer or more remote, can be easily and certainly determined. It can therefore be known whether these parts belong to individuals of the same race or of different races.

It is only the general inference that is unsound, having been drawn too hastily. This I shall have more than one opportunity of proving in the course of the present work.

Affinities are always incomplete when they apply only in an isolated case; that is to say, when they are decided from an examination of a single part taken by itself. But, although incomplete, the value of affinities based upon a single part depends upon the extent to which the part from which they are taken is essential, and *vice versa.*

There are then determinable differences among affinities, and various degrees of importance among the parts which display them; in fact, the knowledge of affinities would have had no application or utility unless the more important parts of living bodies had been distinguished from the less important, and unless a principle had been found for estimating the true values of these important parts.

The most important parts for exhibiting the chief affinities are, among animals, the parts essential to the maintenance of life, and among plants, the parts essential to reproduction.

In animals, therefore, it is always the internal organisation that will guide us in deciding the chief affinities. And in plants, it will be in the parts of fructification that affinities will be sought.

But in both cases the parts most important for seeking out affinities vary. The only principle to be used for determining the importance of any part, without arbitrary assumptions, consists in enquiring either how much use nature makes of it, or else the importance to the animal of the function of that part.

Among animals, whose affinities are mainly determined by their internal organisation, three kinds of special organs have rightly been chosen from among the others as the most suitable for disclosing the most important affinities. They are, in order of importance, as follows:

(1) *The organ of feeling.* The nerves which meet at a centre, either single as in animals with a brain, or multiple as in those with a ganglionic longitudinal cord.

(2) *The organ of respiration.* The lungs, gills and tracheae.

(3) *The organ of circulation.* The arteries and veins, which usually have a centre of action in the *heart.*

The first two of these organs are more widely used by nature, and therefore more important than the third, that is to say, the organ of circulation; for the latter disappears in the series after the crustaceans, while the two former extend to animals of the two classes which follow the crustaceans.

Finally, of the two first, the organ of feeling has the more importance from the point of view of affinities, for it has produced the most exalted of animal faculties, and moreover without that organ muscular activity could not take place.

If I were to refer to plants, among which the reproductive parts alone are of importance in deciding affinities, I should set forth these parts in their order of importance as follows:

(1) The embryo, its accessories (cotyledons, perisperm) and the seed which contains it.

(2) The sexual parts of flowers, such as the pistil and stamens.

(3) The envelopes of the sexual parts; the corolla, calyx, etc.

(4) The pericarp, or envelope of the seed.

(5) The reproductive bodies which do not require fertilisation.

These generally received principles give to natural science a coherence and solidity that it did not previously possess. Affinities are no longer at the mercy of changes of opinion; our general classifications become necessary inferences; and according as we perfect them by this method they approach ever more closely to the actual order of nature.

It was, in fact, due to the perception of the importance of affinities that the attempts of the last few years were originated to determine what is called the natural method; a method which is only a tracing by man of nature's procedure in bringing her productions into existence.

No importance is now attached in France to those artificial systems which ignore the natural affinities among objects; for these systems give rise to divisions and classifications harmful to the progress of natural knowledge.

With regard to animals, there is no longer any doubt that it is purely from their organisation that their natural affinities can be determined. It is, in consequence, chiefly from comparative anatomy that zoology will obtain the data for such determination. But we should pay more attention to the facts collected in the works of anatomists than to the inferences which they draw from them; for too often they hold views which might mislead us and prevent us from grasping the laws and true plan of nature. It seems to be the case that whenever man observes any new fact he is always condemned to rush headlong into error in attempting to explain it; so fertile is his imagination in the creation of ideas. He is not sufficiently careful to guide his judgment by the general principles derived from other facts and observations.

When we consider the natural affinities between objects, and make a sound estimate of them, we can combine species on this principle, and associate groups with definite boundaries forming what are called genera. Genera can be similarly associated on the principle of affinities, and united into higher groups forming what are called families. These families, associated in the same way and on the same principle, make up orders. These again are the primary divisions of classes, while classes are the chief divisions of each kingdom.

We must then be guided everywhere by natural affinities in composing the groups which result by dividing each kingdom into classes, each class into orders, each order into sections or families, each family into genera, and each genus into different species if there is occasion for it.

There is thorough justification for the belief that the complete series of beings making up a kingdom represents the actual order of nature, when it is classified with direct reference to affinities ; but, as I have already pointed out, the different kinds of divisions which have to be set up in that series to help us to distinguish objects with greater ease do not belong to nature at all. They are truly artificial although they exhibit natural portions of the actual order instituted by nature.

It should be added that in the animal kingdom, affinities should be decided mainly from a study of organisation. The principles employed for settling these affinities should not admit of the smallest doubt. We shall thus obtain a solid basis for *zoological philosophy*.

It is known that every science must have its philosophy, and that it cannot make real progress in any other way. It is in vain that naturalists fill their time in describing new species, in grasping all the shades and small details of their varieties, in enlarging the immense list of catalogued species, in establishing genera, and in making incessant changes in the principles which they use. If the philosophy of science is neglected her progress will be unreal, and the entire work will remain imperfect.

It is indeed only since the attempt has been made to fix the extent of affinity between the productions of nature that natural science has obtained any coherence in its principles, and a philosophy to make it really a science.

What progress towards perfection is made every day in our classifications since they were founded upon the study of affinities !

It was through the study of affinities that I recognised that infusorian animals could no longer be put in the same class as polyps ; that radiarians also should not be confused with polyps ; and that soft creatures, such as medusae and neighbouring genera, which Linnæus and even Bruguière placed among the molluscs, were essentially allied to the echinoderms, and should form a special class with them.

It was again the study of affinities which convinced me that worms were a separate group, comprising animals very different from radiarians, and still more from polyps ; that arachnids could no longer be classed with insects, and that cirrhipedes were neither annelids nor molluscs.

Finally, it was through the study of affinities that I succeeded in effecting a number of necessary alterations even in the classification of molluscs, and that I recognised that the pteropods, which are closely allied to but distinct from the gastropods, should not be placed between the gastropods and the cephalapods, but between the acephalic molluscs and the gastropods; since these pteropods, like all acephalic animals, have no eyes and are almost without a head, not even excepting *Hyalea*. (*V.* the special classification of molluscs in Chap. VIII., at the end of Part I.)

When the study of affinities among the different known families of plants has made us better acquainted with the rank held by each in the general series, then the classification of these living bodies will leave nothing more to arbitrary judgment, but will come more closely into conformity with the actual order of nature.

The study of the affinities among observed objects is thus clearly so important that it should now be regarded as the chief instrument for the progress of natural science.

CHAPTER III.

OF SPECIES AMONG LIVING BODIES AND THE IDEA THAT WE SHOULD ATTACH TO THAT WORD.

IT is not a futile purpose to decide definitely what we mean by the so-called *species* among living bodies, and to enquire if it is true that species are of absolute constancy, as old as nature, and have all existed from the beginning just as we see them to-day; or if, as a result of changes in their environment, albeit extremely slow, they have not in course of time changed their characters and shape.

The solution of this question is of importance not only for our knowledge of zoology and botany, but also for the history of the world.

I shall show in one of the following chapters that every species has derived from the action of the environment in which it has long been placed the *habits* which we find in it. These habits have themselves influenced the parts of every individual in the species, to the extent of modifying those parts and bringing them into relation with the acquired habits. Let us first see what is meant by the name of species.

Any collection of like individuals which were produced by others similar to themselves is called a species.

This definition is exact; for every individual possessing life always resembles very closely those from which it sprang; but to this definition is added the allegation that the individuals composing a species never vary in their specific characters, and consequently that species have an absolute constancy in nature.

It is just this allegation that I propose to attack, since clear proofs drawn from observation show that it is ill-founded.

The almost universally received belief is that living bodies constitute species distinguished from one another by unchangeable characteristics, and that the existence of these species is as old as nature herself. This belief became established at a time when no sufficient observations had been taken, and when natural science was still almost negligible. It is continually being discredited for those who have seen much, who have long watched nature, and who have consulted with profit the rich collections of our museums.

Moreover, all those who are much occupied with the study of natural history, know that naturalists now find it extremely difficult to decide what objects should be regarded as species.

They are in fact not aware that species have really only a constancy relative to the duration of the conditions in which are placed the individuals composing it; nor that some of these individuals have varied, and constitute races which shade gradually into some other neighbouring species. Hence, naturalists come to arbitrary decisions about individuals observed in various countries and diverse conditions, sometimes calling them varieties and sometimes species. The work connected with the determination of species therefore becomes daily more defective, that is to say, more complicated and confused.

It has indeed long been observed that collections of individuals exist which resemble one another in their organisation and in the sum total of their parts, and which have kept in the same condition from generation to generation, ever since they have been known. So much so that there seemed a justification for regarding any collection of like individuals as constituting so many invariable species. Now attention was not paid to the fact that the individuals of the species perpetuate themselves without variation only so long as the conditions of their existence do not vary in essential particulars. Since existing prejudices harmonise well with these successive regenerations of like individuals, it has been imagined that every species is invariable and as old as nature, and that it was specially created by the Supreme Author of all existing things.

Doubtless, nothing exists but by the will of the Sublime Author of all things, but can we set rules for him in the execution of his will, or fix the routine for him to observe? Could not his infinite power create an *order of things* which gave existence successively to all that we see as well as to all that exists but that we do not see?

Assuredly, whatever his will may have been, the immensity of his power is always the same, and in whatever manner that supreme will may have asserted itself, nothing can diminish its grandeur.

I shall then respect the decrees of that infinite wisdom and confine myself to the sphere of a pure observer of nature. If I succeed in unravelling anything in her methods, I shall say without fear of error that it has pleased the Author of nature to endow her with that faculty and power.

The idea formed of species among living bodies was quite simple, easy to understand, and seemed confirmed by the constancy in the shapes of individuals, perpetuated by reproduction or generation. Such are a great number of these alleged species that we see every day.

Meanwhile, the farther we advance in our knowledge of the various organised bodies which cover almost every part of the earth's surface, the greater becomes our difficulty in determining what should be regarded as a species, and still more in finding the boundaries and distinctions of genera.

According as the productions of nature are collected and our museums grow richer, we see nearly all the gaps filled up and the lines of demarcation effaced. We find ourselves reduced to an arbitrary decision which sometimes leads us to take the smallest differences of varieties and erect them into what we call species, and sometimes leads us to describe as a variety of some species slightly differing individuals which others regard as constituting a separate species.

Let me repeat that the richer our collections grow, the more proofs do we find that everything is more or less merged into everything else, that noticeable differences disappear, and that nature usually leaves us nothing but minute, nay puerile, details on which to found our distinctions.

How many genera there are both among animals and plants, among which the number of species referred to them is so great that the study and determination of these species are well nigh impracticable! The species of these genera, arranged in series according to their natural affinities, exhibit such slight differences from those next them as to coalesce with them. These species merge more or less into one another, so that there is no means of stating the small differences that distinguish them.

It is only those who have long and diligently studied the question of species, and who have examined rich collections, that are in a position to know to what extent species among living bodies merge into one another. And no one else can know that species only appear to be isolated, because others are lacking which are close to them but have not yet been collected.

I do not mean that existing animals form a very simple series, regularly graded throughout; but I do mean that they form a branching series, irregularly graded and free from discontinuity, or at least once free from it. For it is alleged that there is now occasional discontinuity, owing to some species having been lost. It follows that the species terminating each branch of the general series are connected on one side at least with other neighbouring species which merge into them. This I am now able to prove by means of well-known facts.

I require no hypothesis or supposition; I call all observing naturalists to witness.

Not only many genera but entire orders, and sometimes even classes, furnish instances of almost complete portions of the series which I have just indicated.

When in these cases the species have been arranged in series, and are all properly placed according to their natural affinities, if you choose one, and then, jumping over several others, take another a little way off, these two species when compared will exhibit great differences. It is thus in the first instance that we began to see such of nature's productions as lay nearest to us. Generic and specific distinctions were then quite easy to establish ; but now that our collections are very rich, if you follow the above-mentioned series from the first species chosen to the second, which is very different from it, you reach it by slow gradations without having observed any noticeable distinctions.

I ask, where is the experienced zoologist or botanist who is not convinced of the truth of what I state ?

How great the difficulty now is of studying and satisfactorily deciding on species among that multitude of every kind of polyps, radiarians, worms, and especially insects, such as butterflies, *Phalaena*, *Noctua*, *Tinea*, flies, *Ichneumon*, *Curculio*, *Cerambix*, chafers, rose-chafers, etc. ! These genera alone possess so many species which merge indefinably into one another.

What a swarm of mollusc shells are furnished by every country and every sea, eluding our means of distinction and draining our resources.

Consider again, fishes, reptiles, birds and even mammals ; you will see that except for gaps still to be filled, neighbouring species and even genera are separated by the finest differences, so that we have scarcely any foothold for setting up sound distinctions.

Is there not an exactly similar state of affairs in the case of botany, which deals with the other series, consisting of plants ?

How great indeed are the difficulties of the study and determination of species in the genera *Lichen*, *Fucus*, *Carex*, *Poa*, *Piper*, *Euphorbia*, *Erica*, *Hieracium*, *Solanum*, *Geranium*, *Mimosa*, etc., etc.

When these genera were constituted only a small number of species belonging to them were known, and it was then easy to distinguish them ; but now that nearly all the gaps are filled, our specific differences are necessarily minute and usually inadequate.

Let us see what are the causes which have given rise to this undoubted state of affairs ; let us see if nature affords any explanation, and whether observation can help us.

We learn from a number of facts that, according as the individuals of one of our species change their abode, climate, habits, or manner of life, they become subject to influences which little by little alter the consistency and proportions of their parts, their shape, properties and even their organisation; so that in course of time everything in them shares in these mutations.

In the same climate, very different habitats and conditions at first merely cause variations in the individuals exposed to them; but in course of time the continued change of habitat in the individuals of which I speak, living and reproducing in these new conditions, induces alterations in them which become more or less essential to their being; thus, after a long succession of generations these individuals, originally belonging to one species, become at length transformed into a new species distinct from the first.

Suppose, for example, that the seeds of a grass or any other plant that grows normally in a damp meadow, are somehow conveyed first to the slope of a neighbouring hill where the ground although higher is still rich enough to allow the plant to maintain its existence. Suppose that then, after living there and reproducing itself many times, it reaches little by little the dry and almost barren ground of a mountain side. If the plant succeeds in living there and perpetuating itself for a number of generations, it will have become so altered that botanists who come across it will erect it into a separate species.

The same thing happens in the case of animals that are forced by circumstances to change their climate, habits, and manner of life: but in their case more time is required to work any noticeable change than in the case of plants.

The idea of bringing together under the name of species a collection of like individuals, which perpetuate themselves unchanged by reproduction and are as old as nature, involved the assumption that the individuals of one species could not unite in reproductive acts with individuals of a different species.

Unfortunately, observation has proved and continues every day to prove that this assumption is unwarranted; for the hybrids so common among plants, and the copulations so often noticed between animals of very different species, disclose the fact that the boundaries between these alleged constant species are not so impassable as had been imagined.

It is true that often nothing results from these strange copulations, especially when the animals are very disparate; and when anything does happen the resulting individuals are usually infertile; but we also know that when there is less disparity these defects do not occur. Now this cause is by itself sufficient gradually to create varieties, which then become races, and in course of time constitute what we call species.

To assist us to a judgment as to whether the idea of species has any real foundation, let us revert to the principles already set forth; they show:

(1) That all the organised bodies of our earth are true productions of nature, wrought successively throughout long periods of time.

(2) That in her procedure, nature began and still begins by fashioning the simplest of organised bodies, and that it is these alone which she fashions immediately, that is to say, only the rudiments of organisation indicated in the term *spontaneous generation*.

(3) That, since the rudiments of the animal and plant were fashioned in suitable places and conditions, the properties of a commencing life and established organic movement necessarily caused a gradual development of the organs, and in course of time produced diversity in them as in the limbs.

(4) That the property of growth is inherent in every part of the organised body, from the earliest manifestations of life; and then gave rise to different kinds of multiplication and reproduction, so that the increase of complexity of organisation, and of the shape and variety of the parts, has been preserved.

(5) That with the help of time, of conditions that necessarily were favourable, of the changes successively undergone by every part of the earth's surface, and, finally, of the power of new conditions and habits to modify the organs of living bodies, all those which now exist have imperceptibly been fashioned such as we see them.

(6) That, finally, in this state of affairs every living body underwent greater or smaller changes in its organisation and its parts; so that what we call species were imperceptibly fashioned among them one after another and have only a relative constancy, and are not as old as nature.

But objections may be raised to the allegation that nature has little by little fashioned the various animals known to us by the aid of much time and an infinite variation of environment. It may be asked whether this allegation is not refuted by the single fact of the wonderful variety observed in the *instinct* of various animals, and in the marvellous *skill* of all kinds which they exhibit.

Will anyone, it may be asked, venture to carry his love of system so far as to say that nature has created single-handed that astonishing diversity of powers, artifice, cunning, foresight, patience and skill, of which we find so many examples among animals? Is not what we see in the single class of insects far more than enough to convince us that nature cannot herself produce so many wonders; and to compel the most obstinate philosopher to recognise that the will of the Supreme Author of all things must be here invoked, and could alone suffice for bringing into existence so many wonderful things?

No doubt he would be a bold man, or rather a complete lunatic, who should propose to set limits to the power of the first Author of all things; but for this very reason no one can venture to deny that this infinite power may have willed what nature herself shows us it has willed.

This being so, if I find that nature herself works all the wonders just mentioned; that she has created organisation, life and even feeling, that she has multiplied and diversified within unknown limits the organs and faculties of the organised bodies whose existence she subserves or propagates; that by the sole instrumentality of *needs*, establishing and controlling habits, she has created in animals the fountain of all their acts and all their faculties, from the simplest to instinct, to skill, and finally to reason; if I find all this, should I not recognise in this power of nature, that is to say in the order of existing things, the execution of the will of her Sublime Author, who was able to will that she should have this power?

Shall I admire the greatness of the power of this first cause of everything any the less if it has pleased him that things should be so, than if his will by separate acts had occupied itself and still continued to occupy itself with the details of all the special creations, variations, developments, destructions and renewals, in short, with all the mutations which take place at large among existing things?

Now I hope to prove that nature possesses the necessary powers and faculties for producing herself that so much excite our wonder.

The objection is still raised however that everything we see in living bodies indicates an unchangeable constancy in the preservation of their form. It is held that all animals whose history has come down to us for two or three thousand years have always been the same, and neither lost nor acquired anything in the perfection of their organs and the shape of their parts.

Not only had this apparent stability passed for an undoubted fact, but an attempt has recently been made to find special proofs of it in a report on the natural history collections brought from Egypt by M. Geoffroy. The authors of the report express themselves as follows:

"The collection has in the first place this peculiarity, that it may be said to contain animals of all periods. It has long been asked whether species change their shape in the course of time. This question, apparently so futile, is none the less necessary for the history of the world, and consequently for the solution of innumerable other questions which are not foreign to the gravest subjects of human worship.

"We have never been in so good a position to settle this question, in so far as concerns a large number of remarkable species and some thousands that are not remarkable. It appears as though the superstition of the ancient Egyptians were inspired by nature for the purpose of leaving a record of her history."

. . . .

"It is impossible," continue the authors of the report, "to control our flights of imagination, on seeing still preserved with its smallest bones and hair, perfectly recognisable, an animal which two or three thousand years ago had in Thebes or Memphis its priests and altars. But without giving rein to all the ideas suggested by this approach to antiquity, we shall confine ourselves to the announcement that this part of M. Geoffroy's collection shows that these animals are exactly similar to those of to-day." (*Annales du Muséum d'Hist. natur.*, vol. i. pp. 235 and 236.)

I do not refuse to believe in the close resemblance of these animals with individuals of the same species living to-day. Thus, the birds that were worshipped and embalmed by the Egyptians two or three thousand years ago are still exactly like those which now live in that country.

It would indeed be very odd if it were otherwise; for the position and climate of Egypt are still very nearly what they were in those times. Now the birds which live there, being still in the same conditions as they were formerly, could not possibly have been forced into a change of habits.

Furthermore, it is obvious that birds, since they can travel so easily and choose the places which suit them, are less liable than many other animals to suffer from variations in local conditions, and hence less hindered in their habits.

Indeed there is nothing in the observation now cited that is contrary to the principles which I have set forth on this subject; or which proves that the animals concerned have existed in nature for all time; it proves only that they inhabited Egypt two or three thousand years ago; and every man who has any habit of reflection and at the same time of observing the monuments of nature's antiquity will easily appreciate the import of a duration of two or three thousand years in comparison with it.

Hence we may be sure that this appearance of stability of the things in nature will by the vulgar always be taken for reality; because people in general judge everything with reference to themselves.

For the man who forms his judgment only with reference to the changes that he himself perceives, the eras of these mutations are stationary states which appear to him to be unlimited, on account of the shortness of the existence of individuals of his own species.

Moreover, we must remember that the records of his observations, and the notes of facts which he has been able to register, only extend back a few thousand years; which is a time infinitely great with reference to himself, but very small with reference to the time occupied by the great changes occurring on the surface of the earth. Everything seems to him to be *stable* in the planet which he inhabits; and he is led to repudiate the signs which exist everywhere in the monuments heaped up around him, or buried in the soil which he tramples underfoot.

Magnitudes are relative both in space and time: let man take that truth to heart, and he will then be more reserved in his judgments on the stability which he attributes to the state of things that he observes in nature. (See the Appendix, p. 141, of my *Recherches sur les corps vivants.*)

In order to admit the imperceptible changing of species, and the modifications which their individuals undergo according as they are forced to change their habits and contract new ones, we are not reduced to a mere consideration of the very short spaces of time comprised in our observations; for, in addition to this induction, a number of facts collected many years ago throw enough light on the question to free it from doubt; and I can now affirm that our observations are so far advanced that the solution sought for is patent.

Indeed not only do we know the results of anomalous fertilisations, but we also now know positively that a compulsory and sustained alteration in the habitats and manner of life of animals works after a sufficient time a very remarkable mutation in the individuals exposed to it.

Consider the animal which normally lives in freedom in plains where it habitually exerts itself by swift running; or the bird which is compelled by its needs to pass incessantly through large spaces in the air. When they find themselves imprisoned, the one in the dens of a menagerie or in our stables, the other in our cages or back yards, they undergo in course of time striking alterations, especially after a succession of generations in their new state.

The former loses a great part of his swiftness and agility; his body thickens, the strength and subtleness of his limbs diminish, and his faculties are no longer the same; the latter becomes heavy, can scarcely fly, and takes on more flesh in all his parts.

In Chapter VI. of this Part I., I shall have occasion to prove by well-known facts the power of changes of conditions for giving to animals new needs, and leading them on to new actions; the power of new actions when repeated to induce new habits and inclinations;

finally, the power resulting from the more or less frequent use of any organ to modify that organ either by strengthening, developing and increasing it, or by weakening, reducing, attenuating it, and even making it disappear.

With regard to plants, the same thing may be seen as a result of new conditions on their manner of life and the state of their parts; so that we shall no longer be astonished to see the considerable changes that we have brought about in those that we have long cultivated.

Thus, among living bodies, nature, as I have already said, definitely contains nothing but individuals which succeed one another by reproduction and spring from one another; but the species among them have only a relative constancy and are only invariable temporarily.

Nevertheless, to facilitate the study and knowledge of so many different bodies it is useful to give the name of species to any collection of like individuals perpetuated by reproduction without change, so long as their environment does not alter enough to cause variations in their habits, character and shape.

Of the Species Alleged to be Lost.

I am still doubtful whether the means adopted by nature to ensure the preservation of species or races have been so inadequate that entire races are now extinct or lost.

Yet the fossil remains that we find buried in the soil in so many different places show us the remains of a multitude of different animals which have existed, and among which are found only a very small number of which we now know any living analogues exactly alike.

Does this fact really furnish any grounds for inferring that the species which we find in the fossil state, and of which no living individual completely similar is known to us, no longer exist in nature? There are many parts of the earth's surface to which we have never penetrated, many others that men capable of observing have merely passed through, and many others again, like the various parts of the sea-bottom, in which we have few means of discovering the animals living there. The species that we do not know might well remain hidden in these various places.

If there really are lost species, it can doubtless only be among the large animals which live on the dry parts of the earth; where man exercises absolute sway, and has compassed the destruction of all the individuals of some species which he has not wished to preserve or domesticate. Hence arises the possibility that animals of the genera *Palaeotherium*, *Anoplotherium*, *Megalonix*, *Megatherium*, *Mastodon*, of M. Cuvier, and some other species of genera previously known, are no longer extant in nature: this however is nothing more than a possibility.

But animals living in the waters, especially the sea waters, and in addition all the races of small sizes living on the surface of the earth and breathing air, are protected from the destruction of their species by man. Their multiplication is so rapid and their means of evading pursuit or traps are so great, that there is no likelihood of his being able to destroy the entire species of any of these animals.

It is then only the large terrestrial animals that are liable to extermination by man. This extermination may actually have occurred; but its existence is not yet completely proved.

Nevertheless, among the fossil remains found of animals which existed in the past, there are a very large number belonging to animals of which no living and exactly similar analogue is known; and among these the majority belong to molluscs with shells, since it is only the shells of these animals which remain to us.

Now, if a quantity of these fossil shells exhibit differences which prevent us, in accordance with prevailing opinion, from regarding them as the representatives of similar species that we know, does it necessarily follow that these shells belong to species actually lost? Why, moreover, should they be lost, since man cannot have compassed their destruction? May it not be possible on the other hand, that the fossils in question belonged to species still existing, but which have changed since that time and become converted into the similar species that we now actually find. The following consideration, and our observations throughout this work, will give much probability to such an assumption.

Every qualified observer knows that nothing on the surface of the earth remains permanently in the same state. Everything in time undergoes various mutations, more or less rapid according to the nature of the objects and the conditions; elevated ground is constantly being denuded by the combined action of the sun, rain-waters and yet other causes; everything detached from it is carried to lower ground; the beds of streams, of rivers, even of seas change in shape and depth, and shift imperceptibly; in short, everything on the surface of the earth changes its situation, shape, nature and appearance, and even climates are not more stable.

Now I shall endeavour to show that variations in the environment induce changes in the needs, habits and mode of life of living beings, and especially of animals; and that these changes give rise to modifications or developments in their organs and the shape of their parts. If this is so, it is difficult to deny that the shape or external characters of every living body whatever must vary imperceptibly, although that variation only becomes perceptible after a considerable time.

Let us then no longer be astonished that among the numerous fossils found in all the dry parts of the world, and constituting the remains of so many animals which formerly existed, there are so few of which we recognise the living representatives.

What we should wonder at, on the contrary, is finding amongst these numerous fossil remains of once living bodies, any of which the still existing analogues are known to us. This fact, proved by our collections of fossils, suggests that the fossil remains of animals whose living analogues we know are the least ancient fossils. The species to which each of them belongs doubtless has not had time to undergo variation.

Naturalists who did not perceive the changes undergone by most animals in course of time tried to explain the facts connected with fossils, as well as the commotions known to have occurred in different parts of the earth's surface, by the supposition of a universal catastrophe which took place on our globe. They imagined that everything had been displaced by it, and that a great number of the species then existing had been destroyed.

Unfortunately this facile method of explaining the operations of nature, when we cannot see their causes, has no basis beyond the imagination which created it, and cannot be supported by proof.

Local catastrophes, it is true, such as those produced by earthquakes, volcanoes and other special causes are well known, and we can observe the disorder ensuing from them.

But why are we to assume without proof a universal catastrophe, when the better known procedure of nature suffices to account for all the facts which we can observe?

Consider on the one hand that in all nature's works nothing is done abruptly, but that she acts everywhere slowly and by successive stages; and on the other hand that the special or local causes of disorders, commotions, displacements, etc., can account for everything that we observe on the surface of the earth, while still remaining subject to nature's laws and general procedure. It will then be recognised that there is no necessity whatever to imagine that a universal catastrophe came to overthrow everything, and destroy a great part of nature's own works.

I have said enough on a subject which presents no difficulty. Let us now consider the general principles and essential characters of animals.

CHAPTER IV.

GENERAL PRINCIPLES CONCERNING ANIMALS.

ANIMALS in general are living beings with very curious properties, well calculated to astonish us and excite our study. These beings, infinitely varied in shape, organisation, and faculties, are capable of moving themselves or some of their parts without the impulse of any movement from without. Their irritability is due to an *exciting cause* which in some originates from within, while in others it comes entirely from without. Most of them possess the property of locomotion, and all have parts that are highly irritable.

We find that in their movements some crawl, walk, run or leap; others fly, rising into the atmosphere and passing through wide spaces; others again live in the waters and swim about there freely.

Animals are not, like plants, able to find close by within their reach the material on which they feed; and the predatory animals are actually obliged to go forth and to hunt, chase and seize their prey. It was necessary therefore that they should have the power of motion and even of locomotion, in order to procure the food which they require.

Moreover, those among animals which multiply by sexual reproduction are not hermaphrodite enough to be sufficient to themselves. Hence it was farther necessary that they should be able to travel about for the purpose of effecting acts of fertilisation, and that the environment should provide facilities for it to those which, like oysters, cannot change their position.

Thus the needs of animals have endowed them with the property of moving parts of their bodies, and of carrying out locomotion which subserves their own survival and that of their races.

In Part II. we shall enquire into the origin of this extraordinary faculty, as of the other important faculties found among them; but it suffices at present to draw attention to certain obvious points.

(1) Some only move themselves or their parts when their irritability has been stimulated; but they experience no feeling: these are the most imperfect animals;

(2) Others, in addition to the movements that their parts can undergo through stimulated irritability are capable of experiencing sensations, and possess a very vague inner feeling of their existence; but they only act by the internal impulse of an inclination which leads them towards some or other object; so that their will is always dependent and controlled;

(3) Others again not only exhibit in some of their parts movements resulting from their stimulated irritability; not only are they capable of receiving sensations, and possess an inner feeling of their existence, but they have besides the faculty of forming ideas, although confused ones, and of acting by a free will, subject however to inclinations which lead them exclusively towards certain special objects;

(4) Finally, others (and these are the most perfect) possess in a high degree all the aforementioned faculties; in addition they are able to form clear and precise ideas of the objects which affect their senses and attract their attention; to compare and combine their ideas up to a certain point; to form judgments and complex ideas; in short to think, and to have a will that is less bound down and permits them to introduce more or less variation into their activities.

Life in the most imperfect animals has no energy of movement; and irritability alone suffices for the execution of vital movements. But since vital energy increases in proportion to complexity of organisation, there arrives a time when nature has to improve her methods in order to provide for the necessary activity of vital movements; for this purpose she has utilised muscular activity in establishing the circulatory system, so that the fluids can move with greater rapidity. This rapidity itself is increased in proportion to the increase of the muscular power which works it. Finally, since no muscular activity can take place without nervous influence, this has become everywhere necessary for the acceleration of the fluids in question.

Thus nature has been able to add muscular activity and nervous influence to an irritability which was no longer adequate. But this nervous influence, which gives rise to muscular activity, never does so by means of feeling, as I hope to show in Part II. I shall then prove that sensibility is by no means necessary to the execution of vital movements, even in the most perfect animals.

The various animals which exist are thus clearly distinguished from one another, not only by peculiarities of external shape, consistency of body, size, etc., but, in addition, by the faculties which they possess. Some, such as the most imperfect, are extremely limited in that respect, having no other faculty but those of life in general, and being unable to move except by a power outside them; while others have faculties, progressively more numerous and important, up to the most perfect animals, which exhibit a capacity calculated to excite our wonder.

These remarkable facts no longer surprise us, when we recognise that every faculty is based upon some special organ or system of organs, and when we observe that organisation gradually becomes more complex as we pass from the most imperfect animal, which has no special organ whatever and consequently no faculty but those of life in general, to the most perfect and richly endowed animal. Thus all the organs, even the most important, arise one after the other in the animal scale, and afterwards become successively more perfect through the modifications impressed on them, by which these organs come to harmonise with the state of organisation of which they are part. Hence, by their combination in the most perfect animals, they constitute the highest degree of organisation, giving rise to the most numerous and important faculties.

The examination of the internal organisation of animals; of the various systems presented by that organisation in the animal scale; and, finally, of the special organs, is then the subject of study most deserving of our attention.

If animals, considered as productions of nature, are rendered extremely remarkable by their faculty of locomotion, a great many of them are still more so by their faculty of feeling.

I have said that this faculty of locomotion is very limited in the most imperfect animals, among which it is not voluntary and is only carried out by external stimuli. It then becomes gradually more perfect and ultimately takes its source within the animal itself, and becomes at length subject to its will. In just the same way, the faculty of feeling is still very obscure and limited in the animals among which it begins to exist; but it then develops gradually, and when it has reached its highest development it ultimately gives rise in the animal to the faculties which constitute intelligence.

Indeed the most perfect among animals have simple and even complex ideas; they have passions and memory and they dream, that is to say, they experience involuntary recurrences of their ideas and even of their thoughts; and they are up to a certain point capable of learning. How wonderful is this result of the power of nature!

Nature thus succeeds in endowing a living body with the faculty of locomotion, without the impulse of an external force; of perceiving objects external to it; of forming ideas by comparison of impressions received from one object with those received from others; of comparing or combining these ideas, and of forming judgments which are merely ideas of another order; in short, of thinking. Not only is this the greatest marvel that the power of nature has attained, but it is besides a proof of the lapse of a considerable time; since nature has done nothing but by slow degrees.

As compared to the periods which we look upon as great in our ordinary calculations, an enormous time and wide variation in successive conditions must doubtless have been required to enable nature to bring the organisation of animals to that degree of complexity and development in which we see it at its perfection. If an inspection of the numerous diverse strata composing the external crust of the earth furnishes unimpeachable testimony of its great antiquity; if the very slow but continuous displacement of the sea-bottom,[1] certified by the numerous monuments left everywhere about, gives further confirmation of its prodigious antiquity; then the belief is justified that the state of perfection at which the organisation of the most perfect animals has arrived, contributes to exhibit that truth in the strongest possible light.

But in order that this new proof may be securely based, it will first be necessary to bring into evidence the facts concerning the actual progress of organisation; it will be necessary to verify if possible the reality of that progress; finally, it will be necessary to collect the best established facts and to identify nature's methods in bringing her productions into existence.

Meanwhile, let us note that although the term *productions of nature* is generally accepted for indicating the beings which constitute each kingdom, it seems none the less that no definite idea is attached to that expression. Apparently, prejudices of special origin prevent the recognition of the fact that nature possesses the faculty of herself bringing so many different beings into existence; of causing incessant though very slow variations in living races; and of maintaining everywhere the general order that we observe.

Let us leave aside all opinion whatever on these great subjects; and to avoid any error of imagination let us everywhere consult nature's own works.

In order to be able to bring under our attention the totality of existing animals, and to place these animals under an aspect easily understood, we must remember that all the natural productions that we can observe have long been divided by naturalists into three kingdoms, under the name of animal kingdom, vegetable kingdom and mineral kingdom. By this division, the existences comprised in each of these kingdoms are compared together under a common standard; although some have a very different origin from others.

[1] *Hydrogéologie*, p. 41 *et seq.*

For some time past I have found it more convenient to employ another primary division which is better calculated to give an idea of the beings dealt with. Thus, I distinguish the natural productions comprised in the three aforementioned kingdoms into two main branches:

1. Organised living bodies;
2. Crude bodies without life.

Living beings, such as animals and plants, constitute the first of these two branches of the productions of nature. They possess, as everyone knows, the faculties of alimentation, development, reproduction, and they are subject to death.

But what is not known so well, since the fashionable hypotheses do not permit of the belief, is that living bodies form for themselves their own substances and secretions, as a result of the activity and functions of their organs and of the mutations wrought in them by organic movements (*Hydrogéologie*, p. 112). What is still less known is that the exuviae of these living bodies give rise to all the composite matters, crude or inorganic, that are to be found in nature, matters of which the various kinds increase in course of time and according to the conditions, by reason of the disintegration which they imperceptibly undergo. For this disintegration simplifies them more and more, and after a long period leads to the complete separation of their constituent principles.

These are the various crude and lifeless matters, both solid and liquid, which compose the second branch of the productions of nature, and most of which are known under the name of *minerals*.

It may be said that an immense hiatus exists between crude matters and living bodies, and that this hiatus does not permit of a linear arrangement of these two kinds of bodies, nor of any attempt to unite them by a link, as has been vainly attempted.

All known living bodies are sharply divided into two special kingdoms, based on the essential differences which distinguish animals from plants; and in spite of what has been said I am convinced that these two kingdoms do not really merge into one another at any point, and consequently that there are no animal-plants, as implied by the word *zoophyte*, nor plant-animals.

Irritability in all or some of their parts is the most general characteristic of animals; it is more general than the faculty of voluntary movements and of feeling, more even than that of digestion. Now all plants, as I have elsewhere shown, are completely destitute of irritability, not even excepting the so-called sensitive plants nor those which move certain of their parts on being touched or brought into contact with the air.

It is known that irritability is a faculty essential to the parts or to certain parts of animals, and that it is never suspended or annihilated so long as the animal is alive and the part possessing it has suffered no injury. Its effect is seen in a contraction which takes place instantly throughout the irritable part on contact with a foreign body; a contraction which ends with its cause, and which is renewed whenever the part after relaxation is irritated by new contacts. Now nothing of this kind has ever been observed in any other part of plants.

When I touch the extended branches of the sensitive plant (*Mimosa pudica*), instead of a contraction I observe in the joints of the disturbed branches and petioles a relaxation, which permits these branches and petioles of the leaves to droop, and causes the leaflets themselves to sink down upon one another. When once that sinking has been produced it is useless to touch again the branches and leaves of this plant; no effect follows. A longish time is required, unless it is very hot, for the distension of the joints of the small branches and leaves of the sensitive plant; when all these parts will again be raised and spread out, ready to fall together once more upon a contact or slight shaking.

I cannot see in this phenomenon any relation to the irritability of animals. I reflected however that during growth, especially when it is hot, there are produced in plants many elastic fluids, part of which are incessantly being exhaled. Hence I conceived that in leguminous plants these elastic fluids might accumulate, especially in the joints of the leaves, before being dispelled, and that they might then distend these joints and keep the leaves or leaflets spread out.

In this case, the slow dissipation of the elastic fluids in question set up in leguminous plants by the approach of night; or the sudden dissipation of the same fluid set up in *Mimosa pudica* by a slight shaking, will give rise for leguminous plants in general to the phenomenon known under the name of sleep, and for the sensitive plant to that wrongly attributed to irritability.[1]

It follows from the observations which I shall set forth below, and from the inferences which I have drawn from them, that in general

[1] I have developed in another work (*Hist. Nat. des végétaux*, èdition Déterville, vol. i. p. 202) other analogous phenomena observed in plants such as *Hedysarum girans*, *Dionaea muscipula*, the stamens of the flowers of *Berberis*, etc.; and I have shown that the curious movements observed in the parts of certain plants chiefly in hot weather are never the result of a real irritability essential to any of their fibres; but that they are sometimes hygrometric or pyrometric effects, sometimes the results of elastic relaxations which take place under certain circumstances, and sometimes of a swelling and drooping of parts by the local accumulations and more or less rapid dissipations of elastic and invisible fluids which are being exhaled.

it is not true that animals are sensitive creatures endowed without exception with the power of producing acts of will, and consequently with the faculty of voluntary locomotion. Hence the definition of animals hitherto given to distinguish them from plants is altogether unsuitable; in consequence, I have already proposed to substitute the following as more in harmony with the facts, and more suitable to characterise the beings which compose the two kingdoms of living bodies.

Definition of Animals.

Animals are organised living bodies, which have irritable parts at all times of their lives; which nearly all digest the food on which they live; and which move, some by acts of will, either free or dependent, and others by stimulated irritability.

Definition of Plants.

Plants are organised living bodies whose parts are never irritable, which do not digest or move either by will or true irritability.

We see from these definitions, which are much sounder and more accurate than those hitherto received, that animals are primarily distinguished from plants by the irritability which all or some of their parts possess, and by the movements that they can produce in these parts, or which are set up by external causes as a sequence of their irritability.

It would doubtless be wrong to agree to these new ideas merely on authority; but I think that every unprejudiced reader who takes into consideration the facts and observations which I shall set forth in the course of the present work will be unable to deny them preference over the ancient ones for which I am substituting them; since the latter are obviously contrary to all observation.

We shall terminate this general outlook upon animals by two somewhat curious considerations: one concerning the extreme multiplicity of animals on the surface of the earth and in the waters, the other concerning the means adopted by nature to ensure that their number shall never become injurious to the preservation of her productions and of the general order which should exist.

Of the two kingdoms of living bodies that comprising the animals appears much richer and more varied than the other: at the same time it exhibits more wonderful phenomena in its organisation.

The surface of the earth, the waters, and to some extent even the air are populated by an infinite multitude of diverse animals, the races of which are so varied and numerous that a large proportion of them will probably always evade our researches. This is rendered the more likely since the enormous extent of water, its depth in many places, and the prodigious fertility of nature in the smallest species will doubtless be for all time an almost insuperable obstacle to the progress of knowledge.

A single class of the invertebrate animals, such as insects for instance, equals the entire vegetable kingdom in the number and diversity of its contained objects. The class of polyps is apparently much more numerous still, but we shall never be able to flatter ourselves that we know all the animals which make it up.

As a result of the rapid multiplication of the small species, and particularly of the more imperfect animals, the multiplicity of individuals might have injurious effects upon the preservation of races, upon the progress made in perfection of organisation, in short, upon the general order, if nature had not taken precautions to restrain that multiplication within limits that can never be exceeded.

Animals eat each other, except those which live only on plants; but these are liable to be devoured by carnivorous animals.

We know that it is the stronger and the better equipped that eat the weaker, and that the larger species devour the smaller. Nevertheless, individuals rarely eat others of the same race as themselves; they make war on different races.

The multiplication of the small species of animals is so great, and the succession of generations is so rapid, that these small species would render the globe uninhabitable to any others, if nature had not set a limit to their prodigious multiplication. But since they serve as prey to a multitude of other animals, and since the duration of their life is very short and they are killed by any fall of temperature, their numbers are always maintained in the proper proportions for the preservation of their own and other races.

As to the larger and stronger animals, they might well become dominant and have bad effects upon the preservation of many other races if they could multiply in too large proportions; but their races devour one another, and they only multiply slowly and few at a time; and this maintains in their case also the kind of equilibrium that should exist.

Lastly, man alone, considered apart from all that is special to him, seems to be able to multiply indefinitely, for his intelligence and powers protect him from any limit of multiplication due to the voracity of any animal. He exercises a supremacy over them, so that instead of having to fear the larger and stronger races of animals, he is capable rather of extinguishing them, and he is continually keeping down their numbers.

But nature has given him numerous passions which unfortunately develop with his intelligence, and thus set up a great obstacle to the extreme multiplication of individuals of his species.

It seems, in fact, that man is himself responsible for continually keeping down the numbers of his kind; for I have no hesitation in saying that the earth will never be covered by the population that it might support; several of its habitable regions will always be sparsely populated in turns, although the period of these fluctuations are, so far as we are concerned, immeasurable.

By these wise precautions, everything is thus preserved in the established order; the continual changes and renewals which are observed in that order are kept within limits that they cannot pass; all the races of living bodies continue to exist in spite of their variations; none of the progress made towards perfection of organisation is lost; what appears to be disorder, confusion, anomaly, incessantly passes again into the general order, and even contributes to it; everywhere and always the will of the Sublime Author of nature and of everything that exists is invariably carried out.

Before devoting ourselves to showing the degradation and simplification existing in the organisation of animals, when we proceed according to custom from the most complex to the simplest, let us examine their true arrangement and classification, as well as the principles employed for this purpose. It will then be easier for us to recognise the proofs of the degradation in question.

CHAPTER V.

ON THE TRUE ARRANGEMENT AND CLASSIFICATION OF ANIMALS.

For the progress of zoological philosophy and the object that we have in view, it is necessary to enquire into the true arrangement and classification of animals; to consider how such an arrangement came about; to ascertain what principles should be observed in setting up that general arrangement; and, finally, to investigate what remains to be done in order to bring that arrangement into the closest harmony with the actual order of nature.

But in order that our studies may be profitable, we must first determine the essential aims of an arrangement and of a classification of animals; for these two aims are very different in nature.

The aim of a general arrangement of animals is not only to possess a convenient list for consulting, but it is more particularly to have an order in that list which represents as nearly as possible the actual order followed by nature in the production of animals; an order conspicuously indicated by the affinities which she has set between them.

The aim of a classification of animals, on the other hand, is to furnish points of rest for our imagination, by means of lines of demarcation drawn at intervals in the general series; so that we may be able more easily to identify each race already discovered, to grasp its affinities with other known animals, and to place newly discovered species in their proper position. This device makes up for our own shortcomings, facilitates our studies and our knowledge, and is absolutely necessary for us; but I have already shown that it is a produce of artifice, and that despite appearances it corresponds to nothing real in nature.

An accurate determination of affinities between objects will always begin by fixing in our general arrangements the place of the large groups or primary divisions; then that of the lesser groups, and lastly that of the species or special races that have been observed. Now here is the inestimable advantage accruing to science from a knowledge of affinities. Since these affinities are the actual work of nature, no naturalist will ever be able or indeed desire to alter the consequences of a recognised affinity. The general arrangement will thus become ever more perfect and less arbitrary, according as our knowledge of affinities becomes greater.

The case is different with classifications: that is to say, with the various lines of demarcation that we have to draw at intervals in the general list both of animals and plants. In truth, so long as there are gaps remaining to be filled in our list owing to many animals and plants not having yet been observed, we shall always find these lines of demarcation, which appear to be drawn by nature herself; but this illusion will vanish as our observations accumulate. Have we not already witnessed the effacement of a great number, at least in the smaller divisions, by reason of the numerous discoveries of naturalists during the last half century? Except for the lines of demarcation resulting from gaps to be filled, those which we shall always have to draw will be arbitrary and therefore changeable, so long as naturalists do not adopt some conventional principle for their guidance.

In the animal kingdom such a principle is that *every class should comprise animals distinguished by a special system of organisation.* The strict execution of this principle is quite easy, and attended only with minor inconveniences.

In short, although nature does not pass abruptly from one system of organisation to another, it is possible to draw boundaries between each system, in such a way that there is only a small number of animals near those boundaries and admitting of doubt as to their true class.

The other lines of demarcation which sub-divide classes are usually more difficult to establish, since they depend on less important characters; and for this reason are more arbitrary.

Before examining the true classification of animals, let me endeavour to show that the list of living bodies should form a series, at least as regards the main groups; and not a branching net-work.

Classes should form a Series in the Arrangement of Animals.

Man is condemned to exhaust all possible errors when he examines any set of facts before he recognises the truth. Thus it has been denied that the productions of nature in each kingdom of living bodies can really be arranged in a true series according to their affinities; and that there exists any scale in the general arrangement either of animals or plants.

Naturalists, for instance, have noticed that many species, certain genera and even some families appear to a certain extent isolated in their characters; and several have imagined that the affinities among living beings may be represented something after the manner of the different points of a compass. They regard the small well-marked series, called natural families, as being arranged in the form of a reticulation. This idea, which some modern writers think sublime, is clearly a mistake, and is certain to be dispelled when we have a deeper and wider knowledge of organisation; and especially when the distinction is recognised between what is due to the influence of environment and habits and what is due to the greater or less progress in the complexity or perfection of organisation.

Meanwhile I shall show that nature, by giving existence in the course of long periods of time to all the animals and plants, has really formed a true scale in each of these kingdoms as regards the increasing complexity of organisation; but that the gradations in this scale, which we are bound to recognise when we deal with objects according to their natural affinities, are only perceptible in the main groups of the general series, and not in the species or even in the genera. This fact arises from the extreme diversity of conditions in which the various races of animals and plants exist; for these conditions have no relation to the increasing complexity of organisation, as I shall show; but they produce anomalies or deviations in the external shape and characters which could not have been brought about solely by the growing complexity of organisation.

We have then only to prove that the series constituting the animal scale resides essentially in the arrangement of the main groups composing it, and not in that of species, nor always even of genera.

The series to which I have alluded can then only be made out among the larger groups; since each of these groups, constituting the classes and bigger families, comprises beings whose organisation is dependent on some special system of essential organs.

Thus each distinct group has its special system of essential organs; and it is these special systems which undergo a degradation as we pass from the most complex to the simplest. But each organ taken by itself does not proceed so regularly in its degradations: and less so in proportion to its lesser importance and greater susceptibility to modification by environment.

In fact, the organs that have little importance or are not essential to life are not always at the same stage of perfection or degradation; so that if we follow all the species of a class we shall see that some one organ of any species reaches its highest degree of perfection, while some other organ, which in that same species is quite undeveloped or imperfect, reaches in some other species a high state of perfection.

These irregularities in the perfection and degradation of inessential organs are found in those organs which are the most exposed to the influence of the environment; this influence involves similar irregularities in the shape and condition of the external parts, and gives rise to so great and singular a diversity of species that, instead of being arranged like the main groups in a single linear series as a regularly graduated scale, these species often constitute lateral ramifications around the groups to which they belong, and their extremities are in reality isolated points.

A much more powerful and lasting set of conditions is necessary to modify any internal system of organisation than to alter the external organs.

I observe, however, that in cases of necessity nature passes from one system to another without a break, if they are closely allied; it is indeed by this faculty that she succeeded in fashioning them all in turn, passing from the simplest to the most complex.

So true is it that she has this faculty, that she even passes from one system to another not merely in two different allied families but in one individual.

Those systems of organisation in which respiration is carried on by true lungs are nearer to the systems requiring gills than to those requiring tracheae; thus, nature not only passes from gills to lungs in allied classes and families, as is seen among fishes and reptiles, but she does so even during the existence of one individual: which possesses in turn first one and then the other system. It is known that the frog, in its imperfect condition of tadpole, breathes by gills; while in its more perfect condition of frog it breathes by lungs. But nowhere does nature pass from the system of tracheae to the pulmonary system.

It may then be truly said that in each kingdom of living bodies the groups are arranged in a single graduated series, in conformity with the increasing complexity of organisation and the affinities of the object. This series in the animal and vegetable kingdoms should contain the simplest and least organised of living bodies at its anterior extremity, and ends with those whose organisation and faculties are most perfect.

Such appears to be the true order of nature, and such indeed is the order clearly disclosed to us by the most careful observation and an extended study of all her modes of procedure.

We have seen the necessity of paying attention to the question of affinities, in drawing up our arrangements of the productions of nature; hence we are no longer able to arrange the general series in any way we like. Our knowledge of nature's methods continues to increase in proportion to our studies of the affinities between objects or various groups of objects; and this knowledge compels us to conform to her order.

The first result obtained from the use of affinities in placing the groups in a general scheme is that the two extremities of the order must be occupied by the most dissimilar beings, since they are the most distant from one another from the point of view of affinities, and consequently of organisation. Hence it follows that if one of the extremities of the order is occupied by the most perfect of living bodies, having the most complex organisation, the other extremity of the order must necessarily be occupied by the most imperfect of living bodies, namely, those whose organisation is the simplest.

In the general arrangement of known plants according to the natural methods, that is according to affinities, only one extremity is thoroughly known; and that is occupied by the cryptogams. If the other extremity is not determined with equal certainty, it is due to the fact that our knowledge of plant organisation is much less advanced than our knowledge of the organisation of a great number of known animals. Hence it follows that in the case of plants we have as yet no certain guide to the affinities between the large groups, as we have to those among genera and families.

The same difficulty does not exist in the case of animals, and both extremities of their general series are thus definitely fixed; for as long as importance is attached to the natural method, and hence to affinities, the mammals will of necessity occupy one extremity of the order, while the infusorians will be placed at the other.

For animals then, as well as for plants, there exists in nature an order arising, like the objects which it calls into existence, from powers conferred by the Supreme Author of all things. Nature is herself only the general and immutable order created everywhere by this Sublime Author; she is the sum total of the general and special laws to which that order is subject. By these powers, which she continues unchangeably to make use of, she has given and still continues to give existence to her productions; she is incessantly varying and renewing them, and thus maintains everywhere the entire order which results.

We were obliged to recognise this order of nature in each kingdom of living bodies; and we are already in possession of various parts of it, in our better constituted families and genera. We shall now see that in the animal kingdom it is established in its outlines in a way that leaves no scope for arbitrary opinion.

But the great number of divers animals that we have come to know, and the brilliant light shed by comparative anatomy on their organisation, now place it in our power definitely to draw up the general list of all known animals, and to assign definitely the rank of the main divisions that may be established in the series which they constitute. This it behoves us to recognise; it would indeed be difficult to dispute.

Let us now pass to the actual arrangement and classification of animals.

The True Arrangement and Classification of Animals.

Since the purpose and principles both of a general arrangement and of a classification of living animals were not at first perceived when these subjects were studied, the works of naturalists long suffered from this imperfection of our ideas. The same thing happened in the science of natural history as has happened in all others to which much attention was given, before any principles had been thought out to constitute a basis and to guide their labours.

Instead of subjecting the classification which had to be made in each kingdom of living bodies to an arrangement which should be quite unfettered, attention was entirely devoted to disposing objects in convenient classes, so that their arrangement was thus abandoned to arbitrary opinion.

The affinities among the larger groups in the vegetable kingdom, for example, were very difficult to grasp; and artificial systems were long made use of in botany. They facilitated the making of convenient classifications based upon arbitrary principles, so that every author drew up a new one according to his fancy. Thus the proper arrangement of plants according to the natural method was then always sacrificed. It is only since we have recognised the importance of the parts concerned with fruiting, and the greater importance of some than others that the general arrangement of plants began to make progress towards perfection.

As the case of animals is different, the general affinities which characterise the main groups are much easier to perceive: so that several of these groups were identified at the very beginning of the study of natural history.

Aristotle indeed divided animals primarily into two main divisions or, as he called it, two classes, viz.:

1. Animals that have blood:
   - Viviparous quadrupeds.
   - Oviparous quadrupeds.
   - Fishes.
   - Birds.
2. Animals that have no blood:
   - Molluscs.
   - Crustaceans.
   - Testaceans.
   - Insects.

This primary division of animals into two main groups was fairly good, but the character taken by Aristotle for discrimination was bad. That philosopher gave the name of *blood* to the chief fluid in animals which has a red colour. He imagined that all animals which he placed in his second class only possessed white or whitish fluids; and he thereupon regarded them as having no blood.

Such apparently was the first outline of a classification of animals; it is at any rate the oldest of which we have any knowledge. But this classification also furnishes the earliest example of an arrangement, though in the opposite direction from the order of nature; since we may notice in it a progression, though a very imperfect one, from the most complex to the simplest.

That erroneous direction has been generally followed ever since in the arrangement of animals; and this has clearly retarded our knowledge of nature's procedure.

Modern naturalists have endeavoured to improve upon Aristotle's division by giving to the animals in the first class the name of red-blooded animals, and to those in his second class that of white-blooded animals. It is now well known how defective is this character; since there are some invertebrate animals (many annelids) which have red blood.

In my opinion the essential fluids of animals do not deserve the name of blood, except when they circulate in arteries and veins; for the other fluids are so degraded, and the combination of their principles so imperfect, that it would be wrong to assimilate them to fluids which have a true circulation. One might as well attribute blood to a plant as to a radiarian or polyp.

In order to avoid ambiguity and hypothesis, I divided the entire known animal world in my first course of lectures at the Museum in the spring of 1794 (the year II. of the republic) into two perfectly distinct groups, viz.:

Animals that have vertebrae.
Animals without vertebrae.

I called the attention of my pupils to the fact that the vertebral column, among animals provided with it, indicates the possession of a more or less perfect skeleton and of a plan of organisation on the same plane; whereas its absence among other animals not only distinguishes them sharply from the first, but shows that their whole plan of organisation is very different from those of vertebrate animals.

From Aristotle to Linnæus nothing of note appeared with regard to the general arrangement of animals; but in the course of last century naturalists of the highest distinction made a large number of special observations on animals, and especially on many invertebrate animals. Some recorded their anatomy with greater or less fulness, while others gave an accurate and detailed history of the metamorphoses and habits of a great number of these animals; as a result of their valuable observations, we have become acquainted with many facts of the greatest importance.

At length Linnæus, a man of high genius and one of the greatest of naturalists, after having marshalled the facts and taught us the necessity for great accuracy in the determination of all kinds of characters, gave us the following classification for animals.

He divided known animals into six classes, based upon three stages or characters of organisation.

CLASSIFICATION OF ANIMALS, ESTABLISHED BY LINNÆUS.

Classes.	
I. Mammals.	*First Stage.* Heart with two ventricles: blood red and warm.
II. Birds.	
III. Amphibians (Reptiles).	*Second Stage.* Heart with one ventricle: blood red and cold.
IV. Fishes.	
V. Insects.	*Third Stage.* A cold serum (in place of blood).
VI. Worms.	

Except for the inversion displayed by this arrangement as by all others the four first divisions proposed are now definitely established, and will henceforth always obtain the assent of zoologists as to their position in the general series. For this we are primarily indebted to the illustrious Swedish naturalist.

The case is different with regard to the two final divisions of the arrangement in question; they are wrong and very badly disposed. Since they comprise the greater number of known animals of the most varied characters, they should be more numerous. Hence it has been necessary to re-constitute them and substitute others.

We have seen that Linnæus, and the naturalists who succeeded him, gave very little attention to the necessity for increasing the number of divisions among animals which have a cold serum in place of blood (invertebrate animals), and whose characters and organisation are so greatly varied. Hence they have divided these numerous animals into two classes only, viz. insects and worms; so that everything which was not regarded as an insect, that is to say all invertebrate animals that have not jointed legs, were referred without exception to the class of worms. They placed the class of insects after the fishes, and the worms after the insects. According to this arrangement of Linnæus, the worms constituted the final class of the animal kingdom.

These two classes are still maintained in the same order in all the editions of the *Systema Naturae* published subsequently to Linnæus. The essential vice of this arrangement, as regards the natural order of animals, is obvious; it cannot be denied that Linnæus's class of worms is a sort of chaos in which the most disparate objects are included. Yet the authority of that savant carried so much weight among naturalists, that no one dared to change this monstrous class of worms.

With a view to bringing about some useful reform in this respect, I suggested in my first course the following arrangement for invertebrate animals, which I divided not into two classes, but into five in the following order.

Arrangement of Invertebrate Animals set forth in my First Course.

1. Molluscs;
2. Insects;
3. Worms;
4. Echinoderms;
5. Polyps.

These classes were then identical with some of the orders which Bruguière had suggested in his arrangement of worms (which I did not adopt), and with the class of insects as defined by Linnæus.

The arrival of M. Cuvier in Paris however, towards the middle of the year III. (1795), drew the attention of zoologists to the organisation of animals. I then saw with much satisfaction the conclusive evidence which he produced in favour of the priority of rank accorded to molluscs over insects in the general series. This I had already impressed in my lessons; but it had not been favourably received by the naturalists of this capital.

The change which I had thus instituted, from a consciousness of the inadequacy of the prevailing arrangement of Linnæus, was thoroughly consolidated by M. Cuvier by the most definite facts, several of which, it is true, were already known but had not attracted our attention in Paris.

I took advantage of the light shed since his arrival by this savant over every section of zoology, and particularly over invertebrate animals which he called white-blooded animals. I then added in turn new classes to my arrangement; I was the first to establish them; but, as we shall see, such of those classes as were adopted were only adopted reluctantly.

The personal interests of authors are doubtless a matter of complete indifference to science, and also apparently to those who study it. Nevertheless, a knowledge of the history of the changes undergone during the last fifteen years by the classification of animals is not without its uses: the following are those which I have instituted.

First, I changed the name of my class of echinoderms to *radiarians*, in order to unite with them the jelly-fishes and neighbouring genera.

This class, notwithstanding its utility and inevitableness, has not yet been adopted by naturalists.

In my course in the year VII. (1799) I established the class of *crustaceans*. At that time M. Cuvier, in his *Tableau des animaux*, p. 451, still included crustaceans with insects; and although this class is essentially distinct, yet it was not till six or seven years later that a few naturalists consented to adopt it.

The following year, that is to say, in my course of the year VIII. (1800) I suggested the *arachnids* as a class by itself, easy and necessary to distinguish. From that time its characters have constituted a sure indication of an organisation peculiar to these animals; for it is impossible to believe that they arose from an organisation exactly similar to the insects. Insects undergo metamorphosis, propagate only once in the course of their life, and have only two antennae, two eyes with facets and six jointed legs; while the arachnids never undergo metamorphosis, and exhibit various characters besides which differentiate them from insects. This fact has since been partly confirmed by observation. Yet this class of arachnids is still not admitted into any other work than my own.

M. Cuvier had discovered the existence of arterial and venous vessels in various animals, which used to be confused under the name of worms with other animals of very different organisation. I immediately took this new fact into consideration for the improvement of my classification; and in my course in the year X. (1802) I established the class of *annelids*, placing them after the molluscs and before the crustaceans, as required by their organisation.

By giving a special name to this new class I was able to keep the old name of worms for the animals which have always borne it, and whose organisation was remote from the annelids. So I continued to place the worms after the insects, and to distinguish them from the radiarians and polyps with which they can never again be united.

My class of annelids, published in my lectures and in my *Recherches sur les corps vivants* (p. 24), was several years before being admitted by naturalists. For the last two years however this class has begun to gain recognition; but since it is held desirable to change the name of it and to call it by the name of worms, they do not know what to do with the worms properly so-called which have no nerves or circulatory system. In this difficulty they combine them with the class of polyps, although their organisation is very different.

These instances of perfection at first attained in a classification, then destroyed and subsequently re-established by the necessity of things, are not rare in natural science.

Linnæus in fact united several genera of plants which Tournefort had formerly distinguished as in the case of *Polygonum*, *Mimosa*,

Justicia, *Convallaria*, and many others ; and now botanists are re-establishing the genera which Linnæus had destroyed.

Finally last year (in my course of 1807) I established among invertebrate animals a new class—the tenth—that of *infusorians* ; because after a careful examination of the characters of these imperfect animals, I was convinced that I had been wrong to place them with the polyps.

Thus, by continuing to collect facts from observation and from the rapid progress of comparative anatomy, I instituted successively the various classes which now compose my arrangement of invertebrate animals. These classes, to the number of ten, are arranged in order from the most complex to the simplest as usual, viz. :

Classes of Invertebrate Animals.

Molluscs.	Insects.
Cirrhipedes.	Worms.
Annelids.	Radiarians.
Crustaceans.	Polyps.
Arachnids.	Infusorians.

I shall show, when I come to deal with each of these classes, that they constitute necessary groups, since they are based upon a study of organisation ; and that although races may, nay must, exist near the boundaries, half way between two classes, yet these groups are the best attainable by artifice. They will therefore have to be recognised, so long as the interest of science is our chief concern.

By adding to these ten classes into which the invertebrates are divided, the four classes of vertebrate animals identified and determined by Linnæus, we shall have a classification of all known animals into the following fourteen classes, set out once more in the opposite order to that of nature.

1. Mammals. 2. Birds. 3. Reptiles. 4. Fishes.	Vertebrate animals.
5. Molluscs. 6. Cirrhipedes. 7. Annelids. 8. Crustaceans. 9. Arachnids. 10. Insects. 11. Worms. 12. Radiarians. 13. Polyps. 14. Infusorians.	Invertebrate animals.

The above represents the true arrangement of animals, and also the classes established among them.

We now have to examine a very important problem, which appears never to have been fathomed nor discussed ; but the solution of which is necessary ; it is this :

All the classes, into which the animal kingdom is divided, necessarily form a series of groups arranged according to the increasing or decreasing complexity of their organisation. In drawing up this series, ought we to proceed from the most complex to the simplest, or from the simplest to the most complex ?

We shall endeavour to give the solution of this problem in Chap. VIII. which concludes this part ; but we must first examine a very remarkable fact, most worthy of our attention, which may lead us to a perception of nature's procedure, when bringing her diverse productions into existence. I refer to that remarkable degradation of organisation which is found on traversing the natural series of animals, starting from the most perfect or the most complex towards the simplest and most imperfect.

Although this degradation neither is nor can be finely graduated as I shall show, it so obviously and universally exists in the main groups, including even the variations, that it doubtless depends on some general law which it behoves us to discover and consequently to search for.

CHAPTER VI.

DEGRADATION AND SIMPLIFICATION OF ORGANISATION FROM ONE EXTREMITY TO THE OTHER OF THE ANIMAL CHAIN, PROCEEDING FROM THE MOST COMPLEX TO THE SIMPLEST.

AMONG the problems of interest for zoological philosophy, one of the most important is that which concerns the degradation and simplification observed in animal organisation on passing from one extreme to the other of the animal chain, from the most perfect animals to those whose organisations are the simplest.

Now the question arises whether this is a fact that can be established; for, if so, it will greatly enlighten us as to nature's plan and will set us on the way to discover some of her most important laws.

I here propose to prove that the fact in question is true, and that it is the result of a constant law of nature which always acts with uniformity; but that a certain special and easily recognised cause produces variations now and again in the results which that law achieves throughout the animal chain.

We must first recognise that the general series of animals arranged according to their natural affinities is a series of special groups which result from the different systems of organisation employed by nature; and that these groups are themselves arranged according to the decreasing complexity of organisation, so as to form a real chain.

We notice then that except for the anomalies, of which we shall ascertain the cause, there exists from one end to the other of this chain a striking degradation in the organisation of the animals composing it, and a proportionate diminution in the numbers of these animals' faculties. Thus if the most perfect animals are at one extremity of the chain, the opposite extremity will necessarily be occupied by the simplest and most imperfect animals found in nature.

This examination at length convinces us that all the special organs are progressively simplified from class to class, that they become altered, reduced and attenuated little by little, that they lose their local concentration if they are of the first importance, and that finally they are completely and definitely extinguished before the opposite end of the chain is reached.

As a matter of fact, the degradation of which I speak is not always gradual and regular in its progress, for often some organ disappears or changes abruptly, and these changes sometimes involve it in peculiar shapes not related with any other by recognisable steps.

Often again some organ disappears and re-appears several times before it is definitely extinguished. But we shall see that this could not have been otherwise; for the factor which brings about the progressive complexity of organisation must have had varied effects, owing to its liability to modification by a certain other factor acting with great power. We shall however see that the degradation in question is none the less real and progressive, wherever its effects can be seen.

If the factor which is incessantly working towards complicating organisation were the only one which had any influence on the shape and organs of animals, the growing complexity of organisation would everywhere be very regular. But it is not; nature is forced to submit her works to the influence of their environment, and this environment everywhere produces variations in them. This is the special factor which occasionally produces in the course of the degradation that we are about to exemplify, the often curious deviations that may be observed in the progression.

We shall attempt to set forth in full both the progressive degradation of animal organisation and the cause of the anomalies in the progress of that degradation, in the course of the animal series.

It is obvious that, if nature had given existence to none but aquatic animals and if all these animals had always lived in the same climate, the same kind of water, the same depth, etc., etc., we should then no doubt have found a regular and even continuous gradation in the organisation of these animals.

But the power of nature is not confined within such limits.

It first has to be observed that even in the waters she has established considerable diversity of conditions: fresh-water, sea water, still or stagnant water, running water, the water of hot climates, of cold climates, and lastly shallow water and very deep water; these provide as many special conditions which each act differently on the animals living in them. Now the races of animals exposed to any of these conditions have undergone special influences from them and have been varied by them all the while that their complexity of organisation has been advancing.

After having produced aquatic animals of all ranks and having caused extensive variations in them by the different environments provided by the waters, nature led them little by little to the habit of living in the air, first by the water's edge and afterwards on all the dry parts of the globe. These animals have in course of time been profoundly altered by such novel conditions ; which so greatly influenced their habits and organs that the regular gradation which they should have exhibited in complexity of organisation is often scarcely recognisable.

These results which I have long studied, and shall definitely prove, lead me to state the following zoological principle, the truth of which appears to me beyond question.

Progress in complexity of organisation exhibits anomalies here and there in the general series of animals, due to the influence of environment and of acquired habits.

An examination of these anomalies has led some to reject the obvious progress in complexity of animal organisation and to refuse to recognise the procedure of nature in the production of living bodies.

Nevertheless, in spite of the apparent digressions that I have just mentioned, the general plan of nature and the uniformity of her procedure, however much she varies her methods, are still quite easily distinguished. We have only to examine the general series of known animals and to consider it first in its totality and then in its larger groups ; the most unequivocal proofs will then be perceived of the gradation which she has followed in complexity of organisation ; a gradation which should never be lost sight of by reason of the aforementioned anomalies. Finally, it will be noticed that whenever there have been no extreme changes of conditions, that gradation is found to be perfectly regular in various portions of the general series to which we have given the name of families. This truth becomes still more striking in the study of species ; for the more we observe, the more difficult, complicated and minute become our specific distinctions.

The gradation in complexity of animal organisation can no longer be called in doubt, when once we have given positive and detailed proof of what we have just stated. Now since we are taking the general series of animals in the opposite direction from nature's actual order when she brought them successively into existence, this gradation becomes for us a remarkable degradation which prevails from one end to the other of the animal chain, except for the gaps arising from objects which are not yet discovered and those which arise from anomalies caused by extreme environmental conditions.

Let us now cast an eye over the complexity and totality of the animal series, in order to establish positively the degradation of organisation from one extremity to the other ; let us consider the facts presented and let us then pass rapidly in review the fourteen classes of which it is primarily composed.

The general arrangement of animals set forth above is unanimously accepted as a whole by zoologists: who dispute only as to the boundaries of certain classes. In examining it I notice a very obvious fact which would in itself be decisive for my purpose; it is as follows:

At one extremity of the series (that namely which we are accustomed to consider as the anterior) we find the animals that are most perfect from all points of view, and have the most complex organisation; while at the opposite extremity of the same series we find the most imperfect that exist in nature—those with the simplest organisation and to all appearances hardly endowed with animality.

This accepted fact, which indeed cannot be questioned, becomes the first proof of the degradation which I propose to establish; for it is a necessary condition of it.

Another fact brought forward by an examination of the general series of animals and furnishing a second proof of the degradation prevailing in their organisation from one extremity to the other of their chain, is the following:

The first four classes of the animal kingdom contain animals that are in general provided with a vertebral column, while the animals of all the other classes are absolutely destitute of it.

It is known that the vertebral column is the essential basis of the skeleton, which cannot exist without it; and that wherever there is a vertebral column there is a more or less complete and perfect skeleton.

It is also known that perfection of faculties is a proof of perfection of the organs on which they rest.

Now although man may be above his rank on account of the extreme superiority of his intelligence as compared with his organisation, he assuredly presents the type of the highest perfection that nature could attain to: hence the more an animal organisation approaches his, the more perfect it is.

Admitting this, I observe that the human body not only possesses a jointed skeleton but one that is above all others the most complete and perfect in all its parts. This skeleton stiffens his body, provides numerous points of attachment for his muscles and allows him an almost endless variation of movement.

Since the skeleton is a main feature in the plan of organisation of the human body, it is obvious that every animal possessed of a skeleton has a more perfect organisation than those without it.

Hence the invertebrate animals are more imperfect than the vertebrate animals; hence, too, if we place the most perfect animals at the head of the animal kingdom, the general series exhibits a real degradation in organisation; since after the first four classes all the animals of the following classes are without a skeleton and consequently have a less perfect organisation.

But this is not all: Degradation may be observed even among the vertebrates themselves; and we shall see finally that it is found also among the invertebrates. Hence this degradation follows from the fixed plan of nature, and is at the same time a result of our following her order in the inverse direction; for if we followed her actual order, if, that is to say, we passed along the general series of animals from the most imperfect to the most perfect, instead of a degradation in organisation we should find a growing complexity and we should see animal faculties successively increasing in number and perfection. In order to prove the universal existence of the alleged degradation, let us now rapidly run through the various classes of the animal kingdom.

MAMMALS.

Animals with mammae, four jointed limbs, and all the organs essential to the most perfect animals. Hair on certain parts of the body.

Mammals (*Mammalia*, Lin.) should obviously be at one extremity of the animal chain, viz. that which contains the most perfect animals and the richest in organisation and faculties; for among them alone are found those with the most developed intelligence.

If perfection of faculties is a proof of that of the organs they are based upon as I said above, all mammals (and they alone are truly viviparous) must have the most perfect organisation, since it is agreed that these animals have more intelligence, more faculties and a more perfect set of senses than any others; moreover their organisation approaches most nearly to that of man.

Their organisation exhibits a body whose parts are stiffened by a jointed skeleton, which is generally more complete in these animals than in the three other classes of vertebrates. Most of them have four articulated limbs appended to the skeleton; and all have a diaphragm between the chest and abdomen; a heart with two ventricles and two auricles; red warm blood; free lungs, enclosed within the chest, through which the blood passes before being driven to the other parts of the body; lastly, they are the only viviparous animals, for they are the only animals in which the foetus although enclosed within its membranes is always in communication with its mother and develops at the expense of her substance, and in which the young feed for some time after their birth on the milk of her mammae.

It is then the mammals that must occupy the first rank in the animal kingdom by virtue of their perfection of organisation and greatest number of faculties (*Recherches sur les corps vivants*, p. 15). After the mammals we no longer find a definitely viviparous reproduction, nor lungs limited by a diaphragm to the chest and receiving all the blood which has to be driven to the rest of the body, etc., etc.

Among the mammals themselves it is in truth not easy to distinguish what is really due to degradation from what is the effect of environment, manner of life and long-established habits.

Nevertheless, traces of the general degradation of organisation may be found even among them; for those whose limbs are adapted for grasping objects have a higher perfection than those whose limbs are adapted only for walking. It is among the former that man is placed in respect of his organisation. Now it is clear that since the organisation of man is the most perfect, it should be regarded as the standard for judging of the perfection or degradation of the other animal organisations.

Thus the three divisions, into which the class of mammals is unequally broken up, exhibit among themselves, as we shall see, a conspicuous degradation in the organisation of the animals they contain.

First division: *unguiculate mammals*; they have four limbs, flat or pointed claws at the end of their digits but not investing them. These limbs are in general adapted for grasping objects or at least for hooking on to them. It is among these that the animals with the most perfect organisation are found.

Second division: *ungulate mammals*; they have four limbs and the extremity of their digits is completely invested by a rounded horn called a hoof. Their feet serve no other purpose than that of walking or running on the ground, and cannot be employed either for climbing trees, or for grasping any object or prey, or for attacking and rending other animals. They feed exclusively on vegetable substances.

Third division: *exungulate mammals*; they have only two limbs and these limbs are very short, flat and shaped like fins. Their digits are invested by skin and have no claws or horn. Their organisation is the least perfect of all mammals. They have no pelvis, nor hind feet; they swallow without previous mastication; finally they habitually live in the water; but they come to the surface to breathe air. They have received the name of *cetaceans*.

Although the *amphibians* also live in the water, coming out of it occasionally to crawl upon the shore, they really belong to the first division in the natural order, and not to that which comprises the cetaceans.

Henceforth we have to distinguish the degradation of organisation which arises from the influence of environment and acquired habits, from that which results from the smaller progress in the perfection or complexity of organisation. We must be careful therefore about going into too much detail in this respect; because as I shall show the environment in which animals habitually live, their special habitats, the habits which circumstances have forced upon them, their manner of life, etc., have a great power to modify organs; so that the shapes of parts might be attributed to degradation when they are really due to other causes.

It is obvious for example that the amphibians and cetaceans must have greatly shortened limbs, since they live habitually in a dense medium where well-developed limbs would only impede their movements. It is obvious that the influence of the water alone must have made them such as they are, by interfering with the movements of very long limbs with solid internal parts; and that consequently these animals owe their general shape to the influence of the medium they inhabit. But with regard to that degradation which we are seeking among the mammals themselves, the amphibians must be far removed from the cetaceans because their organisation is much less degraded in its essential parts. Amphibians then have to be joined to the unguiculate mammals, while the cetaceans should form the last order of the class, as being the most imperfect mammals.

We now pass to the birds; but I must first note that there is no gradation between mammals and birds. There exists a gap to be filled, and no doubt nature has produced animals which practically fill this gap, and which must form a special class if they cannot be comprised either among the mammals or among the birds.

This fact has just been realised, by the recent discovery in Australia of two genera of animals, viz.:

Ornithorhyncus }
Echidna } Monotremes (Geoff.).

These animals are quadrupeds with no mammae, with no teeth inserted and no lips; and they have only one orifice for the genital organs, the excrements and the urine (a cloaca). Their body is covered with hair or bristles.

They are not mammals, for they have no mammae and are most likely oviparous.

They are not birds; for their lungs are not pierced through and they have no limbs shaped as wings.

Finally, they are not reptiles; for their heart with only two ventricles removes them from that category.

They belong then to a special class.

BIRDS.

Animals without mammae, with two feet and two arms shaped as wings; the body covered with feathers.

The second rank clearly belongs to the birds; for while we do not find among these animals so many faculties or so much intelligence as among the animals of the first rank, they are the only ones except the monotremes which have like mammals a heart with two ventricles and two auricles, warm blood, the cavity of the cranium completely filled by the brain, and the trunk always enclosed by ribs. They have, then, qualities common to mammals, but not found elsewhere; and consequently affinities with them that are not to be found in any animals of the posterior classes.

But the birds when compared with the mammals display an obvious degradation of organisation which has nothing to do with the influence of the environment. They are for instance naturally devoid of mammae, organs with which only animals of the highest rank are provided and which belong to a system of reproduction that is no longer found in the birds nor in any of the animals of subsequent ranks. In short they are essentially oviparous; for the system of truly viviparous animals, which is adapted to animals of the first rank, is not found in the second nor does it again re-appear. Their foetus is enclosed in an inorganic envelope (the egg-shell) and soon ceases communication with the mother and can develop without feeding on her substance.

The diaphragm, which among mammals completely separates somewhat obliquely the chest from the abdomen, here ceases to exist, or becomes very incomplete.

The vertebrae of the neck and tail are the only mobile parts in the vertebral column of birds. Since movements of the other vertebrae of that column are not necessary to the animal, they are not performed and they thus place no obstacle to the large development of the sternum which now makes such movement almost impossible.

The sternum of birds indeed gives attachment to the pectoral muscles, which have become very thick and strong by reason of their energetic and almost continuous movements. The sternum has thus become extremely large and carinate in the middle. This, however, is due to the habits of these animals and not to the general degradation that we are investigating. The truth of this is exemplified by the fact that the mammal called a bat has also a carinate sternum.

All the blood of birds passes through their lungs before reaching the other parts of the body. Thus they breathe exclusively by lungs like the animals of the first rank; and this is not the case with any known animal after them.

We now come to a very strange peculiarity which is connected with the environment of these animals. They live more than other vertebrates in the air, and are almost continually rising into it and passing through it in every direction. They have adopted a habit of swelling their lungs with air in order to increase their volume and make themselves lighter; and this habit has caused the organ to adhere to the sides of the chest so that the air within, being rarefied by the heat of the place, has had to pierce through the lung with its investing membranes and to penetrate every part of the body even to the inside of the great bones which are hollow, and to the quills of the large feathers.[1] It is, however, only in the lungs that the blood of birds undergoes the necessary influence of the air; for the air which penetrates to the other parts of the body has another use than that of respiration.

Thus the birds, which have been rightly placed after the mammals, exhibit an obvious degradation in their general organisation: not because their lung has a peculiarity not found among the former, for this is due like their feathers only to their acquired habit of launching themselves into the air; but because they no longer have the system of reproduction proper to the most perfect animals, but only that which characterises most of the animals of the posterior classes.

It is very difficult to ascertain among the birds themselves the degradation of organisation which we are now studying; our knowledge of their organisation is still too vague. Hence it has hitherto been a matter of convention which order should be placed at the head of this class and which at the end.

We may reflect however that aquatic birds (like the palmipeds), as also the waders and gallinaceans, have this advantage over all other birds that their young on coming out of the egg can walk and feed. We may pay special attention to the fact that among the palmipeds, the penguins and king-penguins, whose almost featherless wings are merely oars for swimming and of no use for flight,

[1] If it is true that in the case of birds the lungs are pierced through and the hair changed into feathers as a result of their habit of rising into the air, I may be asked why bats have not also feathers and pierced lungs. I reply that it seems to me probable that bats, which have a more perfect organisation than birds, and hence a complete diaphragm to impede the swelling of their lungs, have not been able to pierce them through nor to swell themselves out with air sufficiently for that fluid even by an effort to reach the skin and so to give to the horny matter of the hair the faculty of branching out into feathers. Among birds, in fact, air is introduced as far as the hair bulbs; changing their bases into quills and compelling this same hair to break up into feathers; an event which cannot occur in the bat, where the air does not penetrate beyond the lung.

thus approximate in some ways to the monotremes and cetaceans. We shall then recognise that the palmipeds, waders, and gallinaceans should constitute the first three orders of birds, and that the doves, passerines, birds of prey and climbers should form the last four orders of the class. Now, from what we know of the habits of the birds of these last four orders, we find that their young on coming out of the egg can neither walk nor feed by themselves.

On this principle the climbers are the last order of birds; moreover, they are the only ones which have two posterior digits and two anterior. This character, which they possess in common with the chameleon, appears to justify us in placing them near the reptiles.

REPTILES.

Animals with only one ventricle in the heart and still possessing a pulmonary respiration though incomplete. Their skin is smooth or provided with scales.

In the third rank are naturally and necessarily placed the reptiles; and they will furnish us with new and stronger proofs of the degradation of organisation from one extremity of the animal chain to the other, starting from the most perfect animals. In fact, their heart, which has only one ventricle, no longer displays that conformation which belongs essentially to animals of the first and second ranks, and their blood is cold, almost like that of the animals of the posterior ranks.

We find another proof of the degradation of the organisation of reptiles in their respiration. In the first place they are the last animals to breathe by true lungs; for after them we find no respiratory organ of this nature in any of the succeeding classes, as I shall endeavour to show when speaking of molluscs. Next, the lung has in their case usually very large chambers, proportionally less numerous, and is already much simplified. In many species this organ is absent in youth and is then replaced by gills, a respiratory organ which is never found in animals of the anterior ranks. Sometimes the two kinds of respiratory organs are present together in the same individual.

But the strongest proof of degradation in the respiration of reptiles is that only part of their blood passes through the lungs, while the rest reaches the parts of the body without having undergone the influence of respiration.

Finally, among reptiles the four limbs essential to the most perfect animals begin to be lost, and indeed many of them (nearly all the snakes) lack them altogether.

Independently of the degradation of organisation indicated by the shape of the heart, by the temperature of the blood which scarcely arises above the level of the environment, by the incomplete respiration and by the almost regular simplification of the lung, it is found that reptiles differ considerably among themselves ; so that there are greater differences of organisation and external shape among the animals of the various orders of this class than among those of the two preceding classes. Some habitually live in the air, and of these, such as have no legs can only crawl ; others live in the water or on its banks, sometimes withdrawing into the water and sometimes going into open places. There are some that are clothed in scales and others that have a naked skin. Lastly, although they all have a heart with one ventricle, in some there are two auricles, while in others there is only one. All these differences are due to environment, manner of life, etc. ; conditions which doubtless act more strongly upon an organisation that is still remote from the goal to which nature is tending, than they could do on one more advanced towards perfection.

Reptiles are oviparous animals (including even those in which the eggs are hatched in the body of the mother) ; their skeleton is modified and usually very degraded ; their respiration and circulation are less perfect than those of mammals and birds ; and they all have a small brain which does not fill the cavity of the cranium. Hence they are less perfect than the animals of the two preceding classes, and in their turn confirm the fact that the degradation of organisation increases, according as we approach the most imperfect animals.

Within this class of animals themselves, independently of the modifications in their parts due to environment, we find in addition traces of the general degradation of organisation ; for in the last of their orders (the batrachians) the individuals, when they are first born, breathe by gills.

If the absence of legs observed among snakes were regarded as a result of degradation, the ophidians ought to be the last order of reptiles ; but it would be a mistake to suppose this. The fact is that snakes are animals which for purposes of concealment have adopted the habit of crawling directly on the ground, and their body has thus acquired a considerable length, out of proportion to its size. Now elongated legs would have impeded their efforts in crawling and concealing themselves ; while very short legs, of which there could only be four since these animals are vertebrates, would have been incapable of moving their body. Thus the habits of these animals have caused the disappearance of their legs ; although the batrachians, which have legs, are more degraded in organisation and nearer to the fishes.

The proofs of the important principle which I am stating will be based upon positive facts; they will consequently always hold good in contact with the arguments that are brought against them.

FISHES.

Animals breathing by gills, with a smooth or scaly skin; the body provided with fins.

On following the course of that degradation undergone by organisation as a whole and of the diminution in the number of animal faculties, we see that the fishes must of necessity be placed in the fourth rank, that is, after the reptiles. Their organisation in fact is even less advanced towards perfection than is that of reptiles, and is consequently more remote from that of the most perfect animals.

It is true no doubt that their general shape, the absence of a constriction between the head and body to form a neck, and the various fins which for them take the place of limbs, are results of the influence of the dense medium they inhabit, and not of the degradation of organisation. But that degradation is none the less real and very great, as we may convince ourselves by an examination of their internal organs; so that we are forced to assign to fishes a lower rank than to reptiles.

We no longer find in them the respiratory organ of the most perfect animals; for they have no true lung, and in its place have only gills or vascular pectinate folds arranged on both sides of the neck or head, four altogether on each side. The water which these animals breathe goes in by the mouth, passes between the folds of the gills, and bathes the numerous vessels which run there. Now since the water is mixed with air or contains it in solution, that air although small in quantity acts upon the blood of the gills and there achieves the function of respiration. The water then issues through open holes on either side of the neck.

Note that this is the last time that the respired fluid enters by the animal's mouth in order to reach the organ of respiration.

These animals, like those of the posterior ranks, have no trachea or larynx or true voice (including even those called *grondeurs*[1]) or eyelids, etc. These organs and faculties are here lost and are not again found throughout the animal kingdom.

Yet the fishes are still part of the division of vertebrate animals;

[1] [The Grey Gurnard. H.E.]

but they are the last of them and they terminate the fifth stage of organisation, being in common with reptiles the only animals which have:

A vertebral column;

Nerves, terminating in a brain, which does not fill the cranium;

A heart with one ventricle;

Warm blood;

Lastly, a completely internal ear.

Fishes thus display an oviparous reproduction; a body without mammae, of a shape adapted for swimming; fins which are not all invariably analogous with the four limbs of the most perfect animals; a very incomplete skeleton curiously modified and rudimentary in the last animals of this class; only one ventricle in the heart and cold blood; gills instead of lungs; a very small brain; the sense of touch incapable of giving knowledge of the shapes of bodies; and apparently without any sense of smell, for odours are only conveyed by air. It is clear that these animals strongly confirm in their turn also the degradation of organisation that we have undertaken to follow throughout the animal kingdom.

We shall now see that fishes are primarily divided into what are called bony fishes, which are the most perfect of them, and cartilaginous fishes, which are the least perfect. These two facts confirm the degradation of organisation within the class itself; for among the cartilaginous fishes the softness and cartilaginous condition of the parts intended to stiffen their bodies and aid their movements indicate that it is among them that the skeleton ends or rather that nature has sketched its first rudiments.

By continually following the order of nature in the inverse direction, the eight last genera of this class should include the fishes whose branchial apertures have no operculum or membrane and are nothing but holes at the sides or under the throat; finally the lampreys and hag-fishes should terminate the class, for these fishes differ greatly from all others by the imperfection of their skeleton and in having a naked slimy body without lateral fins, etc.

Observations on the Vertebrates.

The vertebrate animals, although differing greatly from one another as regards their organs, appear to be all formed on a common plan of organisation. On passing from the fishes to the mammals, we find that this plan becomes more perfect from class to class and that it only reaches completion in the most perfect mammals; but we may also notice that this plan while approaching perfection has undergone numerous modifications, some of them very large, through the influence of the environment of the animals and of the habits which each race has been forced to contract by the conditions in which it is placed.

Hence we see, on the one hand, that if vertebrates differ markedly from one another in their organisation, it is because nature only started to carry out her plan in their respect with the fishes; that she made further advances with the reptiles; that she carried it still nearer perfection with the birds, and that finally she only attained the end with the most perfect mammals.

On the other hand, we cannot fail to recognise that if the perfection of the plan of organisation of the vertebrates does not everywhere show a regular and even gradation from the most imperfect fishes to the most perfect mammals, the reason is that nature's work has often been modified, thwarted and even reversed by the influence exercised by very different and indeed conflicting conditions of life upon animals exposed to them throughout a long succession of generations.

Annihilation of the Vertebral Column.

On reaching this point in the animal scale the vertebral column becomes entirely annihilated. Since this column is the basis of every true skeleton, and since this bony framework is an important part of the organisation of the most perfect animals, it follows that all the invertebrate animals, which we are about to investigate in turn, must have an organisation still more degraded than that of the four classes that we have just passed in review. Henceforth, therefore, the supports for muscular activity will no longer reside in any internal parts.

Moreover, none of the invertebrate animals breathes by cellular lungs; none of them has any voice nor consequently any organ for this faculty; finally they mostly appear devoid of true blood, that is to say, of that fluid which in the vertebrate is essentially red, but which only owes its colour to the intensity of their animalisation, and proves especially a real circulation. How grave an abuse of words it would be to give the name of blood to the thin and colourless fluid which moves slowly through the cellular substance of the polyps! We might as well apply the name to the sap of plants.

Besides the vertebral column, we also lose here the iris which is characteristic of the eyes of the most perfect animals; for such of the invertebrates as have eyes have no distinct irises.

Kidneys in the same way are only found among the vertebrates, and fishes are the last animals where this organ is met with. Henceforward there is no more spinal cord, no more great sympathetic nerve.

A final very important observation is that among vertebrates, and especially in the neighbourhood of that extremity of the animal scale where the most perfect animals are found, all the essential organs are isolated or have each an isolated seat in as many special places. We shall soon see that the complete contrary holds good according as we approach the other extremity of the scale.

It is then obvious that all the invertebrate animals have a less perfect organisation than any of those which possess a vertebral column; while the organisation of mammals is that which from all aspects includes the most perfect animals and is beyond question the true type of the highest perfection.

Let us now enquire whether the classes and large families into which the long series of invertebrate animals is divided also exhibit, when we compare them together, an increasing degradation in the complexity and perfection of their organisation.

INVERTEBRATE ANIMALS.

On reaching invertebrate animals we enter upon an immense series of diverse creatures, the most numerous of any existing in nature, the most curious and interesting with regard to the variations observed in their organisation and faculties.

On observing their condition, we are convinced that in bringing them successively into existence, nature has proceeded gradually from the simplest to the most complex. Now since the purpose in view has been to attain a plan of organisation which should admit of the highest perfection (that of the vertebrates)—a plan very different from those which nature had hitherto used to reach this point —we may be sure that among these numerous animals we shall not meet with a single system of organisation progressively perfected, but with various quite distinct systems, each one taking its start at the point where each organ of highest importance began to exist.

For instance, when nature attained to the creation of a special organ for digestion (as in the polyps) she then gave for the first time a special constant shape to the animals provided with it; seeing that the infusorians with which she began everything could not possess either the faculty endowed by this organ, or the kind of shape and organisation favourable to its functions.

She subsequently established a special organ for respiration, and in proportion as she varied this organ in order to perfect it and to accommodate it to the animal's environment, she diversified their organisation, in so far as the existence and development of the other special organs rendered it necessary.

When afterwards she succeeded in producing the nervous system, it then immediately became possible to create the muscular system. Thereupon, fixed points for the attachments of the muscles became necessary, and also paired parts so as to constitute a symmetrical shape. Hence have resulted various schemes of organisation due to the environment and to the parts acquired, which could not previously have come about.

When finally she secured sufficient movement in the contained fluids of the animal to permit a circulation to be organised, there again resulted important peculiarities of organisation which distinguished it from the organic systems in which there is no circulation.

In order to perceive the truth of what I have stated and to furnish evidence of the degradation and simplification of organisation (since we are following the order of nature in the inverse direction) let us rapidly run through the various classes of invertebrate animals.

MOLLUSCS.

Soft unjointed animals which breathe by gills and have a mantle. No ganglionic longitudinal cord; no spinal cord.

The fifth rank, as we descend the graduated scale of the animal series, necessarily belongs to the molluscs; for they have to be placed a stage lower than the fishes since they have no vertebral column, but they are yet the most highly organised of invertebrate animals. They breathe by gills, which vary greatly not only in their shape and size, but in their position within or without the animal according to the genera, and the habits of the races comprised in these genera. They all have a brain; nerves without nodes, that is to say, without a row of ganglia stretching down a longitudinal cord. They have arteries and veins and one or several single-chambered hearts. They are the only known animals which, although possessing a nervous system, have neither a spinal cord, nor a ganglionic longitudinal cord.

Gills, which are essentially intended by nature to carry out respiration during immersion in the water, have been subjected to modification both in function and shape in those aquatic animals which have been constantly exposed for generations to contact with the air, and even in some cases have stayed in it altogether.

The respiratory organ of these animals has imperceptibly become accustomed to the air; and this is no mere supposition: for it is known that all the crustaceans have gills and yet there are crabs (*Cancer ruricola*) which habitually live on land and breathe air quite naturally with their gills. Eventually this habit of breathing air with gills became a necessity to many molluscs which acquired it: it even modified the organ in such a way that the gills of these animals, having no further need for so many points of contact with the respired fluid, became adherent to the walls of the cavity which contains them.

As a result we may distinguish among molluscs two kinds of gills.

The first kind consist of networks of vessels running through the skin of an internal cavity which is not protruded and can only breathe air: these may be called aerial gills.

The second kind are organs nearly always protruded either within or without the animal and forming fringes or pectinate lamellae or edgings, etc.: these can only achieve respiration by means of the contact of fluid water, and may be called aquatic gills.

If the differences in the habits of animals produce differences in their organs, it will be useful in describing the special characters of certain orders of molluscs to distinguish those which have aerial gills from those whose gills can only breathe water; but in any case they are always gills and it appears to us quite improper to say that the molluscs which breathe air possess a lung. How often the abuse of words and wrong applications of names have served to distort objects and lead us into error!

After all, is the difference so great between the respiratory organ of *Pneumoderma*, which consists in a vascular network running over an external skin, and the vascular network of snails, which runs over an internal skin? Yet *Pneumoderma* appears to breathe nothing but water.

Let us further enquire for a moment if there are any affinities between the respiratory organ of air-breathing molluscs and the lung of vertebrates.

A lung is essentially a peculiar spongy mass composed of more or less numerous cells into which air is always entering in nature. The entrance is effected through the animal's mouth and thence by a more or less cartilaginous canal called the trachea, which usually sub-divides into branches known as bronchi, culminating in the cells. The cells and bronchi are alternately filled and emptied of air by successive swellings and shrinkings of the cavity of the body containing the mass; so that distinct alternate inspirations and expirations are characteristic of a lung. This organ can only tolerate the contact of air and is highly irritated by water or any other material. It is therefore different in character from the branchial cavity of certain molluscs, which is quite peculiar, exhibits no alternate swelling and shrinking, never has a trachea or bronchi and in which the respired fluid never enters by the animal's mouth.

A respiratory cavity which has neither trachea nor bronchi nor alternate swelling and shrinking, and in which the respired fluid does not enter by the mouth, and which is adapted either for air or water, cannot be a lung. To confuse such different things by the same name is not to advance science but to retard it.

The lung is the only respiratory organ that can give the animal the faculty of having a voice. After the reptiles no animal has a lung; nor therefore a voice.

I conclude that it is not true that there are molluscs which breathe by lungs. If some in nature breathe air, so also do certain crustaceans and all insects; but none of these animals has true lungs, unless the same name is to be given to very different objects.

The molluscs also furnish proof of the progressive degradation that we are investigating in the animal chain; for their general organisation is less perfect than that of fishes. But it is not so easy to recognise the same degradation among the molluscs themselves; for it is difficult to distinguish in so numerous and varied a class what is due to the degradation in question, from what is caused by the environment and habits of these animals.

The only two orders into which the large class of molluscs is divided, are strongly contrasted by the importance of their distinctive characters. The animals of the first of these orders (cephalic molluscs) have a very distinct head, eyes, jaws or a proboscis and reproduce by copulation.

All the molluscs of the second order (acephalic molluscs) on the contrary are destitute of a head, eyes, jaws, proboscis; and they never copulate for the purpose of reproduction.

Now it can hardly be denied that the second order of molluscs is inferior to the first as regards perfection of organisation.

It is important, however, to remember that the absence of head, eyes, etc., in the acephalic molluscs is not wholly due to the general degradation of organisation, since we find again at lower stages of the animal chain, animals which have a head, eyes, etc. We have here again apparently one of those deviations in the progress of perfection of organisation that are produced by environment, and consequently by causes foreign to those which make for a gradual increase of complexity in animal organisation.

When we come to consider the influence of the use of organs and of an absolute and permanent disuse, we shall see that a head, eyes, etc., would in fact have been of very little use to molluscs of the second order, because the large development of their mantle would have prevented the functioning of these organs.

In conformity with that law of nature which requires that every organ permanently disused should imperceptibly deteriorate, become reduced and finally disappear, the head, eyes, jaws, etc., have in fact become extinct in the acephalic molluscs: we shall see elsewhere many other examples of the same thing.

In the invertebrates nature no longer finds in the internal parts any support for muscular movement; she has therefore supplied the molluscs with a mantle for that purpose. Now the strength and compactness of this mantle of the molluscs is proportional to the necessity entailed by their locomotion and means of support.

Thus in the cephalic molluscs, where there is more locomotion than in those which have no head, the mantle is closer, thicker and stronger; and among the cephalic molluscs, those which are naked (without shells) have in addition a cuirass in their mantle which is stronger than the mantle itself and greatly facilitates the locomotion and contraction of the animal (slugs).

But if, instead of following the animal chain in the opposite direction from the actual order of nature, we followed it from the most imperfect animals to the most perfect, we should then easily perceive that nature when she was about to start the plan of organisation of the vertebrates, was forced in the molluscs to abandon the use of a crustaceous or horny skin as a support for muscular action, and to prepare to transfer these fulcra into the interior of the animal. In this way the molluscs are to some extent in the midst of this change of system of organisation; they have in consequence only feeble powers of locomotive movements and they all carry out such movements with remarkable slowness.

CIRRHIPEDES.

Animals without eyes which breathe by gills and have a mantle and jointed arms with a horny skin.

The cirrhipedes, of which only four genera [1] are yet known, should be considered as a special class, since these animals cannot belong to any other class of invertebrate animals.

They approach the molluscs by their mantle and should be placed immediately after the acephalic molluscs, since like them they have neither head nor eyes.

Yet the cirrhipedes cannot be a part of the class of molluscs; for

[1] *Anatifa*, *Balanus*, *Coronula*, and *Tubicinella*.

their nervous system is characterised like the animals of the three following classes by a ganglionic longitudinal cord. They have moreover jointed arms with a horny skin and several pairs of transverse jaws. They are therefore of lower rank than molluscs. Their fluids move by a true circulation with arteries and veins.

These animals are fixed on marine bodies and in consequence carry out no locomotion; their principal movements are those of their arms. Now although they have a mantle like the molluscs, nature could not obtain from it any assistance for the movements of their arms, and was forced to create in the skin of those arms fulcra for their muscles. Hence the skin is coriaceous and almost horny like that of crustaceans and insects.

ANNELIDS.

Animals with elongated annulated bodies without jointed legs, breathing by gills and having a circulatory system and a ganglionic longitudinal cord.

The class of annelids necessarily comes after that of cirrhipedes, because no annelid has a mantle. We are moreover compelled to place them before the crustaceans, because they have no jointed legs and it would not do to interpose them in the series of those which have; nor does their organisation permit us to place them lower than the insects.

Although these animals in general are still very little known, the rank to which their organisation entitles them proves that in their case again the degradation of organisation is continued; for from this aspect they are inferior to the molluscs in that they have a ganglionic longitudinal cord; they are inferior also to the cirrhipedes, which have a mantle like molluscs; and the fact that they have not jointed legs prevents us from interposing them in the series of those which are so organised.

Annelids owe their elongated form to their habits of life, for they either live buried in damp earth or in mud or actually in the water, mostly in tubes of various materials which they enter and leave at will. Thus they are so like worms that all naturalists hitherto have confused the two.

Their internal organisation shows a very small brain, a ganglionic longitudinal cord, arteries and veins in which circulates blood that is usually coloured red; they breathe by gills, sometimes external and protruding, and sometimes internal and hidden or invisible.

CRUSTACEANS.

Animals with a jointed body and limbs, crustaceous skin, a circulatory system, and breathing by gills.

We now enter upon the long series of animals, whose body and limbs are jointed, and whose integuments are hard, crustaceous, horny or coriaceous.

The solid or hard parts of these animals are all on the exterior. Since nature created the muscular system very little in advance of the earlier animals of this series, and since she had need of solid support to endow it with energy, she was obliged to establish the method of articulation in order to secure the possibility of movement.

All the animals that exhibit this method of articulation were held by Linnæus and subsequently as forming only a single class, to which was given the name of insects; but it was at length recognised that this large series of animals has several important divisions which must be distinguished.

The class of crustaceans, which had thus been confused with that of insects, although all the ancient naturalists had always kept it apart, is a division indicated by nature and must be maintained. It should follow immediately upon the annelids and occupy the eighth rank in the general series of animals; this is required by their organisation and is not a matter of arbitrary opinion.

The crustaceans indeed have a heart, arteries and veins; a transparent and almost colourless circulating fluid, and they all breathe by true gills. This is unquestionable and will always constitute a difficulty in the way of those who persist in placing them among the insects on account of their having jointed legs.

If the crustaceans are completely distinguished from the arachnids and insects by their circulation and respiratory organ, and if their rank is therefore obviously superior, they yet share one trait of inferiority of organisation with the arachnids and insects as compared with the annelids; that, namely, of being a part of the series of animals with jointed limbs: a series in the course of which the circulatory system and consequently the heart, arteries and veins are seen to diminish and disappear, and in which again the branchial system of respiration is likewise lost. The crustaceans therefore again confirm the continuous degradation of organisation in the direction in which we are following the animal scale. The transparency and extreme thinness of the fluid which circulates in their vessels, like that of insects, is a further proof of their degradation.

As to their nervous system, it consists of a very small brain and a ganglionic longitudinal cord. This is a sign of poverty of that system observed among the animals of the two preceding and the two following classes; for the animals of these classes are the last in which a nervous system is still to be seen.

It is in the crustaceans that the last traces of an organ of hearing have been identified; after them it is no more found in any animal.

Observations.

Here ends the existence of a true circulatory system, that is to say, of a system of arteries and veins, which is part of the organisation of the most perfect animals and with which those of all the preceding classes are provided. The organisation of the animals of which we shall now speak is still more imperfect than that of the crustaceans, which are the last in which a circulation is actually to be found. The degradation of organisation is thus clearly in progress; since according as we advance along the series of animals, all features of resemblance are successively lost between the organisation of those we come to and that of the most perfect animals.

Whatever may be the nature of the movement of the fluids in the animals of the classes that we are about to traverse, that move ment is secured by less active methods, and constantly tends to become slower.

ARACHNIDS.

Animals breathing by limited tracheae, undergoing no metamorphosis, and having throughout their lives jointed legs and eyes in their head.

On continuing the order that we have hitherto followed, the ninth rank in the animal kingdom necessarily belongs to the arachnids; they have so much affinity with the crustaceans that we shall always have to bring them together, immediately following one another. They are however entirely distinct; for the arachnids furnish us with the first example of a respiratory organ lower than gills,—one never met with in animals which have a heart, arteries and veins.

Arachnids in fact breathe only by stigmata and air-carrying tracheae, which are respiratory organs analogous to those of insects. But these tracheae, instead of extending throughout the body as in the insects, are limited to a small number of sacs: this shows that nature is bringing to an end in the arachnids the method of respiration which she had to employ before the establishment of gills, just as she brought to an end in the fishes or later reptiles that which she had to make use of, before she could form a true lung.

If the arachnids are quite distinct from the crustaceans, through not breathing by gills but by very limited air-carrying tracheae, they are also to be distinguished from the insects. It would be quite as improper to combine them with the insects of which they lack the classic character and from which they differ even in internal organisation, as it would be to confuse the crustaceans with the insects.

The arachnids, indeed, although having strong affinities with the insects are essentally distinct from them:

1. In that they never undergo metamorphosis, that they have at birth the shape and all the parts of an adult and that, consequently, they have eyes in their head and jointed legs throughout their lives. This is an order of things that follows from the nature of their internal organisation and therein differs greatly from that of insects;

2. In that in the arachnids of the first order (pedipalp-arachnids) we begin to see the outlines of a circulatory system; [1]

3. In that their respiratory system, although of the same order as that of insects, is nevertheless very different; since their tracheae are limited to a small number of sacs, and do not constitute the very numerous air-canals extending throughout the animal's body that are witnessed in the tracheae of insects;

4. Lastly, in that the arachnids procreate several times in the course of their life; a faculty which the insects do not possess.

These considerations suffice to show how faulty are those arrangements in which the arachnids and insects are combined into one class, through paying exclusive regard to the joints in these animals' legs, and the more or less crustaceous skin which covers them. It is almost as if we were to consider only the more or less scaly integuments of reptiles and fishes, and thus to combine them into one class.

The general degradation of organisation that we are seeking throughout the entire animal scale is extremely obvious in the arachnids: these animals indeed breathe by an organ inferior in organic perfection to lungs and even to gills, and have only the rudiments of a circulation apparently not yet finished off. They thus confirm in their turn the continuous degradation in question.

This degradation may even be observed in the series of species belonging to that class; for the arachnids with antennae, making up the second order, are sharply distinguished from the others, are very inferior to them in progress of organisation, and come close to the insects; they differ from the latter however in undergoing no meta-

[1] "It is especially in the spiders that this heart may be easily observed: it may be seen beating through the skin of the abdomen in species that are not hairy. On removing this skin, a hollow oblong organ is seen, pointed at the two ends, with the anterior end directed towards the thorax and from the sides of which there issue visibly two or three pairs of vessels" (Cuvier, *Anatom. comp.* vol. iv. p. 419).

morphosis; and as they never launch themselves into the air, it is very probable that their tracheae do not generally extend throughout all parts of their bodies.

INSECTS.

Animals which undergo metamorphoses, and have in the perfect state two eyes and two antennae in their head, six jointed legs and wo tracheae which extend throughout their body.

As we continue to follow the inverse order from that of nature, the insects necessarily succeed the arachnids. They constitute that immense series of imperfect animals which have no arteries or veins; which breathe by air-carrying tracheae not limited to special parts; lastly, which are born in a state less perfect than that in which they reproduce; and which consequently undergo metamorphoses.

In their perfect state all insects without exception have six jointed legs, two antennae and two eyes in their head, and most of them also have wings.

The insects necessarily occupy the tenth rank of the animal kingdom in the order that we are following; for they are inferior to the arachnids in perfection of organisation since they are not born like these latter in their perfect state, and they only procreate once in the course of their life. It is particularly in the insects that we begin to observe that the organs essential to maintenance of life are almost equally distributed, and in most cases situated throughout their bodies instead of being isolated in special places, as is the case in the most perfect animals. The exceptions to this rule gradually disappear, so that it becomes ever more striking in the lower classes of animals.

Nowhere hitherto has the general degradation of organisation been more manifest than in the insects, whose organisation is less perfect than that of the animals of any of the preceding classes. This degradation comes out even within the various orders into which insects are naturally divided; for those of the three first orders (Coleoptera, Orthoptera, Neuroptera) have mandibles and maxillae in their mouths; those of the fourth order (Hymenoptera) begin to possess a sort of proboscis; finally, those of the four last orders (Lepidoptera, Hemiptera, Diptera and Aptera) have really nothing more than a proboscis. Now paired maxillae are nowhere found again in the animal kingdom, after the insects of the three first orders. With regard to wings, the insects of the six first orders have four, all of which or only two serve for flight. Those of the seventh and eighth have only two wings or else they are quite aborted. The larvae of the insects of the two last orders have no legs and are like worms.

It appears that the insects are the last animals which have a quite distinct sexual reproduction and are probably oviparous.

Lastly, we shall see that insects are rendered highly remarkable by what is called their skill; but this alleged skill is far from being the product of any thought or any combination of ideas on their part.

Observation.

Just as among the vertebrates the fishes display in their general conformation and anomalies of organisation the product of the action of their environment; so the insects among the invertebrates exhibit, in their shape, organisation and metamorphoses, the obvious effects of the action of the air in which they live; for most of them launch themselves into it and habitually maintain themselves there like birds.

If the insects had had lungs, if they had been able to swell themselves out with air, and if the air which penetrates into every part of their body could have there become rarefied like that which is introduced into the body of birds, their hair would no doubt have changed into feathers.

Lastly, if among invertebrate animals we are surprised to find so few affinities between the insects which undergo remarkable metamorphoses and other classes of invertebrates, let us remember that these are the only invertebrate animals which launch themselves into the air and there execute movements of progression; we shall then no longer be surprised that such peculiar conditions and habits must have produced peculiar results.

The insects are allied only to the arachnids by their affinities, and in fact these two are in general the only invertebrate animals that live in the air; but no arachnid has the faculty of flight; none therefore undergoes metamorphosis; and when I come to treat of the influence of habits, I shall show that these animals, being accustomed to remain on the surface of the earth and to live in retreats, must have lost a part of the faculties of insects and acquired characters which conspicuously distinguish the two groups.

Extinction of Several Organs essential to the Most Perfect Animals.

After the insects it appears that there is a rather large gap in the series remaining to be filled by animals not yet observed; for in this part of the series several organs essential to the most perfect animals suddenly drop out and are really annihilated, since they are not found again in the classes which remain to be considered.

Disappearance of the Nervous System.

Here for instance the nervous system (the nerves and their centre of communication) completely disappears, and is no more found in any of the animals of the succeeding classes.

In the most perfect animals, this system consists of a brain which appears to serve for carrying out acts of intelligence. At its base is the nucleus of sensations from which issue nerves and also a dorsal spinal cord which sends out other nerves to various parts.

Among the vertebrates the brain becomes regularly reduced, and as its volume diminishes the spinal cord becomes larger and seems to take its place.

Among the molluscs which constitute the first class of the invertebrates the brain still exists, but there is no spinal cord nor a ganglionic longitudinal cord, and as ganglia are rare the nerves do not appear to have nodes.

Lastly, in the five following classes the nervous system is approaching its end and is reduced to the very small rudiments of a brain and to a longitudinal cord from which issue nerves. Thereafter there is no longer a separate nucleus for sensations, but a multitude of small nuclei scattered throughout the length of the animal's body.

Hence among insects, the important system of feeling comes to an end; a system which at a certain stage of development gives birth to ideas and which in its highest perfection can produce all the acts of intelligence; which, lastly, is the source whence muscular action derives its power and without which sexual reproduction apparently could not exist.

The centre of communication of the nervous system is situated in the brain or at its base or in a ganglionic longitudinal cord. There is still a longitudinal cord when there is no longer any obvious brain; but when there is neither a brain nor a longitudinal cord, the nervous system ceases to exist.

Disappearance of the Sexual Organs.

Here again all traces of sexual reproduction disappear; indeed among the animals about to be cited it is no longer possible to recognise organs for true impregnation. We shall, however, still find among the animals of the two following classes kinds of ovaries, abounding in oviform corpuscles that are alleged to be eggs; but I look upon these supposed eggs which can develop without previous fertilisation as buds or internal gemmules; they establish a connection between internal gemmiparous reproduction and sexual oviparous reproduction.

The strength of habit is so great that man always perseveres in the same view of things, even when it is contrary to the evidence.

Thus botanists, accustomed to observing the sexual organs of a great number of plants, affirm that all without exception have such organs. Consequently several botanists have made every conceivable effort to discover stamens and pistils in cryptogamic or agamic plants; and they have preferred to attribute arbitrarily and without proof these functions to parts of whose use they are ignorant, rather than admit that nature may attain the same end by different means.

It was believed that every reproductive body is a seed or egg, that is to say, a body which must undergo the influence of sexual fertilisation in order to be reproductive. This is what caused Linnæus to say: *Omne vivum ex ovo.* But we now know well plants and animals which reproduce entirely by means of bodies that are neither seeds nor eggs, and which consequently do not require sexual fertilisation. These bodies are therefore differently fashioned and develop in another manner.

The following is the principle to be observed in judging of the method of reproduction in any living body.

Any reproductive corpuscle which *without having any investment to break through* lengthens, grows and becomes a plant or animal similar to that from which it sprang, is not a seed nor an egg; it undergoes no germination and does not hatch after beginning to grow, and its formation requires no sexual fertilisation: thus it does not contain an embryo in an investment which has to be broken through, as does the seed or egg.

Now, if you follow attentively the development of the reproductive corpuscles of algae, fungi, etc., you will see that the result of the lengthening and growth of these corpuscles is to take imperceptibly the shape of the plant from which they spring; that they do not break through any investment as does the embryo in the seed or egg.

Similarly, if you follow the gemma or bud of a polyp like a hydra, you will be convinced that this reproductive body does nothing but lengthen out and grow; that it breaks through no investment; in short, that it does not hatch like a chicken or silkworm coming out of its egg.

It is then clear that all reproduction of individuals does not come about by means of sexual fertilisation; and that when sexual fertilisation does not occur there is not really a true sexual organ. Now as no organ for fertilisation is to be distinguished in the four classes following the insects it appears that this is the point in the animal chain at which sexual reproduction ceases to exist.

Disappearance of the Organ of Sight.

Here again the organ of sight, so useful to the most perfect animals, is entirely extinguished. This organ began to be deficient in some of the molluscs and cirrhipedes and in most of the annelids, and is only found afterwards in the crustaceans, arachnids and insects in a very imperfect state and of little or no use; after the insects it does not re-appear in any animal.

Here again, finally, the head altogether ceases to exist,—an essential part of the body of the most perfect animals and the seat of the brain and nearly all the senses; for the swelling at the anterior extremity of the body of some worms like *Taenia* is caused by the arrangement of their suckers and is not the seat of a brain nor of any organ of hearing, sight, etc., since there are no such organs in the animals of the neighbouring classes. Hence this swelling cannot be considered as a true head.

We see that at this part of the animal scale the degradation of organisation becomes extremely rapid, and strongly foreshadows the greatest simplification of animal organisation.

WORMS.

Animals with soft elongated bodies, without head, eyes or jointed legs, and no longitudinal cord or circulatory system.

We now come to worms, which have no vessels for circulation; including those known under the name of intestinal worms, and some others not intestinal whose organisation is quite as imperfect. They are animals with soft more or less elongated bodies, which undergo no metamorphosis and are all destitute of a head, eyes and jointed legs.

The worms should be placed immediately after the insects and before the radiarians, and occupy the eleventh rank in the animal kingdom. It is among them that we note the origin of the tendency of nature to establish the system of articulations, a system that she subsequently carried to completion in the insects, arachnids and crustaceans. But the organisation of the worms is less perfect than that of the insects, since they have no longitudinal cord, head, eyes or true legs, so that we are forced to place them after the insects; lastly, the new kind of shape, which nature initiates in them on passing from a radiating arrangement of the parts to the system of articulations, shows that the worms should be placed even before the radiarians. After the insects, moreover, the plan followed by nature in the animals of preceding classes is lost sight of, viz. that general shape of the animal which consists of a bilateral symmetry of the parts, so that each part is opposite to another exactly like it.

In the worms we no longer find this bilateral symmetry, nor do we yet witness the radiating arrangement of the organs both internal and external which characterises the radiarians.

After I established the annelids, some naturalists called them by the name of worms; and as they did not then know what to do with the animals now under discussion, they united them with the polyps. I leave the reader to imagine what may be the affinities and classic characters that justify the union in one class of *Taenia* or *Ascaris* with a hydra or any other polyp.

Several worms still appear to breathe like insects by tracheae of which the external openings are kinds of stigmata; but there is reason to believe that these limited or imperfect tracheae are water-carrying and not air-carrying like those of insects; because these animals never live in the open air, but are continuously in the water or bathed by fluids which contain water.

As no organ for fertilisation is distinguished in them, I suppose that sexual reproduction does not occur in these animals. It may be however that, just as there exists a primitive circulation in arachnids so there may exist a sexual reproduction in the worms, as is suggested by the various shapes of the tail of *Strongylus*; but observation has not yet fully established such reproduction in these animals.

Objects which are found in some of them and supposed to be ovaries (as in *Taenia*) appear to be merely clusters of reproductive corpuscles which do not require fertilisation. These oviform corpuscles are internal, like those of sea-urchins, and not external like those of *Coryne*, etc. Polyps exhibit similar differences in the situation of their gemmules; it is therefore probable that the worms are internally gemmiparous.

Animals like the worms, which have no head, eyes, legs or perhaps sexual reproduction, provide further evidence in their turn of the continuous degradation of organisation that we are seeking throughout the animal scale.

RADIARIANS.

Animals with regenerating bodies, destitute of a head, eyes or jointed legs; with the mouth on the inferior surface and a radiating arrangement of the parts both internal and external.

In the usual order the radiarians occupy the twelfth rank in the lengthy series of known animals, and constitute one of the three last classes of invertebrates.

When we reach this class we find animals with a general shape and arrangement of the parts and organs, both internal and external, that nature has not employed in any of the animals of the anterior classes.

The radiarians indeed conspicuously exhibit in their internal and external parts that radiating arrangement around a centre or axis, which constitutes a special shape not hitherto used by nature. Its rudiments are found in the polyps, which accordingly come next.

Nevertheless, the radiarians form a stage in the animal scale quite distinct from the polyps; so that we can no more confuse radiarians with polyps than we can class crustaceans with insects or reptiles with fishes. Among the radiarians indeed, not only do we find again organs apparently intended for respiration (tubes or kinds of water-bearing tracheae), but we discover in addition special organs for reproduction, such as kinds of ovaries of various shapes to which there is nothing analogous in the polyps. Moreover, the intestinal canal of the radiarians is not generally a cul-de-sac with a single opening as in all the polyps; their mouth is always on the inferior surface and displays a special arrangement which is quite different from that commonly found in polyps.

Although the radiarians are very remarkable animals, and as yet little known, what we do know of their organisation plainly points to the rank which I am assigning to them. Like the worms, radiarians have no head, eyes, jointed legs, circulatory system or perhaps nerves. Yet the radiarians necessarily come next to the worms, for the latter have nothing in the arrangement of their internal organs that suggests a radiating shape, and it is among them that the system of articulations begins.

If the radiarians are destitute of nerves, they cannot have the faculty of feeling, but are simply irritable; this fact seems to be confirmed by observations made on living star-fishes, whose arms have been cut off without their showing any sign of pain.

In many radiarians fibres may still be distinguished; but can we call these fibres muscles? Not unless we are justified in saying that a muscle can function without nerves. Do not plants show us that cellular tissue may be reduced to fibres? Yet we cannot possibly regard these fibres as muscular. In my opinion it does not follow that because a living being has distinguishable fibres, it must therefore have muscles; I hold that where there are no nerves, there is no muscular system. There is reason to believe that in animals without nerves the fibres which are still to be found possess the faculty by mere irritability of producing movements which replace those of the muscles, although less energetically.

Not only does it appear that the muscular system has ceased to exist in the radiarians, but also there seems to be no sexual reproduction. There is indeed nothing to show or even to suggest that the little oviform bodies, the clusters of which are called ovaries in these animals, undergo any fertilisation or are true eggs: this is rendered still less probable by the fact that they are found in all individuals alike. Hence I regard these little oviform bodies as already perfected internal gemmules, and the clusters of them in special places as nature's preliminary step towards sexual reproduction.

The radiarians in their turn contribute to prove the general degradation of animal organisation; for on reaching this class of animals we find a shape and a new arrangement of the parts and organs that are far removed from the animals of the preceding classes. Furthermore they appear to be destitute of feeling, muscular movement and sexual reproduction; among them the intestinal canal no longer has two exits, the clusters of oviform corpuscles disappear, and the body becomes completely gelatinous.

Observation.

It appears that in very imperfect animals such as the polyps and radiarians, the centre of movement of the fluids does not exist except in the alimentary canal; it is here that it is first established, and it is through this canal that the subtle surrounding fluids enter, mainly for the purpose of stimulating the movement of the fluids which belong to these animals themselves. What would plant life be without external stimuli? What indeed would be the life of the most imperfect animals without this factor, that is, without the caloric and electricity of the environment?

The radiating form has no doubt been acquired as a consequence of this method, which nature employs feebly at first in the polyps and afterwards with greater vigour in the radiarians; for the subtle surrounding fluids which enter the alimentary canal are expansive and must by incessant repulsion from the centre towards every point of the circumference give rise to this radiating arrangement of the parts.

This is the reason why in the radiarians the intestinal canal, although still very imperfect, since it has usually only one opening, is none the less provided with numerous radiating vasculiform and often branched appendages.

This again is no doubt the reason why in the soft radiarians, such as jelly-fishes, etc., we may observe a continual isochronous movement, a movement which very probably results from the alternative movements of the masses of subtle fluids, which penetrate into the interior of these animals and escape again, after having spread throughout all their parts.

Let it not be said that the isochronous movements of the soft radiarians are signs of respiration; for nature does not exhibit in any animal after the vertebrates those alternate and measured movements of inspiration and expiration; whatever the respiration of radiarians may be, it is extremely slow and involves no appreciable movement.

POLYPS.

Animals with sub-gelatinous and regenerating bodies, with no special organ but an alimentary canal with only one opening. Terminal mouth supplied with radiating tentacles or a ciliated and rotatory organ.

With the polyps we reach the penultimate stage of the animal scale, that is to say, the last but one of the classes which have to be established among animals.

Here the imperfection and simplicity of organisation are very striking, so that the animals of this group have scarcely any faculties left, and their animal nature has long been doubted.

They are gemmiparous animals with homogeneous bodies, usually gelatinous, and with very regenerative parts, not displaying the radiating shape (for it is only here that nature began it) except in radiating tentacles around their mouth, and having no special organ but an intestinal canal, which has only one opening and is therefore incomplete.

The polyps may be described as much more imperfect animals than any of those which make up the preceding classes, for they have no brain, longitudinal cord, nerves, special respiratory organs, vessels for the circulation of fluids nor ovary for reproduction. The substance of their body is to a great extent homogeneous and composed of a gelatinous and irritable cellular tissue in which fluids move slowly. Lastly, their viscera are entirely reduced to an imperfect alimentary canal, rarely folded on itself or provided with appendages, and usually resembling a mere elongated sac, always with a single opening which serves at once for mouth and anus.

There can be no justification for the statement that, although we find in these animals no nervous system, respiratory organ or muscle, etc., yet these organs still exist infinitely reduced and distributed or dissolved throughout the general substance of the body, and equally divided up in every molecule instead of being collected in special places; consequently that every point in their body can experience every kind of sensation, muscular movement, will, ideas and thought; this would be an altogether gratuitous, baseless and improbable supposition. On such a supposition a hydra must have in every part of its body all the organs of the most perfect animals, and hence every point in the body of this polyp must see, hear, distinguish odours, tastes, etc.; and also must have ideas, form judgments, think and, in short, reason; each molecule of the body of a hydra or any other polyp would in itself be a perfect animal; and the whole hydra would be a more perfect animal even than man. since each of its molecules would be equivalent, as regards the completion of organisation and faculties, to an entire individual of the human species.

There is no reason why such an argument should not be extended to the *Monas*—the most imperfect of known animals—and even to plants themselves, which also possess life. We should then attribute to each molecule of a plant all the aforementioned faculties, though restricted within limits set by the nature of the living body of which it is part.

Assuredly it is not to this that the study of nature leads us. This study teaches us, on the contrary, that wherever an organ ceases to exist, the function depending on it ceases likewise. Any animal which has no eyes or whose eyes have been destroyed cannot see; and although in the last analysis the various senses derive their origin from touch, of which they are only special modifications, yet no animal which is without nerves, the special organ of feeling, could experience any kind of sensation; for it has not the intimate feeling of its existence, it has not the central nucleus to which sensation has to be conveyed, and consequently it could not feel.

Thus the sense of touch, which is the basis of the other senses and is spread throughout every part of the bodies of animals which have nerves, no longer exists in those which, like the polyps, have no nerves. Among the latter, the parts are nothing more than merely irritable; they are so in a very high degree, but they are devoid of feeling and hence of every kind of sensation. In order that a sensation may arise, an organ is first necessary to receive it (nerves); and then some central nucleus must exist (a brain or ganglionic longitudinal cord) to which this sensation may be conveyed.

A sensation is always the sequel of an impression received and immediately conveyed to an internal nucleus, where the sensation is formed. Interrupt the communication between the organ which receives the impression and the nucleus where the sensation is formed, and all feeling will immediately cease. This principle can never be disputed.

No polyp can really be oviparous; for it has no special organ for reproduction. Now in order to produce true eggs, it is necessary not only that the animal should have an ovary, but in addition that it or some other individual of its species should have a special organ for fertilisation, and it cannot be shown that the polyps have such organs; in place of them we are well aware of the buds which some of them produce for purposes of multiplication; and on paying them a little attention we note that these buds are themselves nothing more than somewhat isolated portions of the animal's body,—portions less simple than those employed by nature for the multiplication of the animalcules which compose the last class of the animal kingdom.

Polyps, being highly irritable, only move by external stimuli foreign to themselves. All their movements are necessary results of impressions received, and are in general carried out without any act of will; they are thus without any possibility of choice, since they cannot have any will.

They invariably and inevitably move towards the light, just like the branches and leaves or flowers of plants, although in their case the movement is slower. No polyp pursues its prey, nor seeks for it with its tentacles; but when some foreign body touches these same tentacles, they hold it and carry it to the mouth, and the polyp swallows it without making any distinction as to its suitability or the reverse. It digests and feeds on the body if it is capable of being digested, but rejects it entirely if it remains some time untouched in the alimentary canal; finally, it brings up such of the débris as can be no more broken up; but in all this, there is the same necessity in the action and never any possibility of choice to vary it.

The distinction of the polyps from the radiarians is very wide and glaring; nowhere in the interior of the polyps is the radiating arrangement to be found: their tentacles alone have this arrangement, thus resembling the arms of the cephalapod molluscs, with which however they certainly cannot be confused. Moreover the polyps have a superior terminal mouth, while the mouth of the radiarians is otherwise situated.

It is altogether improper to give the polyps the name of zoophytes, which means animal-plants; because they are entirely and completely animals. They have faculties absent in plants, that, for instance, of true irritablity and generally of digestion; and, lastly, their nature has nothing essentially in common with that of a plant.

The only affinities existing between polyps and plants are: (1) A similar simplification of organisation; (2) the faculty possessed by many polyps of adhering to one another with a common conmunication by their alimentary canal, and of forming compound animals; (3) the external shape of the groups formed by these combined polyps, a shape which has long caused these groups to be taken for true plants, since they are often branched almost in the same way.

Whether polyps have one or several mouths, there is always an alimentary canal to which they lead and consequently an organ for digestion, of which all plants are destitute.

If the degradation of organisation that we have observed in all classes starting from the mammals is anywhere obvious, it is assuredly among the polyps, whose organisation is reduced to an extreme simplification.

INFUSORIANS.

Infinitely small animals with gelatinous, transparent, homogeneous and very contractile bodies ; with no distinct special organ internally, but often oviform gemmules ; and having externally no radiating tentacles nor rotatory organs.

At length we reach the last class of the animal kingdom, comprising the most imperfect animals from all points of view ; that is, those which have the simplest organisation, possess the fewest faculties and seem all to be mere rudiments of animal nature.

Hitherto I have placed these small animals in the class of polyps, of which they constituted the last order under the name of amorphous polyps, since they have no constant shape peculiar to them all ; but I have recognised the necessity of separating them to form a class apart, though this in no wise changes the rank that I had assigned to them. The only result of this change is to establish a line of demarcation which appears to be called for, on account of the greater simplicity of their organisation and their lack of radiating tentacles and rotatory organs.

Since the organisation of the infusorians becomes ever more simple as we pass down their genera, the last of these genera shows us in some degree the limit of animality, the limit at all events of what we can reach. It is especially in the animals of the second order of this class that we can verify the entire disappearance of any trace of an intestinal canal and mouth ; so that they have no special organ whatever nor any digestion.

They are only very tiny gelatinous, transparent, contractile and homogeneous bodies, consisting of cellular tissue, with very slight cohesion and yet irritable throughout. These tiny bodies, which look like animated or moving points, feed by absorption and continual imbibition ; and they are doubtless animated by the influence of the subtle surrounding fluids, such as caloric and electricity, which stimulate in them the movements constituting life.

If we were to imagine that such animals possess all the organs known in other animals, but that these organs are dissolved throughout their bodies, how absurd such a supposition would be!

The extremely slight cohesion between the parts of these tiny gelatinous bodies is an indication that such organs cannot exist, since they could not possibly carry on their functions. It is clear that in order that any organs may have the power of reacting on fluids and of carrying on their appropriate functions, their parts must have enough cohesion and firmness to give them strength; now this is not to be imagined in the case of these fragile animalcules. It is exclusively among the animals of this class that nature appears to carry out direct or spontaneous generations, which are incessantly renewed whenever conditions are favourable; and we shall endeavour to show that it is through this means that she acquired the power after an enormous lapse of time to produce indirectly all the other races of animals that we know.

Justification for the belief that the infusorians or most of them owe their existence exclusively to spontaneous generation is found in the fact that all these fragile animals perish during the reduction of temperature in bad seasons; and it surely will not be suggested that such delicate bodies could leave any bud sufficiently hardy to be preserved and to reproduce them in warm weather.

Infusorians are found in stagnant waters and infusions of plant or animal substances, and even in the seminal fluid of the most perfect animals. They are found just the same in all parts of the world, but only in conditions suitable for their existence.

Thus on examining in turn the various systems of organisation of animals from the most complex to the simplest, we have seen the degradation of animal organisation beginning even in the class that comprises the most perfect animals and thence advancing progressively from class to class, although with anomalies due to environment, and finally ending with the infusorians. These last are the most imperfect animals and the simplest in organisation,—the animals in which the degradation that we have traced reaches its limit. Animal organisation is then reduced to a simple homogeneous gelatinous body with very slight cohesion, destitute of special organs, and entirely formed of a very delicate and primitive cellular tissue, which appears to be vivified by subtle surrounding fluids incessantly penetrating it and exhaling from it.

We have seen each special organ in turn, including even the most essential, become slowly degraded till it is less special, less isolated, and finally completely lost and gone, long before reaching the other extremity of the order we are tracing; and we have noticed that it is chiefly among the invertebrate animals that the extinction of special organs occurs.

It is true that even before leaving the division of vertebrates, we already witness great changes in the perfection of organs ; while some even disappear altogether, such as the urinary bladder, the diaphragm, the organ of voice, the eyelids, etc. The lung, for instance, which is the most perfect respiratory organ, begins its degradation in the reptiles and ceases to exist in the fishes, not to reappear again in any invertebrate animal. Finally, the skeleton, the appendages of which constitute the basis of the four extremities or limbs possessed by most vertebrates, begins its deterioration mainly in the reptiles and comes entirely to an end with the fishes.

But it is in the division of invertebrates that the extinction takes place of the heart, brain, gills, conglomerate glands, vessels for circulation, the organs of hearing and sight, that of sexual reproduction and even that of feeling, as also of movement.

As I have already said, we should vainly seek in a polyp, such as the hydra or most animals of that class, the slightest vestiges either of nerves (organs of feeling) or of muscles (organs of movement)! Irritability, with which every polyp is highly endowed, alone replaces the faculty of feeling which no polyp possesses, since it has not the essential organ for it.

It also replaces the faculty of voluntary movement, since all will is an act of the organ of intelligence, and this animal is absolutely destitute of any such organ. All its movements are necessary results of the impressions on its irritable parts of external stimuli ; and they are carried out without any scope for choice.

Put a hydra in a glass of water, and set this glass in a room where daylight only enters by one window and therefore only from one side. When this hydra has fixed on some point of the sides of the glass, turn the glass so that the light strikes it on the opposite side to that where the animal is. You will then always see the hydra go with a slow movement and place itself where the light strikes, and stay there so long as you do not change this point. This is the same in those parts of plants which, without any act of will, lean towards the side from which the light comes.

Doubtless wherever a special organ no longer exists, the function which it supports also ceases to exist ; and we may furthermore clearly observe that according as an organ is degraded and reduced, the function resulting from it becomes proportionally more vague and imperfect. Thus, we find that on descending from the most complex towards the simplest, the insects are the last animals which have eyes ; but there is sound reason for the belief that they see very dimly and make little use of them.

Thus on traversing the chain of animals from the most perfect to the most imperfect, and on examining in turn the various systems of organisation distinguished in the course of this chain, the degradation of organisation and of each organ up to their complete disappearance is seen to be a positive fact which we have now verified.

This degradation comes out even in the nature and consistency of the essential fluids and flesh of animals. For the flesh and blood of mammals and birds are the most complex and animalised materials that can be obtained from the soft parts of animals. Hence after the fishes these materials are progressively degraded until in the soft radiarians, the polyps and the infusorians, the essential fluid has only the consistency and colour of water and the flesh is nothing more than a gelatinous scarcely animalised material. The bouillon made from such flesh would scarcely be found very nourishing or strengthening by any one who tried to live upon it.

Whether or no we recognise these interesting truths, they will nevertheless always be forced upon the attention of those who closely observe facts, and who, overcoming prevailing prejudices, consult the phenomena of nature and study her laws and regular procedure.

We shall now pass to the examination of another kind of subject, and shall endeavour to prove that the environment exercises a great influence over the activities of animals, and that as a result of this influence the increased and sustained use or disuse of any organ are causes of modification of the organisation and shape of animals and give rise to the anomalies observed in the progress of the complexity of animal organisation.

CHAPTER VII.

OF THE INFLUENCE OF THE ENVIRONMENT ON THE ACTIVITIES AND HABITS OF ANIMALS, AND THE INFLUENCE OF THE ACTIVITIES AND HABITS OF THESE LIVING BODIES IN MODIFYING THEIR ORGANISATION AND STRUCTURE.

WE are not here concerned with an argument, but with the examination of a positive fact—a fact which is of more general application than is supposed, and which has not received the attention that it deserves, no doubt because it is usually very difficult to recognise. This fact consists in the influence that is exerted by the environment on the various living bodies exposed to it.

It is indeed long since the influence of the various states of our organisation on our character, inclinations, activities and even ideas has been recognised; but I do not think that anyone has yet drawn attention to the influence of our activities and habits even on our organisation. Now since these activities and habits depend entirely on the environment in which we are habitually placed, I shall endeavour to show how great is the influence exerted by that environment on the general shape, state of the parts and even organisation of living bodies. It is, then, with this very positive fact that we have to do in the present chapter.

If we had not had many opportunities of clearly recognising the result of this influence on certain living bodies that we have transported into an environment altogether new and very different from that in which they were previously placed, and if we had not seen the resulting effects and alterations take place almost under our very eyes, the important fact in question would have remained for ever unknown to us.

The influence of the environment as a matter of fact is in all times and places operative on living bodies; but what makes this influence difficult to perceive is that its effects only become perceptible or recognisable (especially in animals) after a long period of time.

Before setting forth to examine the proofs of this fact, which deserves our attention and is so important for zoological philosophy, let us sum up the thread of the discussions that we have already begun.

In the preceding chapter we saw that it is now an unquestionable fact that on passing along the animal scale in the opposite direction from that of nature, we discover the existence, in the groups composing this scale, of a continuous but irregular degradation in the organisation of animals, an increasing simplification in their organisation, and, lastly, a corresponding diminution in the number of their faculties.

This well-ascertained fact may throw the strongest light over the actual order followed by nature in the production of all the animals that she has brought into existence, but it does not show us why the increasing complexity of the organisation of animals from the most imperfect to the most perfect exhibits only an *irregular gradation*, in the course of which there occur numerous anomalies or deviations with a variety in which no order is apparent.

Now on seeking the reason of this strange irregularity in the increasing complexity of animal organisation, if we consider the influence that is exerted by the infinitely varied environments of all parts of the world on the general shape, structure and even organisation of these animals, all will then be clearly explained.

It will in fact become clear that the state in which we find any animal, is, on the one hand, the result of the increasing complexity of organisation tending to form a regular gradation; and, on the other hand, of the influence of a multitude of very various conditions ever tending to destroy the regularity in the gradation of the increasing complexity of organisation.

I must now explain what I mean by this statement: *the environment affects the shape and organisation of animals*, that is to say that when the environment becomes very different, it produces in course of time corresponding modifications in the shape and organisation of animals.

It is true if this statement were to be taken literally, I should be convicted of an error; for, whatever the environment may do, it does not work any direct modification whatever in the shape and organisation of animals.

But great alterations in the environment of animals lead to great alterations in their needs, and these alterations in their needs necessarily lead to others in their activities. Now if the new needs become permanent, the animals then adopt new habits which last as long as the needs that evoked them. This is easy to demonstrate, and indeed requires no amplification.

It is then obvious that a great and permanent alteration in the environment of any race of animals induces new habits in these animals.

Now, if a new environment, which has become permanent for some race of animals, induces new habits in these animals, that is to say, leads them to new activities which become habitual, the result will be the use of some one part in preference to some other part, and in some cases the total disuse of some part no longer necessary.

Nothing of all this can be considered as hypothesis or private opinion ; on the contrary, they are truths which, in order to be made clear, only require attention and the observation of facts.

We shall shortly see by the citation of known facts in evidence, in the first place, that new needs which establish a necessity for some part really bring about the existence of that part, as a result of efforts ; and that subsequently its continued use gradually strengthens, develops and finally greatly enlarges it ; in the second place, we shall see that in some cases, when the new environment and the new needs have altogether destroyed the utility of some part, the total disuse of that part has resulted in its gradually ceasing to share in the development of the other parts of the animal ; it shrinks and wastes little by little, and ultimately, when there has been total disuse for a long period, the part in question ends by disappearing. All this is positive ; I propose to furnish the most convincing proofs of it.

In plants, where there are no activities and consequently no habits, properly so-called, great changes of environment none the less lead to great differences in the development of their parts ; so that these differences cause the origin and development of some, and the shrinkage and disappearance of others. But all this is here brought about by the changes sustained in the nutrition of the plant, in its absorption and transpiration, in the quantity of caloric, light, air and moisture that it habitually receives ; lastly, in the dominance that some of the various vital movements acquire over others.

Among individuals of the same species, some of which are continually well fed and in an environment favourable to their development, while others are in an opposite environment, there arises a difference in the state of the individuals which gradually becomes very remarkable. How many examples I might cite both in animals and plants which bear out the truth of this principle ! Now if the environment remains constant, so that the condition of the ill-fed, suffering or sickly individuals becomes permanent, their internal organisation is ultimately modified, and these acquired modifications are preserved by reproduction among the individuals in question, and finally give rise to a race quite distinct from that in which the individuals have been continuously in an environment favourable to their development.

A very dry spring causes the grasses of a meadow to grow very little, and remain lean and puny; so that they flower and fruit after accomplishing very little growth.

A spring intermingled with warm and rainy days causes a strong growth in this same grass, and the crop is then excellent.

But if anything causes a continuance of the unfavourable environment, a corresponding variation takes place in the plants: first in their general appearance and condition, and then in some of their special characters.

Suppose, for instance, that a seed of one of the meadow grasses in question is transported to an elevated place on a dry, barren and stony plot much exposed to the winds, and is there left to germinate; if the plant can live in such a place, it will always be badly nourished, and if the individuals reproduced from it continue to exist in this bad environment, there will result a race fundamentally different from that which lives in the meadows and from which it originated. The individuals of this new race will have small and meagre parts; some of their organs will have developed more than others, and will then be of unusual proportions.

Those who have observed much and studied large collections, have acquired the conviction that according as changes occur in environment, situation, climate, food, habits of life, etc., corresponding changes in the animals likewise occur in size, shape, proportions of the parts, colour, consistency, swiftness and skill.

What nature does in the course of long periods we do every day when we suddenly change the environment in which some species of living plant is situated.

Every botanist knows that plants which are transported from their native places to gardens for purposes of cultivation, gradually undergo changes which ultimately make them unrecognisable. Many plants, by nature hairy, become glabrous or nearly so; a number of those which used to lie and creep on the ground, become erect; others lose their thorns or excrescences; others again whose stem was perennial and woody in their native hot climates, become herbaceous in our own climates and some of them become annuals; lastly, the size of their parts itself undergoes very considerable changes. These effects of alterations of environment are so widely recognised, that botanists do not like to describe garden plants unless they have been recently brought into cultivation.

Is it not the case that cultivated wheat (*Triticum sativum*) is a plant which man has brought to the state in which we now see it? I should like to know in what country such a plant lives in nature, otherwise than as the result of cultivation.

Where in nature do we find our cabbages, lettuces, etc., in the same state as in our kitchen gardens? and is not the case the same with regard to many animals which have been altered or greatly modified by domestication?

How many different races of our domestic fowls and pigeons have we obtained by rearing them in various environments and different countries; birds which we should now vainly seek in nature?

Those which have changed the least, doubtless because their domestication is of shorter standing and because they do not live in a foreign climate, none the less display great differences in some of their parts, as a result of the habits which we have made them contract. Thus our domestic ducks and geese are of the same type as wild ducks and geese; but ours have lost the power of rising into high regions of the air and flying across large tracts of country; moreover, a real change has come about in the state of their parts, as compared with those of the animals of the race from which they come.

Who does not know that if we rear some bird of our own climate in a cage and it lives there for five or six years, and if we then return it to nature by setting it at liberty, it is no longer able to fly like its fellows, which have always been free? The slight change of environment for this individual has indeed only diminished its power of flight, and doubtless has worked no change in its structure; but if a long succession of generations of individuals of the same race had been kept in captivity for a considerable period, there is no doubt that even the structure of these individuals would gradually have undergone notable changes. Still more, if instead of a mere continuous captivity, this environmental factor had been further accompanied by a change to a very different climate; and if these individuals had by degrees been habituated to other kinds of food and other activities for seizing it, these factors when combined together and become permanent would have unquestionably given rise imperceptibly to a new race with quite special characters.

Where in natural conditions do we find that multitude of races of dogs which now actually exist, owing to the domestication to which we have reduced them? Where do we find those bull-dogs, greyhounds, water-spaniels, spaniels, lap-dogs, etc., etc.; races which show wider differences than those which we call specific when they occur among animals of one genus living in natural freedom?

No doubt a single, original race, closely resembling the wolf, if indeed it was not actually the wolf, was at some period reduced by man to domestication. That race, of which all the individuals were then alike, was gradually scattered with man into different countries and climates; and after they had been subjected for some time to the influences of their environment and of the various habits which had been forced upon them in each country, they underwent remarkable alterations and formed various special races. Now man travels about to very great distances, either for trade or any other purpose; and thus brings into thickly populated places, such as a great capital, various races of dogs formed in very distant countries. The crossing of these races by reproduction then gave rise in turn to all those that we now know.

The following fact proves in the case of plants how the change of some important factor leads to alteration in the parts of these living bodies.

So long as *Ranunculus aquatilis* is submerged in the water, all its leaves are finely divided into minute segments; but when the stem of this plant reaches the surface of the water, the leaves which develop in the air are large, round and simply lobed. If several feet of the same plant succeed in growing in a soil that is merely damp without any immersion, their stems are then short, and none of their leaves are broken up into minute divisions, so that we get *Ranunculus hederaceus*, which botanists regard as a separate species.

There is no doubt that in the case of animals, extensive alterations in their customary environment produce corresponding alterations in their parts; but here the transformations take place much more slowly than in the case of plants; and for us therefore they are less perceptible and their cause less readily identified.

As to the conditions which have so much power in modifying the organs of living bodies, the most potent doubtless consist in the diversity of the places where they live, but there are many others as well which exercise considerable influence in producing the effects in question.

It is known that localities differ as to their character and quality, by reason of their position, construction and climate: as is readily perceived on passing through various localities distinguished by special qualities; this is one cause of variation for animals and plants living in these various places. But what is not known so well and indeed what is not generally believed, is that every locality itself changes in time as to exposure, climate, character and quality, although with such extreme slowness, according to our notions, that we ascribe to it complete stability.

Now in both cases these altered localities involve a corresponding alteration in the environment of the living bodies that dwell there, and this again brings a new influence to bear on these same bodies.

Hence it follows that if there are extremes in these alterations, there are also finer differences: that is to say, intermediate stages which fill up the interval. Consequently there are also fine distinctions between what we call species.

It is obvious then that as regards the character and situation of the substances which occupy the various parts of the earth's surface, there exists a variety of environmental factors which induces a corresponding variety in the shapes and structure of animals, independent of that special variety which necessarily results from the progress of the complexity of organisation in each animal.

In every locality where animals can live, the conditions constituting any one order of things remain the same for long periods: indeed they alter so slowly that man cannot directly observe it. It is only by an inspection of ancient monuments that he becomes convinced that in each of these localities the order of things which he now finds has not always been existent; he may thence infer that it will go on changing.

Races of animals living in any of these localities must then retain their habits equally long: hence the apparent constancy of the races that we call species,—a constancy which has raised in us the belief that these races are as old as nature.

But in the various habitable parts of the earth's surface, the character and situation of places and climates constitute both for animals and plants environmental influences of extreme variability. The animals living in these various localities must therefore differ among themselves, not only by reason of the state of complexity of organisation attained in each race, but also by reason of the habits which each race is forced to acquire; thus when the observing naturalist travels over large portions of the earth's surface and sees conspicuous changes occurring in the environment, he invariably finds that the characters of species undergo a corresponding change.

Now the true principle to be noted in all this is as follows:

1. Every fairly considerable and permanent alteration in the environment of any race of animals works a real alteration in the needs of that race.

2. Every change in the needs of animals necessitates new activities on their part for the satisfaction of those needs, and hence new habits.

3. Every new need, necessitating new activities for its satisfaction, requires the animal, either to make more frequent use of some of its parts which it previously used less, and thus greatly to develop and enlarge them; or else to make use of entirely new parts, to which the needs have imperceptibly given birth by efforts of its inner feeling; this I shall shortly prove by means of known facts.

Thus to obtain a knowledge of the true causes of that great diversity of shapes and habits found in the various known animals, we must reflect that the infinitely diversified but slowly changing environment in which the animals of each race have successively been placed, has involved each of them in new needs and corresponding alterations in their habits. This is a truth which, once recognised, cannot be disputed. Now we shall easily discern how the new needs may have been satisfied, and the new habits acquired, if we pay attention to the two following laws of nature, which are always verified by observation.

First Law.

In every animal which has not passed the limit of its development, a more frequent and continuous use of any organ gradually strengthens, develops and enlarges that organ, and gives it a power proportional to the length of time it has been so used; while the permanent disuse of any organ imperceptibly weakens and deteriorates it, and progressively diminishes its functional capacity, until it finally disappears.

Second Law.

All the acquisitions or losses wrought by nature on individuals, through the influence of the environment in which their race has long been placed, and hence through the influence of the predominant use or permanent disuse of any organ; all these are preserved by reproduction to the new individuals which arise, provided that the acquired modifications are common to both sexes, or at least to the individuals which produce the young.

Here we have two permanent truths, which can only be doubted by those who have never observed or followed the operations of nature, or by those who have allowed themselves to be drawn into the error which I shall now proceed to combat.

Naturalists have remarked that the structure of animals is always in perfect adaptation to their functions, and have inferred that the shape and condition of their parts have determined the use of them. Now this is a mistake: for it may be easily proved by observation that it is on the contrary the needs and uses of the parts which have caused the development of these same parts, which have even given birth to them when they did not exist, and which consequently have given rise to the condition that we find in each animal.

If this were not so, nature would have had to create as many different kinds of structure in animals, as there are different kinds of environment in which they have to live; and neither structure nor environment would ever have varied.

This is indeed far from the true order of things. If things were really so, we should not have race-horses shaped like those in England;

we should not have big draught-horses so heavy and so different from the former, for none such are produced in nature; in the same way we should not have basset-hounds with crooked legs, nor grey-hounds so fleet of foot, nor water-spaniels, etc.; we should not have fowls without tails, fantail pigeons, etc.; finally, we should be able to cultivate wild plants as long as we liked in the rich and fertile soil of our gardens, without the fear of seeing them change under long cultivation.

A feeling of the truth in this respect has long existed; since the following maxim has passed into a proverb and is known by all, *Habits form a second nature.*

Assuredly if the habits and nature of each animal could never vary, the proverb would have been false and would not have come into existence, nor been preserved in the event of any one suggesting it.

If we seriously reflect upon all that I have just set forth, it will be seen that I was entirely justified when in my work entitled *Recherches sur les corps vivants* (p. 50), I established the following proposition:

"It is not the organs, that is to say, the nature and shape of the parts of an animal's body, that have given rise to its special habits and faculties; but it is, on the contrary, its habits, mode of life and environment that have in course of time controlled the shape of its body, the number and state of its organs and, lastly, the faculties which it possesses."

If this proposition is carefully weighed and compared with all the observations that nature and circumstances are incessantly throwing in our way, we shall see that its importance and accuracy are substantiated in the highest degree.

Time and a favourable environment are as I have already said nature's two chief methods of bringing all her productions into existence: for her, time has no limits and can be drawn upon to any extent.

As to the various factors which she has required and still constantly uses for introducing variations in everything that she produces, they may be described as practically inexhaustible.

The principal factors consist in the influence of climate, of the varying temperatures of the atmosphere and the whole environment, of the variety of localities and their situation, of habits, the commonest movements, the most frequent activities, and, lastly, of the means of self-preservation, the mode of life and the methods of defence and multiplication.

Now as a result of these various influences, the faculties become extended and strengthened by use, and diversified by new habits that are long kept up. The conformation, consistency and, in short, the character and state of the parts, as well as of the organs, are imperceptibly affected by these influences and are preserved and propagated by reproduction.

These truths, which are merely effects of the two natural laws stated above, receive in every instance striking confirmation from facts; for the facts afford a clear indication of nature's procedure in the diversity of her productions.

But instead of being contented with generalities which might be considered hypothetical, let us investigate the facts directly, and consider the effects in animals of the use or disuse of their organs on these same organs, in accordance with the habits that each race has been forced to contract.

Now I am going to prove that the permanent disuse of any organ first decreases its functional capacity, and then gradually reduces the organ and causes it to disappear or even become extinct, if this disuse lasts for a very long period throughout successive generations of animals of the same race.

I shall then show that the habit of using any organ, on the contrary, in any animal which has not reached the limit of the decline of its functions, not only perfects and increases the functions of that organ, but causes it in addition to take on a size and development which imperceptibly alter it; so that in course of time it becomes very different from the same organ in some other animal which uses it far less.

The permanent disuse of an organ, arising from a change of habits, causes a gradual shrinkage and ultimately the disappearance and even extinction of that organ.

Since such a proposition could only be accepted on proof, and not on mere authority, let us endeavour to make it clear by citing the chief known facts which substantiate it.

The vertebrates, whose plan of organisation is almost the same throughout, though with much variety in their parts, have their jaws armed with teeth; some of them, however, whose environment has induced the habit of swallowing the objects they feed on without any preliminary mastication, are so affected that their teeth do not develop. The teeth then remain hidden in the bony framework of the jaws, without being able to appear outside; or indeed they actually become extinct down to their last rudiments.

In the right-whale, which was supposed to be completely destitute of teeth, M. Geoffroy has nevertheless discovered teeth concealed in the jaws of the foetus of this animal. The professor has moreover discovered in birds the groove in which the teeth should be placed, though they are no longer to be found there.

Even in the class of mammals, comprising the most perfect animals, where the vertebrate plan of organisation is carried to its highest completion, not only is the right-whale devoid of teeth, but the ant-eater (*Myrmecophaga*) is also found to be in the same condition, since it has acquired a habit of carrying out no mastication, and has long preserved this habit in its race.

Eyes in the head are characteristic of a great number of different animals, and essentially constitute a part of the plan of organisation of the vertebrates.

Yet the mole, whose habits require a very small use of sight, has only minute and hardly visible eyes, because it uses that organ so little.

Olivier's *Spalax* (*Voyage en Égypte et en Perse*), which lives underground like the mole, and is apparently exposed to daylight even less than the mole, has altogether lost the use of sight: so that it shows nothing more than vestiges of this organ. Even these vestiges are entirely hidden under the skin and other parts, which cover them up and do not leave the slightest access to light.

The *Proteus*, an aquatic reptile allied to the salamanders, and living in deep dark caves under the water, has, like the *Spalax*, only vestiges of the organ of sight, vestiges which are covered up and hidden in the same way.

The following consideration is decisive on the question which I am now discussing,

Light does not penetrate everywhere; consequently animals which habitually live in places where it does not penetrate, have no opportunity of exercising their organ of sight, if nature has endowed them with one. Now animals belonging to a plan of organisation of which eyes were a necessary part, must have originally had them. Since, however, there are found among them some which have lost the use of this organ and which show nothing more than hidden and covered up vestiges of them, it becomes clear that the shrinkage and even disappearance of the organ in question are the results of a permanent disuse of that organ.

This is proved by the fact that the organ of hearing is never in this condition, but is always found in animals whose organisation is of the kind that includes it: and for the following reason.

The substance of sound,[1] that namely which, when set in motion by the shock or the vibration of bodies, transmits to the organ of hearing

[1] Physicists believe and even affirm that the atmospheric air is the actual substance of sound, that is to say, that it is the substance which, when set in motion by the shocks or vibrations of bodies, transmits to the organ of hearing the impression of the concussions received.

That this is an error is attested by many known facts, showing that it is impossible the impression received, penetrates everywhere and passes through any medium, including even the densest bodies: it follows that every animal, belonging to a plan of organisation of which hearing is an essential part, always has some opportunity for the exercise of this organ wherever it may live. Hence among the vertebrates we do not find any that are destitute of the organ of hearing; and after them, when this same organ has come to an end, it does not subsequently recur in any animal of the posterior classes.

It is not so with the organ of sight; for this organ is found to disappear, re-appear and disappear again according to the use that the animal makes of it.

In the acephalic molluscs, the great development of the mantle would make their eyes and even their head altogether useless. The permanent disuse of these organs has thus brought about their disappearance and extinction, although molluscs belong to a plan of organisation which should comprise them.

Lastly, it was part of the plan of organisation of the reptiles, as of other vertebrates, to have four legs in dependence on their skeleton. Snakes ought consequently to have four legs, especially since they are by no means the last order of the reptiles and are farther from the fishes than are the batrachians (frogs, salamanders, etc.).

Snakes, however, have adopted the habit of crawling on the ground and hiding in the grass; so that their body, as a result of continually repeated efforts at elongation for the purpose of passing through narrow spaces, has acquired a considerable length, quite out of proportion to its size. Now, legs would have been quite useless to these animals and consequently unused. Long legs would have interfered

that the air should penetrate to all places to which the substance producing sound actually does penetrate.

See my memoir *On the Substance of Sound*, printed at the end of my *Hydrogéologie*, p. 225, in which I furnished the proofs of this mistake.

Since the publication of my memoir, which by the way is seldom cited, great efforts have been made to make the known velocity of the propagation of sound in air tally with the elasticity of the air, which would cause the propagation of its oscillations to be too slow for the theory. Now, since the air during oscillation necessarily undergoes alternate compressions and dilatations in its parts, recourse has been had to the effects of the caloric squeezed out during the sudden compressions of the air and of the caloric absorbed during the rarefactions of that fluid. By means of these effects, quantitatively determined by convenient hypotheses, geometricians now account for the velocity with which sound is propagated through air. But this is no answer to the fact that sound is also propagated through bodies which air can neither traverse nor set in motion.

These physicists assume forsooth a vibration in the smallest particles of solid bodies; a vibration of very dubious existence, since it can only be propagated through homogeneous bodies of equal density, and cannot spread from a dense body to a rarefied one or *vice versâ*. Such a hypothesis offers no explanation of the well-known fact that sound is propagated through heterogeneous bodies of very different densities and kinds.

with their need of crawling, and very short legs would have been incapable of moving their body, since they could only have had four. The disuse of these parts thus became permanent in the various races of these animals, and resulted in the complete disappearance of these same parts, although legs really belong to the plan of organisation of the animals of this class.

Many insects, which should have wings according to the natural characteristics of their order and even of their genus, are more or less completely devoid of them through disuse. Instances are furnished by many Coleoptera, Orthoptera, Hymenoptera and Hemiptera, etc., where the habits of these animals never involve them in the necessity of using their wings.

But it is not enough to give an explanation of the cause which has brought about the present condition of the organs of the various animals,—a condition that is always found to be the same in animals of the same species; we have in addition to cite instances of changes wrought in the organs of a single individual during its life, as the exclusive result of a great mutation in the habits of the individuals of its species. The following very remarkable fact will complete the proof of the influence of habits on the condition of the organs, and of the way in which permanent changes in the habits of an individual lead to others in the condition of the organs, which come into action during the exercise of these habits.

M. Tenon, a member of the Institute, has notified to the class of sciences, that he had examined the intestinal canal of several men who had been great drinkers for a large part of their lives, and in every case he had found it shortened to an extraordinary degree, as compared with the same organ in all those who had not adopted the like habit.

It is known that great drinkers, or those who are addicted to drunkenness, take very little solid food, and eat hardly anything; since the drink which they consume so copiously and frequently is sufficient to feed them.

Now since fluid foods, especially spirits, do not long remain either in the stomach or intestine, the stomach and the rest of the intestinal canal lose among drinkers the habit of being distended, just as among sedentary persons, who are continually engaged on mental work and are accustomed to take very little food; for in their case also the stomach slowly shrinks and the intestine shortens.

This has nothing to do with any shrinkage or shortening due to a binding of the parts which would permit of the ordinary extension, if instead of remaining empty these viscera were again filled; we have to do with a real shrinkage and shortening of considerable extent, and such that these organs would burst rather than yield at once to any demand for the ordinary extension.

Compare two men of equal ages, one of whom has contracted the habit of eating very little, since his habitual studies and mental work have made digestion difficult, while the other habitually takes much exercise, is often out-of-doors, and eats well; the stomach of the first will have very little capacity left and will be filled up by a very small quantity of food, while that of the second will have preserved and even increased its capacity.

Here then is an organ which undergoes profound modification in size and capacity, purely on account of a change of habits during the life of the individual.

The frequent use of any organ, when confirmed by habit, increases the functions of that organ, leads to its development and endows it with a size and power that it does not possess in animals which exercise it less.

We have seen that the disuse of any organ modifies, reduces and finally extinguishes it. I shall now prove that the constant use of any organ, accompanied by efforts to get the most out of it, strengthens and enlarges that organ, or creates new ones to carry on functions that have become necessary.

The bird which is drawn to the water by its need of finding there the prey on which it lives, separates the digits of its feet in trying to strike the water and move about on the surface. The skin which unites these digits at their base acquires the habit of being stretched by these continually repeated separations of the digits; thus in course of time there are formed large webs which unite the digits of ducks, geese, etc., as we actually find them. In the same way efforts to swim, that is to push against the water so as to move about in it, have stretched the membranes between the digits of frogs, sea-tortoises, the otter, beaver, etc.

On the other hand, a bird which is accustomed to perch on trees and which springs from individuals all of whom had acquired this habit, necessarily has longer digits on its feet and differently shaped from those of the aquatic animals that I have just named. Its claws in time become lengthened, sharpened and curved into hooks, to clasp the branches on which the animal so often rests.

We find in the same way that the bird of the water-side which does not like swimming and yet is in need of going to the water's edge to secure its prey, is continually liable to sink in the mud. Now this bird tries to act in such a way that its body should not be immersed in the liquid, and hence makes its best efforts to stretch and lengthen its legs. The long-established habit acquired by this bird and all its race of continually stretching and lengthening its legs, results in the individuals of this race becoming raised as though on stilts, and gradually obtaining long, bare legs, denuded of feathers up to the thighs and often higher still. (*Système des Animaux sans vertèbres*, p. 14.)

We note again that this same bird wants to fish without wetting its body, and is thus obliged to make continual efforts to lengthen its neck. Now these habitual efforts in this individual and its race must have resulted in course of time in a remarkable lengthening, as indeed we actually find in the long necks of all water-side birds.

If some swimming birds like the swan and goose have short legs and yet a very long neck, the reason is that these birds while moving about on the water acquire the habit of plunging their head as deeply as they can into it in order to get the aquatic larvae and various animals on which they feed ; whereas they make no effort to lengthen their legs.

If an animal, for the satisfaction of its needs, makes repeated efforts to lengthen its tongue, it will acquire a considerable length (ant-eater, green-woodpecker) ; if it requires to seize anything with this same organ, its tongue will then divide and become forked. Proofs of my statement are found in the humming-birds which use their tongues for grasping things, and in lizards and snakes which use theirs to palpate and identify objects in front of them.

Needs which are always brought about by the environment, and the subsequent continued efforts to satisfy them, are not limited in their results to a mere modification, that is to say, an increase or decrease of the size and capacity of organs ; but they may even go so far as to extinguish organs, when any of these needs make such a course necessary.

Fishes, which habitually swim in large masses of water, have need of lateral vision ; and, as a matter of fact, their eyes are placed on the sides of their head. Their body, which is more or less flattened according to the species, has its edges perpendicular to the plane of the water ; and their eyes are placed so that there is one on each flattened side. But such fishes as are forced by their habits to be constantly approaching the shore, and especially slightly inclined or gently sloping beaches, have been compelled to swim on their flattened surfaces in order to make a close approach to the water's edge. In this position, they receive more light from above than below and stand in special need of paying constant attention to what is passing above them ; this requirement has forced one of their eyes to undergo a sort of displacement, and to assume the very remarkable position found in the soles, turbots, dabs, etc. (*Pleuronectes* and *Achirus*). The position of these eyes is not symmetrical, because it results from an incomplete mutation. Now this mutation is entirely completed in the skates, in which the transverse flattening of the body is altogether horizontal, like the head. Accordingly the eyes of skates are both situated on the upper surface and have become symmetrical.

Snakes, which crawl on the surface of the earth, chiefly need to see objects that are raised or above them. This need must have had its effect on the position of the organ of sight in these animals, and accordingly their eyes are situated in the lateral and upper parts of their head, so as easily to perceive what is above them or at their sides; but they scarcely see at all at a very short distance in front of them. They are, however, compelled to make good the deficiency of sight as regards objects in front of them which might injure them as they move forward. For this purpose they can only use their tongue, which they are obliged to thrust out with all their might. This habit has not only contributed to making their tongue slender and very long and contractile, but it has even forced it to undergo division in the greater number of species, so as to feel several objects at the same time; it has even permitted of the formation of an aperture at the extremity of their snout, to allow the tongue to pass without having to separate the jaws.

Nothing is more remarkable than the effects of habit in herbivorous mammals.

A quadruped, whose environment and consequent needs have for long past inculcated the habit of browsing on grass, does nothing but walk about on the ground; and for the greater part of its life is obliged to stand on its four feet, generally making only few or moderate movements. The large portion of each day that this kind of animal has to pass in filling itself with the only kind of food that it cares for, has the result that it moves but little and only uses its feet for support in walking or running on the ground, and never for holding on, or climbing trees.

From this habit of continually consuming large quantities of food-material, which distend the organs receiving it, and from the habit of making only moderate movements, it has come about that the body of these animals has greatly thickened, become heavy and massive and acquired a very great size: as is seen in elephants, rhinoceroses, oxen, buffaloes, horses, etc.

The habit of standing on their four feet during the greater part of the day, for the purpose of browsing, has brought into existence a thick horn which invests the extremity of their digits; and since these digits have no exercise and are never moved and serve no other purpose than that of support like the rest of the foot, most of them have become shortened, dwindled and, finally, even disappeared.

Thus in the pachyderms, some have five digits on their feet invested in horn, and their hoof is consequently divided into five parts; others have only four, and others again not more than three; but in the ruminants, which are apparently the oldest of the mammals that are permanently confined to the ground, there are not more than two digits on the feet and indeed, in the solipeds, there is only one (horse, donkey).

Nevertheless some of these herbivorous animals, especially the ruminants, are incessantly exposed to the attacks of carnivorous animals in the desert countries that they inhabit, and they can only find safety in headlong flight. Necessity has in these cases forced them to exert themselves in swift running, and from this habit their body has become more slender and their legs much finer; instances are furnished by the antelopes, gazelles, etc.

In our own climates, there are other dangers, such as those constituted by man, with his continual pursuit of red deer, roe deer and fallow deer; this has reduced them to the same necessity, has impelled them into similar habits, and had corresponding effects.

Since ruminants can only use their feet for support, and have little strength in their jaws, which only obtain exercise by cutting and browsing on the grass, they can only fight by blows with their heads, attacking one another with their crowns.

In the frequent fits of anger to which the males especially are subject, the efforts of their inner feeling cause the fluids to flow more strongly towards that part of their head; in some there is hence deposited a secretion of horny matter, and in others of bony matter mixed with horny matter, which gives rise to solid protuberances: thus we have the origin of horns and antlers, with which the head of most of these animals is armed.

It is interesting to observe the result of habit in the peculiar shape and size of the giraffe (*Camelo-pardalis*): this animal, the largest of the mammals, is known to live in the interior of Africa in places where the soil is nearly always arid and barren, so that it is obliged to browse on the leaves of trees and to make constant efforts to reach them. From this habit long maintained in all its race, it has resulted that the animal's fore-legs have become longer than its hind legs, and that its neck is lengthened to such a degree that the giraffe, without standing up on its hind legs, attains a height of six metres (nearly 20 feet).

Among birds, ostriches, which have no power of flight and are raised on very long legs, probably owe their singular shape to analogous circumstances.

The effect of habit is quite as remarkable in the carnivorous mammals as in the herbivores; but it exhibits results of a different kind.

Those carnivores, for instance, which have become accustomed to climbing, or to scratching the ground for digging holes, or to tearing their prey, have been under the necessity of using the digits of their feet: now this habit has promoted the separation of their digits, and given rise to the formation of the claws with which they are armed.

But some of the carnivores are obliged to have recourse to pursuit in order to catch their prey: now some of these animals were compelled by their needs to contract the habit of tearing with their claws, which they are constantly burying deep in the body of another animal in order to lay hold of it, and then make efforts to tear out the part seized. These repeated efforts must have resulted in its claws reaching a size and curvature which would have greatly impeded them in walking or running on stony ground: in such cases the animal has been compelled to make further efforts to draw back its claws, which are so projecting and hooked as to get in its way. From this there has gradually resulted the formation of those peculiar sheaths, into which cats, tigers, lions, etc. withdraw their claws when they are not using them.

Hence we see that efforts in a given direction, when they are long sustained or habitually made by certain parts of a living body, for the satisfaction of needs established by nature or environment, cause an enlargement of these parts and the acquisition of a size and shape that they would never have obtained, if these efforts had not become the normal activities of the animals exerting them. Instances are everywhere furnished by observations on all known animals.

Can there be any more striking instance than that which we find in the kangaroo? This animal, which carries its young in a pouch under the abdomen, has acquired the habit of standing upright, so as to rest only on its hind legs and tail; and of moving only by means of a succession of leaps, during which it maintains its erect attitude in order not to disturb its young. And the following is the result:

1. Its fore legs, which it uses very little and on which it only supports itself for a moment on abandoning its erect attitude, have never acquired a development proportional to that of the other parts, and have remained meagre, very short and with very little strength.

2. The hind legs, on the contrary, which are almost continually in action either for supporting the whole body or for making leaps, have acquired a great development and become very large and strong.

3. Lastly, the tail, which is in this case much used for supporting the animal and carrying out its chief movements, has acquired an extremely remarkable thickness and strength at its base.

These well-known facts are surely quite sufficient to establish the results of habitual use on an organ or any other part of animals. If on observing in an animal any organ particularly well-developed, strong, and powerful, it is alleged that its habitual use has nothing to do with it, that its continued disuse involves it in no loss, and finally, that this organ has always been the same since the creation of the species to which the animal belongs, then I ask, Why can our domestic ducks no longer fly like wild ducks ? I can, in short, cite a multitude of instances among ourselves, which bear witness to the differences that accrue to us from the use or disuse of any of our organs, although these differences are not preserved in the new individuals which arise by reproduction : for if they were their effects would be far greater.

I shall show in Part II., that when the will guides an animal to any action, the organs which have to carry out that action are immediately stimulated to it by the influx of subtle fluids (the nervous fluid), which become the determining factor of the movements required. This fact is verified by many observations, and cannot now be called in question.

Hence it follows that numerous repetitions of these organised activities strengthen, stretch, develop and even create the organs necessary to them. We have only to watch attentively what is happening all around us, to be convinced that this is the true cause of organic development and changes.

Now every change that is wrought in an organ through a habit of frequently using it, is subsequently preserved by reproduction, if it is common to the individuals who unite together in fertilisation for the propagation of their species. Such a change is thus handed on to all succeeding individuals in the same environment, without their having to acquire it in the same way that it was actually created.

Furthermore, in reproductive unions, the crossing of individuals who have different qualities or structures is necessarily opposed to the permanent propagation of these qualities and structures. Hence it is that in man, who is exposed to so great a diversity of environment, the accidental qualities or defects which he acquires are not preserved and propagated by reproduction. If, when certain peculiarities of shape or certain defects have been acquired, two individuals who are both affected were always to unite together, they would hand on the same peculiarities ; and if successive generations were limited to such unions, a special and distinct race would then be formed. But perpetual crossings between individuals, who have not the same peculiarities of shape, cause the disappearance of all peculiarities acquired by special action of the environment. Hence, we may be sure that if men were not kept apart by the distances of their habitations, the crossing in reproduction would soon bring about the disappearance of the general characteristics distinguishing different nations.

If I intended here to pass in review all the classes, orders, genera and species of existing animals, I should be able to show that the conformation and structure of individuals, their organs, faculties, etc., etc., are everywhere a pure result of the environment to which each species is exposed by its nature, and by the habits that the individuals composing it have been compelled to acquire; I should be able to show that they are not the result of a shape which existed from the beginning, and has driven animals into the habits they are known to possess.

It is known that the animal called the *ai* or sloth (*Bradypus tridactylus*) is permanently in a state of such extreme weakness that it only executes very slow and limited movements, and walks on the ground with difficulty. So slow are its movements that it is alleged that it can only take fifty steps in a day. It is known, moreover, that the organisation of this animal is entirely in harmony with its state of feebleness and incapacity for walking; and that if it wished to make other movements than those which it actually does make it could not do so.

Hence on the supposition that this animal had received its organisation from nature, it has been asserted that this organisation forced it into the habits and miserable state in which it exists.

This is very far from being my opinion; for I am convinced that the habits which the ai was originally forced to contract must necessarily have brought its organisation to its present condition.

If continual dangers in former times have led the individuals of this species to take refuge in trees, to live there habitually and feed on their leaves, it is clear that they must have given up a great number of movements which animals living on the ground are in a position to perform. All the needs of the ai will then be reduced to clinging to branches and crawling and dragging themselves among them, in order to reach the leaves, and then to remaining on the tree in a state of inactivity in order to avoid falling off. This kind of inactivity, moreover, must have been continually induced by the heat of the climate; for among warm-blooded animals, heat is more conducive to rest than to movement.

Now the individuals of the race of the ai have long maintained this habit of remaining in the trees, and of performing only those slow and little varied movements which suffice for their needs. Hence their organisation will gradually have come into accordance with their new habits; and from this it must follow:

1. That the arms of these animals, which are making continual efforts to clasp the branches of trees, will be lengthened;

2. That the claws of their digits will have acquired a great length and a hooked shape, through the continued efforts of the animal to hold on;

3. That their digits, which are never used in making independent movements, will have entirely lost their mobility, become united and have preserved only the faculty of flexion or extension all together;

4. That their thighs, which are continually clasping either the trunk or large branches of trees, will have contracted a habit of always being separated, so as to lead to an enlargement of the pelvis and a backward direction of the cotyloid cavities;

5. Lastly, that a great many of their bones will be welded together, and that parts of their skeleton will consequently have assumed an arrangement and form adapted to the habits of these animals, and different from those which they would require for other habits.

This is a fact that can never be disputed; since nature shows us in innumerable other instances the power of environment over habit and that of habit over the shape, arrangement and proportions of the parts of animals.

Since there is no necessity to cite any further examples, we may now turn to the main point elaborated in this discussion.

It is a fact that all animals have special habits corresponding to their genus and species, and always possess an organisation that is completely in harmony with those habits.

It seems from the study of this fact that we may adopt one or other of the two following conclusions, and that neither of them can be verified.

Conclusion adopted hitherto: Nature (or her Author) in creating animals, foresaw all the possible kinds of environment in which they would have to live, and endowed each species with a fixed organisation and with a definite and invariable shape, which compel each species to live in the places and climates where we actually find them, and there to maintain the habits which we know in them.

My individual conclusion: Nature has produced all the species of animals in succession, beginning with the most imperfect or simplest, and ending her work with the most perfect, so as to create a gradually increasing complexity in their organisation; these animals have spread at large throughout all the habitable regions of the globe, and every species has derived from its environment the habits that we find in it and the structural modifications which observation shows us.

The former of these two conclusions is that which has been drawn hitherto, at least by nearly everyone: it attributes to every animal a fixed organisation and structure which never have varied and never do vary; it assumes, moreover, that none of the localities inhabited by animals ever vary; for if they were to vary, the same animals could no longer survive, and the possibility of finding other localities and transporting themselves thither would not be open to them.

The second conclusion is my own : it assumes that by the influence of environment on habit, and thereafter by that of habit on the state of the parts and even on organisation, the structure and organisation of any animal may undergo modifications, possibly very great, and capable of accounting for the actual condition in which all animals are found.

In order to show that this second conclusion is baseless, it must first be proved that no point on the surface of the earth ever undergoes variation as to its nature, exposure, high or low situation, climate, etc., etc. ; it must then be proved that no part of animals undergoes even after long periods of time any modification due to a change of environment or to the necessity which forces them into a different kind of life and activity from what has been customary to them.

Now if a single case is sufficient to prove that an animal which has long been in domestication differs from the wild species whence it sprang, and if in any such domesticated species, great differences of conformation are found between the individuals exposed to such a habit and those which are forced into different habits, it will then be certain that the first conclusion is not consistent with the laws of nature, while the second, on the contrary, is entirely in accordance with them.

Everything then combines to prove my statement, namely : that it is not the shape either of the body or its parts which gives rise to the habits of animals and their mode of life ; but that it is, on the contrary, the habits, mode of life and all the other influences of the environment which have in course of time built up the shape of the body and of the parts of animals. With new shapes, new faculties have been acquired, and little by little nature has succeeded in fashioning animals such as we actually see them.

Can there be any more important conclusion in the range of natural history, or any to which more attention should be paid than that which I have just set forth ?

Let us conclude this Part I. with the principles and exposition of the natural order of animals.

CHAPTER VIII.

OF THE NATURAL ORDER OF ANIMALS, AND THE WAY IN WHICH THEIR CLASSIFICATION SHOULD BE DRAWN UP SO AS TO BE IN CONFORMITY WITH THE ACTUAL ORDER OF NATURE.

I HAVE already observed that the true aim of a classification of animals should not be merely the possession of a list of classes, genera and species, but also the provision of the greatest facilities for the study of nature and for obtaining a knowledge of her procedure, methods and laws.

I do not hesitate to say, however, that our general classifications of animals up to the present have been in the inverse order from that followed by nature when bringing her living productions successively into existence; thus, when we proceed from the most complex to the simplest in the usual way, we increase the difficulty of acquiring a knowledge of the progress in complexity of organisation; and we also find it less easy to grasp both the causes of that progress and of the interruptions in it.

When once we have recognised that a thing is useful and indeed indispensable for the end in view and that it is free from drawbacks, we should hasten to carry it into execution although it is contrary to custom.

This is the case with regard to the way in which a general classification of animals should be drawn up.

We shall see that it is not a matter of indifference from which end we begin this general classification of animals, and that the beginning of the order is not a mere matter of choice.

The existing custom of placing at the head of the animal kingdom the most perfect animals, and of terminating this kingdom with the most imperfect and simplest in organisation, is due, on the one hand, to that natural prejudice towards giving the preference to the objects which strike us most or in which we are most pleased or interested; and, on the other hand, to the preference for passing from the better known to what is less known.

When the study of natural history began to occupy attention, these reasons were no doubt very plausible; but they must now yield to the needs of science and especially to those of facilitating the progress of natural knowledge.

With regard to the numerous and varied animals which nature has produced, if we cannot flatter ourselves that we possess an exact knowledge of the real order which she followed in bringing them successively into existence, it is nevertheless true that the order which I am about to set forth is probably very near it: reason and all our acquired knowledge testify in favour of this probability.

If indeed it is true that all living bodies are productions of nature, we are driven to the belief that she can only have produced them one after another and not all in a moment. Now if she shaped them one after another, there are grounds for thinking that she began exclusively with the simplest, and only produced at the very end the most complex organisations both of the animal and vegetable kingdoms.

The botanists were the first to set an example to the zoologists as to the proper way of drawing up a general classification in order to represent the actual order of nature; for it is with the Acotyledons or agamous plants that they constitute the first class among plants, that is to say, with the simplest in organisation and the most imperfect under every aspect, plants in short which have no cotyledons, no recognisable sex, no vessels in their tissue, and which in fact are composed of nothing but cellular tissue more or less modified according to their various expansions.

What botanists have done in the case of plants, we should now do with regard to the animal kingdom; and we should do it, not only because nature herself indicates it and reason demands it, but also because the natural order of classes in accordance with their growing complexity of organisation is much easier to determine among animals than it is in the case of plants.

While this order represents most closely the order of nature, it also makes the study of objects much easier, advances our knowledge of the organisation of animals with its increasing complexity from class to class, and exhibits still more clearly the affinities existing among the various stages of complexity of animal organisation, and the external differences that we commonly utilise for the characterisation of classes, orders, families, genera and species.

To these two principles, whose validity can scarcely be questioned, I add another, viz.: that if nature, who has not succeeded in endowing organised bodies with eternal existence, had not had the power of giving these bodies the faculty of reproducing others like themselves to carry on and perpetuate the race in the same way, she would have been forced to create directly all races, or rather she would only have been able to create a single race in each organic kingdom, viz. the simplest and most imperfect animals and plants.

Moreover, if nature had not been able to endow the organising activity with the faculty of gradually increasing the complexity of organisation by accelerating the energy of the movement of the fluids and hence that of organic movement, and if she had not preserved by reproduction all the progress made in complexity of organisation and all acquired improvements, she would assuredly never have produced that infinitely varied multitude of animals and plants which differ so greatly from one another both in their organisation and in their faculties.

Finally, she could not create at once the highest faculties of animals, for they are only found in conjunction with highly complex systems of organs: and she had to prepare slowly the methods by which such systems might be brought into existence.

Thus, in order to establish the state of affairs that we now see in living bodies, the only direct production that is required from nature, that is to say, the only production that occurs without the co-operation of any organic activity, is in the case of the simplest organised bodies, both of animals and plants; these she continues to produce every day in the same way at favourable times and places. Now she endows these bodies, which she has herself created, with the faculties of feeding, growing, multiplying, and always preserving the progress made in organisation. She transmits these same faculties to all individuals organically reproduced throughout time and the immense variety of ever-changing conditions. By these means living bodies of all classes and orders have been successively produced.

In the study of the natural order of animals, the very definite gradation existing in the growing complexity of their organisation and in the number and perfection of their faculties is very far from being a new truth for it was known even to the Greeks; [1] but they could not set forth its underlying principles and proofs, because they lacked the necessary knowledge.

Now, in order to facilitate an acquaintance with the principles which have guided me in the exposition that I am about to give, and in order to bring home more closely the gradation observed in the complexity of organisation from the most imperfect animals at the head of the series to the most perfect at the end of it, I have divided into six distinct stages the various modes of organisation recognised throughout the animal scale.

[1] See the *Voyage du jeune Anacharsis*, by J. Barthélemy, vol. v. pp. 353, 354.

Of these six stages of organisation, the first four comprise the invertebrate animals, and consequently the first ten classes of the animal kingdom according to the new order that we are going to follow; the two last stages comprises all the vertebrate animals and consequently the four (or five) last classes of animals.

By this method it will be easier to study and follow the procedure of nature in the production of the animals that she has brought into existence; to recognise throughout the animal scale the progress made in complexity of organisation and everywhere to verify both the accuracy of the classification and the propriety of the rank assigned by examining such characters and facts of organisation as are known.

In lecturing on invertebrates at the Museum, I have for some years past followed this plan of always proceeding from the simplest to the most complex.

In order to bring out more clearly the arrangement and totality of the general series of animals, I shall first present a table of the fourteen classes into which the animal kingdom is divided, confining myself to a very brief account of their characters and of the stages of organisation which they include.

TABLE

OF THE ARRANGEMENT AND CLASSIFICATION OF ANIMALS ACCORDING TO THE ORDER MOST IN CONFORMITY WITH THE ORDER OF NATURE.

INVERTEBRATE ANIMALS.

CLASSES	
I. INFUSORIANS. Amorphous animals, reproducing by fission or budding; with bodies gelatinous, transparent, homogeneous, contractile and microscopic; no radiating tentacles, or rotatory appendage; no special organ, even for digestion. II. POLYPS. Reproducing by budding; bodies gelatinous and regenerating, but with no other internal organ than an alimentary canal with a single aperture. Terminal mouth, surrounded by radiating tentacles or furnished with ciliated or rotatory organs. They mostly form compound animals.	*1st Stage.* No nerves; no vessels; no specialised internal organ except for digestion.

CLASSES	
III. Radiarians. Suboviparous animals, free, with regenerating bodies, destitute of head, eyes, or jointed legs; parts arranged radially. Mouth on inferior surface. IV. Worms. Suboviparous animals, with soft regenerating bodies; undergoing no metamorphosis, and never having eyes, jointed legs, nor a radial arrangement of the internal parts.	*2nd Stage.* No ganglionic longitudinal cord; no vessels for circulation; a few internal organs in addition to those of digestion.
V. Insects. Oviparous animals, which undergo metamorphosis, and have, in the perfect state, eyes in their heads, six jointed legs, and tracheae which spread everywhere; a single fertilisation in the course of their life. VI. Arachnids. Oviparous, having always jointed legs, and eyes in their heads, and undergoing no metamorphosis. Limited trachae for respiration; a primitive circulation; several fertilisations in the course of their life.	*3rd Stage.* Nerves terminating in a ganglionic longitudinal cord; respiration by air-carrying tracheae; circulation absent or imperfect.
VII. Crustaceans. Oviparous, with jointed body and limbs, crustaceous skin, eyes in their head, and usually four antennae; respiration by gills; a ganglionic longitudinal cord. VIII. Annelids. Oviparous, with elongated and ringed bodies; no jointed legs, rarely any eyes; respiration by gills; a ganglionic longitudinal cord. IX. Cirrhipedes. Oviparous, with a mantle and jointed arms, the skin of which is horny; no eyes; respiration by gills; ganglionic longitudinal cord. X. Molluscs. Oviparous, with soft moist bodies, unjointed, and with a variable mantle; respiration by gills of various shapes and situations; no spinal cord, nor ganglionic longitudinal cord, but nerves terminating in a brain.	*4th Stage.* Nerves terminating in a brain or a ganglionic longitudinal cord; respiration by gills; arteries and veins for circulation.

VERTEBRATE ANIMALS.

CLASSES	
XI. FISHES. Oviparous, and without mammae; respiration complete and always by gills; two or four primitive limbs; fins for locomotion; no hair or feathers on the skin. XII. REPTILES. Oviparous, and without mammae; respiration incomplete, usually by lungs which exist either throughout life or during the latter part of it; four limbs, or two, or none; no hair or feathers on the skin.	*5th Stage.* Nerves terminating in a brain which is far from filling the cranial cavity; heart with one ventricle, and the blood cold.
XIII. BIRDS. Oviparous, and without mammae; four jointed limbs, two of which are shaped as wings; respiration complete, by adherent lungs, which are pierced through; feathers on the skin. XIV. MAMMALS. Viviparous, and possessing mammae; four jointed limbs, or two only; respiration complete, by lungs not pierced through; hair on some part of the body.	*6th Stage.* Nerves terminating in a brain which fills the cranial cavity; heart with two ventricles, and the blood warm.

The above is a table of the fourteen classes of known animals arranged in the order most in conformity with that of nature. The arrangement of these classes is such that we shall always be obliged to adhere to it even though we may refuse to adopt the lines of demarcation between them; because this arrangement is based on a study of the organisation of the living bodies concerned, and because this highly important study reveals affinities among the objects comprised in each division, and determines the rank of each division throughout the series.

For these reasons no solid grounds can ever be found for changing the general features of this classification, though changes may be made as to detail, particularly in the divisions that are subordinate to the classes; because the affinities between the objects comprised in the sub-divisions are more difficult to determine and leave more to arbitrary opinion.

Now in order to bring home more closely the conformity of this arrangement of animals with the actual order of nature, I shall set forth the general series of known animals divided into its main groups, proceeding from the simplest to the most complex according to the principles indicated above.

My object in the exposition is to enable the reader to recognise the rank in the general series that is occupied by the animals which I have often had occasion to cite in the course of the present work, and to save him the trouble of having recourse to other works on zoology for this purpose.

I shall, however, here give merely a list of genera and of the main groups; but this list will suffice to show the extent of the general series, the arrangement of it that is most in conformity with nature, and the places necessarily occupied by classes and orders as well perhaps as by families and genera. We must of course refer to the good works on zoology that we possess for a study of the details of all the animals named in this list, for that does not come within the scope of the present work.

GENERAL CLASSIFICATION OF ANIMALS.

Forming a series in conformity with the actual order of nature.

INVERTEBRATE ANIMALS.

They have no vertebral column and consequently no skeleton; those which have fulcra for the movement of their parts have them under the integument. They lack a spinal cord and exhibit great variety in the complexity of their organisation.

FIRST STAGE OF ORGANISATION.

No nerves or ganglionic longitudinal cord; no vessels for circulation; no organs of respiration; no specialised internal organ but that for digestion.

(*Infusorians and Polyps.*)

INFUSORIANS.

(Class I. of the Animal Kingdom.)

Amorphous animals, reproducing by fission; with bodies gelatinous, transparent, homogeneous, contractile, and microscopic; no radiating tentacles, or rotatory appendage; internally no special organ, even for digestion.

Observations.

Of all known animals the infusorians are the most imperfect, the most simply organised and possessed of the fewest faculties; they certainly have not the faculty of feeling.

Infinitely minute, gelatinous, transparent, contractile, almost homogeneous and incapable of the possession of any special organ on account of the very delicate consistency of their parts, the infusorians are in truth mere rudiments of animalisation.

These fragile animals are the only creatures which do not have to carry on any digestion when feeding, and which in fact only feed by absorption through the pores of their skin and by an internal imbibition.

In this they resemble plants, which live entirely by absorption, carry on no digestion and in which the organic movements are only achieved by external stimuli; but the infusorians are irritable and contractile and perform sudden movements which they can repeat several times running; this it is that indicates their animal nature and distinguishes them essentially from plants.

TABLE OF INFUSORIANS.

Order 1.—Naked Infusorians.

They are destitute of external appendages.

Monas.	—
Volvox.	Bursaria.
Proteus [Amoeba. H. E.].	Colpoda.
Vibrio.	

Order 2.—Appendiculate Infusorians.

They have projecting parts, like hair, kinds of horns or a tail.

Cercaria [Trematode. H. E.].
Trichocerca.
Trichoda.

Remarks. The monas, and especially *Monas termo*, is the most imperfect and simplest of the known animals, since its extremely minute body is nothing but a point which is gelatinous and transparent, but contractile. This animal then must be the one with which the animal series begins, when arranged according to the order of nature.

POLYPS.

(Class II. of the Animal Kingdom.)

Gemmiparous animals, with gelatinous, regenerating bodies, and having no internal organ except an alimentary canal with a single aperture.

Terminal mouth, surrounded by radiating tentacles, or furnished with ciliated or rotatory organs. They mostly adhere together, are in communication by their alimentary canal, and then form compound animals.

Observations.

We have seen that the infusorians are infinitely small and fragile animalcules without coherence, without a shape peculiar to their class, without any organs and hence without a distinct mouth and alimentary canal.

In the polyps the simplicity and imperfection of organisation, although very conspicuous, are less than in the infusorians. Organisation has clearly made some progress; for nature has already obtained a permanent and regular shape for the animals for this class; they are all provided with a special organ for digestion, and consequently with a mouth which leads into the alimentary sac.

Imagine a small, elongated gelatinous, highly irritable body, which has at its superior extremity a mouth furnished either with rotatory organs or with radiating tentacles serving as the entrance to an alimentary canal which has no other opening: and we shall have a good idea of a polyp.

Add to this that many of these little bodies become adherent and live together in a common life, and we shall then know the most general and curious fact that concerns them.

The polyps are more imperfect in organisation than the animals of the following classes, since they have no nerves for feeling, no special organs for respiration and no vessels for the circulation of their fluids.

TABLE OF POLYPS.

Order 1.—Rotifer Polyps.

They have ciliated and rotatory organs at their mouths.

Urceolaria.
Brachionus (?).
Vorticella [Infusorian. H. E.].

Order 2.—Polyps which form Polyparies.

[Hydrozoa, Anthozoa, Polyzoa, Sponge, etc. H. E.]

They have radiating tentacles around the mouth, and are fixed in a polypary which does not float upon the waters.

(1) *Polypary membranous or horny, without distinct bark.*

Cristatella.
Plumatella.
Tubularia.
Sertularia.
Cellaria.
Flustra.
Cellepora.
Botryllus [Ascidian. H. E.].

(2) *Polypary with a horny axis, covered with an encrustation.*

Acetabulum [Alga. H. E.].
Corallina [Alga. H. E.].
—
Sponge.
Alcyon.
Antipathes (black coral).
Gorgonia (sea-fan).

(3) *Polypary with an axis partly or wholly stony, and covered with a bark-like encrustation.*

Isis.
Coral.

(4) *Polypary wholly stony, and without encrustation.*

Tubipora (organ-pipe coral).
Lunulites.
Ovulites.
Siderolites.
Orbulites.
Alveolites.
Ocellaria.
Eschara.
Retepora.
Millepora.
Agaricia.
Pavonia.
Meandrina.
Astrea.
Madrepora.
Caryophyllia.
Turbinolia.
Fungia (mushroom-coral).
Cyclolites.
Dactylopora.
Virgularia.

ORDER 3.—FLOATING POLYPS.

A free, elongated, polypary floating in the waters, with a horny or bony axis, covered with flesh that is common to all the polyps; radiating tentacles around the mouth.

Funiculina.
Veretillum.
Pennatula (Sea-pen).
Encrinus [Echinoderm. H. E.].
Umbellularia.

ORDER 4.—NAKED POLYPS.

They have radiating tentacles, often multiple, at the mouth, and form no polypary.

Pedicellaria.
Coryne.
Hydra.
Zoantha.
Actinia (Sea-anemone).

SECOND STAGE OF ORGANISATION.

No ganglionic longitudinal cord; no circulatory vessels; a few special internal organs (either tubes or pores, which draw in water or kinds of ovaries) in addition to those of digestion.

(*Radiarians and worms.*)

RADIARIANS.

(Class III. of the Animal Kingdom.)

Subgemmiparous animals, free or vagrant; with regenerating bodies and a radiating arrangement of the parts both internal and external and a complex digestive organ; mouth underneath, simple or multiple.

No head, eyes or jointed legs; a few internal organs in addition to those of digestion

OBSERVATIONS.

This is the third main line of demarcation which has to be drawn in the natural classification of animals.

We here find altogether new forms which, however, belong in general to one type, viz. the radiating arrangement of the parts both internal and external.

We have no longer to deal with animals with elongated bodies, a superior terminal mouth, usually fixed in a polypary, and living together in great numbers which share a common life, but we have to deal with animals whose organisation is more complex than that of the polyps and which are not compound but always free, which have a conformation peculiar to themselves and assume in general the inverted position.

Nearly all the radiarians have tubes which draw in water and appear to be water-bearing tracheae, and in a great many of them are found peculiar bodies resembling ovaries.

From a memoir which I lately heard read at a meeting of the professors of the Museum, I learn that a skilful observer, Dr. Spix, a Bavarian doctor, has discovered the apparatus of a nervous system in star-fishes and sea-anemones.

Dr. Spix affirms that he has seen in the red star-fish, under a tendinous membrane which is suspended over the stomach like a tent, a plexus consisting of whitish nodules and threads, and in addition, at the origin of each arm, two nodules or ganglia communicating together by a thread and giving rise to other threads which go to the neighbouring parts. Among these are two very long ones which traverse the entire length of the arm and send out branches to the tentacles.[1]

According to the observations of this savant there are in each arm two nodules, a short prolongation of the stomach (*caecum*), two hepatic lobes, two ovaries and tracheal canals.

In sea-anemones Dr. Spix observed at the base of these animals below the stomach several pairs of nodules arranged about a centre and communicating together by cylindrical threads. These give rise to others which pass to the upper parts: he found moreover four ovaries surrounding the stomach, from the base of which issue canals which unite together and open at a point within the alimentary cavity.

It is surprising that the apparatus of such complicated organs should have escaped the notice of all those who have studied the organisation of these animals.

If Dr. Spix is correct in what he describes; if he is not mistaken by

[1] [Tube-feet. H. E.]

attributing to these organs a nature and functions that do not really belong to them (as has happened to so many botanists who imagined they saw male and female organs in nearly all the cryptogams), then the following result ensues:

1. That we must no longer refer the beginning of the nervous system to the insects;

2. That this system must be regarded as existing in a rudimentary form in the worms, radiarians and even in the sea-anemone, the last genus of the polyps;

3. That this however is no reason why all the polyps should possess the rudiments of this system; just as it does not follow that because some reptiles have gills, therefore they must all have them;

4. Finally, that the nervous system is none the less a special organ, not common to all living bodies; for, not only is it absent in plants, but it is absent also in some animals. As I have shown the infusorians cannot possibly have it, nor assuredly can it be possessed by the majority of polyps; thus we should seek it in vain in the hydras which belong nevertheless to the first order of polyps, that, namely, which is nearest to the radiarians, since it comprises the sea-anemones.

Thus, whatever truth there may be in the facts named above, the considerations set forth in this work as to the successive formation of the various special organs hold good at whatever point in the animal scale each of these organs begins; and it remains true that the various faculties of animals only take their origin from the existence of the organs underlying them.

TABLE OF RADIARIANS.

Order 1.—Soft Radiarians.

[Various Coelenterates, exclusive of Anthozoa. H. E.]

Bodies gelatinous; soft, transparent skin without jointed spines; no anus.

Stephanomia.
Lucernaria.
Velella.
Porpita.
Pyrosoma [Tunicate. H. E.].
Beroë.
Physsophora.
Physalia.
Aequorea.
Rhizostoma.
Medusa (jelly-fish).

Order 2.—Echinoderm Radiarians.

Opaque, crustaceous or coriaceous skin, provided with retractile tubercles, or spines articulated on tubercles, and pierced with holes in series.

(1) Stellerides. *Skin not irritable, but mobile; no anus.*

Ophiura (brittle-star).
Asterias (star-fish).

(2) Echinoids. *Skin neither irritable, nor mobile ; an anus.*

Clypeaster (cake-urchin).
Cassidites.
Spatangus (heart-urchin).
Ananchytes.
Calorites.
Nucleolites.
Sea-urchin.

(3) Fistulides. *Elongated body, skin irritable and mobile ; an anus.*

Holothuria (sea-cucumber).
Sipunculus [Gephyrean. H. E.].

Remark. Sipunculus is an animal very similar to the worms, but its recognised affinities with the holothurians have caused it to be placed among the radiarians, although it has not the characters of that group and must therefore be placed at the end of it.

As a rule, in a thoroughly natural classification the first and last genera of the classes are those in which the standard characters are least pronounced. Since the lines of demarcation are artificial, the genera which are close to these lines display the characters of their class less conspicuously than the others.

WORMS.

(Class IV. of the Animal Kingdom.)

Suboviparous animals with soft elongated bodies, no head, eyes, legs, or bundles of setae ; destitute of a circulation, and having a complete, intestinal canal, that is, with two openings.

Mouth consisting of one or several suckers.

Observations.

The general shape of worms is quite different from that of radiarians, and their mouth, which is always formed as a sucker, has no analogy with that of polyps, where there is merely an aperture associated with radiating tentacles or rotatory organs.

The worms in general have an elongated body, very slightly contractile, although quite soft ; and as regards their intestine they are no longer limited to a single aperture.

In the fistulide radiarians nature has begun to abandon the radiating structure and to give an elongated shape to the bodies of animals, the only shape which could conduct towards the end she had in view.

After having fashioned the worms, she will henceforth tend to establish a type that is symmetrical as regards parts in pairs. She could not have attained this type except through the type of articulations ; but in the somewhat ambiguous class of worms, she has merely sketched out the rudiments of certain features of it.

TABLE OF WORMS.

Order 1.—Cylindrical Worms.

[Nematodes, cestodes, and other flat and round worms. H. E.]

Gordius [Nematode. H. E.].
Filaria (guinea-worm).
Proboscidea [Turbellarian. H. E.].
Crino.
Ascaris [Nematode. H. E.].
Fissula.
Trichocephalus [Nematode. H. E.].
Cucullanus.
Strongylus.
Scolex [head of tapeworm. H. E.].
Caryophyllaeus [Cestode. H. E.].
Tentacularia.
Echinorhyncus [Acanthocephala. H. E.].

Order 2.—Bladder Worms.

"Bicorne."
Hydatis.

Order 3.—Flat Worms.

Taenia [Cestode. H. E.].
Linguatula [Arthropod. H. E.].
Lingula [doubtless a misprint for Ligula, a Cestode. H. E.].
Fasciola [*v.* Introd. H. E.].

THIRD STAGE OF ORGANISATION.

Nerves terminating in a ganglionic longitudinal cord; respiration by air-carrying tracheae; circulation absent or imperfect.

(*Insects and Arachnids.*)

INSECTS.

(Class V. of the Animal Kingdom.)

Oviparous animals which undergo metamorphoses, which may have wings, and which have in the perfect state six jointed legs, two antennae, two eyes with facets and a horny skin.

Respiration by air-carrying tracheae which spread everywhere; no circulatory system; two distinct sexes; a single copulation in the course of their life.

Observations.

On reaching the insects we find among the extremely numerous animals comprised in this class a state of affairs very different from what we have met with in the animals of the four preceding classes; so instead of a gradual progress in the complexity of animal organisation we find on reaching the insects that a considerable leap has been made.

Here for the first time, animals from the outward aspect exhibit a distinct head; very remarkable, although still imperfect, eyes; jointed legs arranged in two rows; and that symmetrical form of paired and opposed parts that we shall henceforth find employed by nature up to and including the most perfect animals.

On examining the interior of insects, we also see a complete nervous system, consisting of nerves which terminate in a ganglionic longitudinal cord; but although complete, this nervous system is still very imperfect, since the nucleus to which sensations are conveyed appears much broken up, and the senses themselves are few and ill-developed; lastly, we see a true muscular system, and sexes which are distinct but which as in plants can only provide for a single fertilisation.

It is true that we do not yet find any circulatory system; and we shall have to pass higher up the animal chain before we meet with this improvement in organisation.

It is characteristic of all insects to have wings in their perfect state; so that those which have none owe their condition to the fact that their wings have become habitually and permanently aborted.

Observations.

In the table which I shall now give, the number of genera is greatly reduced from what has been hitherto constituted among the animals of this class. Such a reduction appears to me to be required in the interests of study, and also of simplicity and clearness of method. I have not carried it so far as to be detrimental to our knowledge of the animals. If we were to utilise every appreciable peculiarity in the characters of animals and plants for indefinitely multiplying their genera, we should, as I have already said, merely encumber and darken science instead of serving it; we should make the study of it so complicated and difficult that it would only be practicable for those who were willing to devote their entire life to gaining a knowledge of the immense nomenclature and the minute characters selected for marking the distinctions between these animals.

TABLE OF INSECTS.

(A) Sucking Insects.

Their mouth has a sucking-organ with or without a sheath.

Order 1.—Apterous Insects.

A bivalve, three-jointed, proboscis, enclosing a sucking-organ of two setae.

Wings generally abortive in both sexes; larva without legs; pupa motionless, in a cocoon.

Flea

Order 2.—Dipterous Insects.

An unjointed proboscis, straight or elbowed, sometimes retractile.

Two naked, membranous, veined wings; two balancers; larva worm-like, usually without legs.

Hippobosca (horse-fly).
Oestrus.
—
Stratiomys.
Syrphus (hover-fly).
Anthrax.
Fly.
Tabanus (gad-fly).
Rhagio.
—
Gnat.

—
Stomoxis.
Myopa.
Conops.
Empis.
Bombylus.
Asilus.
Tipula (crane-fly).
Simulium (sand-midge).
Bibio.

Order 3.—Hemipterous Insects.

Sharp, jointed beak, curved under the breast, serving as a sheath for a sucking-tube of three setae.

Two wings hidden under membranous elytra; larva hexapod; the pupa walks and eats.

Dorthesia.
Cochineal insect.
Psylla.
Plant-louse.
Aleyrodes.
Thrips.
—
Cicada.
Fulgora.
Tettigonia.
—
Scutellera.

Pentatoma.
Bed-Bug.
Coraeus.
Reduvius.
Hydrometra.
Gerris.
—
Nepa (water-scorpion).
Notonecta (water-boatman).
Nancoris.
Corixa (water-bug).

Order 4.—Lepidopterous Insects.

Sucking tube in two pieces, without a sheath, resembling a tubular proboscis, and rolled spirally when not in use.

Four membranous wings, covered with coloured and flour-like scales.

Larva with eight to sixteen legs; motionless chrysalis.

(1) *Antennae subulate or setaceous* [moths. H. E.].

Pterophorus.
Orneodes.
Cerastoma.
Tinea.
Noctua.
Phalaena.

Alucita.
Adella [Trichoptera. H. E.].
Pyralis.
—
Hepialus.
Bombyx (silk-worm).

(2) *Antennae swollen at some part of their length.*

Zygoena (burnet-moth).
Butterfly.

Sphinx (hawk-moth).
Sesia (clear-wing).

(B) Biting Insects.

The mouth exhibits mandibles, usually accompanied by maxillae.

Order 5.—Hymenopterous Insects.

Mandibles, and a sucking-tube of three more or less elongated pieces, whose base is enclosed in a short sheath.

Four wings, naked, membranous, veined and unequal ; anus of the females armed with a sting or provided with a boring-apparatus ; pupa motionless.

(1) *Anus of the females armed with a sting.*

Bee.
Monomelites.
Nomada.
Eucera.
Andrena.
—
Wasp.
Polistes.

Ant.
Mutilla (solitary ant).
Scolia.
Tiphia.
Bembex.
Crabro (digging-wasp).
Sphex.

(2) *Anus of the females provided with a boring-apparatus.*

Chrysis.
Oxyurus.
—
Leucopsis [Diptera. H. E.].
—
Evania.
Foenus.
—

Chalcis.
Cinips.
Diplolepis.
Ichneumon.
Wood-wasp.
Oryssus.
Tenthredo (saw-fly).
Cimbex (saw-fly).

Order 6.—Neuropterous Insects.

Mandibles and maxillae.

Four wings, naked, membranous and reticulated ; abdomen elongated, without sting or boring-apparatus ; larva hexapod ; metamorphosis variable.

(1) *Pupa motionless.*

Perla.
Nemoura.
Caddis-fly.

Hemerobius.
Ascalaphus.
Myrmeleon.

(2) *Pupa active.*

Nemoptera.
Panorpa.
Psocus.
Termes (white ant).
—
Corydalis.
Chauliodes.

Raphidia.
Ephemera.
—
Agrion.
Aeshna (dragon-fly).
Libellula (dragon-fly).

ORDER 7.—ORTHOPTEROUS INSECTS.

Mandibles, maxillae and galeae covering the maxillae.

Two straight wings, folded longitudinally, and covered by two almost membranous elytra.

Larva like the perfect insect, but with no wings or elytra ; pupa active.

Grasshopper.
Acheta.

—

Mantis (praying-insect).
Phasma (stick- and leaf-insects).
Spectrum.

Locust.
Truxalis.

—

Cricket.
Cockroach.
Earwig.

ORDER 8.—COLEOPTEROUS INSECTS.

Mandibles and maxillae.

Two membranous wings, folded longitudinally when at rest, under two hard or coriaceous but shorter elytra. Larva hexapod, with a scaly head and no eyes; pupa inactive.

(1) *Two or three segments in all the tarsi.*

Pselaphus.

—

Lady-bird.
Eumorphus.

(2) *Four segments in all the tarsi.*

Erotylus.
Cassida.
Chrysomela.
Galeruca.
Crioceris.
Clythra.
Cryptocephalus.

—

Leptura.
Stencorus.
Saperda.
Necydalis.
Callidium.
Cerambix.

Prionus.
Spondilus.

—

Bostrichus.
Mycetophagus.
Trogossita.
Cucujus.

—

Bruchus (pea-beetle).
Attelabus.
Brentus.
Curculio.
Brachycerus [Diptera. H. E.].

(3) *Five segments in the tarsi in the first pairs of legs, and four in those of the third pair.*

Opatrum.
Tenebrio (meal-worm beetle).
Blaps.
Pimelia.
Sepidium.
Scaurus.
Erodius.
Chiroscelis.

—

Helops.
Diaperis.

—

Cistela.

Mordella.
Rhipiphorus.
Pyrochroa (cardinal-beetle).
Cossiphus.
Notoxus.
Lagria.
Cerocoma.
Apalus.
Horia.
Mylabris.
Cantharis.
Melöe (oil-beetle).

(4) Five segments in all the tarsi.

Lymexylon.
Telephorus.
Malachius.
Melyris.
Lampyris (glow-worm).
Lycus.
Omalysus.
Drilus.

—

Melasis.
Buprestis.
Click-beetle.

—

Ptilinus.
Death-watch.
Ptinus.

—

Staphylinus (rove-beetle).
Ips.
Dermestes.
Anthrenus.
Byrrhus (pill-beetle).
Hister.
Sphoeridinus.

—

Trox.
Cetonia (rose-chafer).
Oxyporus (rove-beetle).
Poederus (rove-beetle).

—

Cicindela (tiger-beetle).
Elaphrus.
Scarites.
Manticora.
Carabus.
Dyticus (water-beetle).

—

Hydrophilus (water-beetle).
Gyrinus (whirligig-beetle).
Dryops.
Clerus.

—

Necrophorus (burying-beetle).
Carrion-beetle.
Nitidula.
Goliathus.
Cockchafer.
Lethrus.
Geotrupes (dor-beetle).
Copris (dung-beetle).
Scarabaeus (chafer).
Passealus.
Lucanus (stag-beetle).

ARACHNIDS.

(Class VI. of the Animal Kingdom.)

Oviparous animals which have jointed legs throughout their lives and eyes in their head ; they undergo no metamorphosis and never have wings or elytra.

Stigmata and limited tracheae for respiration ; a rudimentary circulation ; several fertilisations in the course of their life.

Observations.

The arachnids, which come after the insects in the order that we have established, display obvious progress in the perfection of organisation.

Sexual reproduction, for instance, is found among them, and for the first time in its full capacity, since these animals copulate and procreate several times in the course of their life ; whereas in the insects the sexual organs, like those of plants, can only achieve a single fertilisation. Moreover, the arachnids are the first animals in which we find a rudimentary circulation, for according to M. Cuvier they have a heart, from the sides of which issue two or three pairs of vessels.

Arachnids live in the air like insects which have attained their perfect state; but they undergo no metamorphosis, never have wings or elytra (nor is this due to any mere abortion), and they generally keep hidden or live in solitude, feeding on other animals or sucking blood.

In the arachnids, the method of respiration is the same as in the insects, but this method is on the verge of changing; for the tracheae of arachnids are very limited, and do not extend throughout the body. These tracheae are reduced to a small number of vesicles, as we learn from M. Cuvier again (*Anatom.* vol. iv. p. 419); and after the arachnids this method of respiration does not recur in any of the succeeding classes.

This class of animals should be treated with much caution: many of them are venomous, especially those living in hot climates.

TABLE OF ARACHNIDS.

ORDER 1.—ARACHNIDS WITH PEDIPALPS.

No antennae, but only pedipalps; the head fused with the thorax; eight legs.

Mygale.
Spider.
Scorpion.
—
Chelifer.
Galeodes.
Harvestman.
Trogulus.
Elais.

Phrynus.
Thelyphonus.
Trombidium.
—
Hydrachna.
Bdella.
Mite.
Nymphon.
Pycnogonum.

ORDER 2.—ARACHNIDS WITH ANTENNAE.

[Myriapods and a few insects. H. E.]

Two antennae; the head distinct from the thorax.

Louse [Hemiptera. H. E.].
Ricinus.
—
Silver-fish [Aptera. H. E.].
Spring-tail.

—
Centipede.
Scutigera.
Julus (millipede).

FOURTH STAGE OF ORGANISATION.

Nerves terminating in a ganglionic longitudinal cord, or in a brain without a spinal cord; respiration by gills; arteries and veins for the circulation.

(*Crustaceans, annelids, cirrhipedes and molluscs.*)

CRUSTACEANS.

(Class VII. of the Animal Kingdom.)

Oviparous animals with jointed body and limbs, a crustaceous skin, several pairs of maxillae, eyes and antennae in the head.

Respiration by gills; a heart and vessels for circulation.

Observations.

The great changes that we find in the organisation of the animals of this class, indicate that in forming the crustaceans, nature has succeeded in making great progress in animal organisation.

In the first place, the method of respiration is altogether different from that employed in the arachnids and insects; and this method, which is characterised by the organs called gills, continues as far as the fishes. Tracheae will appear no more, and gills themselves disappear as soon as nature can form a cellular lung.

Then again the circulation, of which only rudiments are found in the arachnids, is thoroughly established in the crustaceans; for in them we find a heart and arteries for the dispatch of blood to the various parts of the body, and veins which bring back this fluid to the chief organ which sets it in motion.

We still find in the crustaceans the type of articulations, always used by nature in the insects and arachnids, to facilitate muscular movement by means of the induration of the skin; but hereafter nature abandons this type to establish a system of organisation in which it is no longer required.

Most crustaceans live either in brackish or salt water. Some, however, keep on land and breathe air with their gills: they all feed on animal substances.

TABLE OF CRUSTACEANS.

ORDER 1.—SESSILE-EYED CRUSTACEANS.

Eyes sessile and immovable.

Wood-louse.	Cephaloculus.
Ligia.	Amymone.
Asellus.	Daphnia.
Cyamus (whale-louse).	Lynceus.
Shrimp.	Osole.
Caprella.	Limulus [Apus, not the modern Limulus. H. E.].
—	Caligus.
Cyclops (water-flea).	Polyphemus.
Zoea [Decapod larva. H. E.].	

ORDER 2.—STALK-EYED CRUSTACEANS.

Two distinct eyes, raised upon movable stalks.

(1) *Elongated tail, furnished with swimming blades, or hooks or setae.*

Branchiopod.	Pagurus (hermit-crab).
Squilla.	—
Palaemon.	Ranina.
Crangon.	Albunea.
Palinurus (rock-lobster).	Hippa (sand-crab).
Scyllarus.	Corystes.
Galathea.	Porcellana.
Cray-fish.	

(2) *Tail short, without appendages, and applied to the lower surface of the abdomen.*

Pinnotheres.	Dorippe.
Leucosia.	Plagusia.
Arctopsis.	Grapsus.
Maia.	Ocypode.
—	Calappa.
Matuta.	Hepatus.
Orithyia.	Dromia.
Podophthalmus.	Cancer.
Portunus.	

ANNELIDS.

(Class VIII. of the Animal Kingdom.)

Oviparous animals with soft elongated bodies, with transverse rings; they rarely have eyes or a distinct head and are destitute of jointed legs.

Arteries and veins for circulation; respiration by gills; a ganglionic longitudinal cord.

OBSERVATIONS.

We find in the annelids that nature is striving to abandon the type of articulations which she always used in the insects, arachnids and crustaceans. Their soft elongated body, which in most of them simply consists of rings, makes these animals appear as imperfect as the worms with which they used to be confused. Since, however, they have arteries and veins and breathe by gills, these animals are quite distinct from the worms and should be placed with the cirrhipedes between the crustaceans and the molluscs.

They have no jointed legs,[1] and most of them have on their sides setae or bundles of setae which take the place of legs: they nearly all have suckers and feed only on fluid substances.

TABLE OF ANNELIDS.

Order 1.—Cryptobranch Annelids.

Planaria [Triclad. H. E.].
Leech.
Lernea [Copepod. H. E.].
Clavella [Copepod. H. E.].
—
Furia (?).
Nais.
Lumbricus.
Thalassema.

Order 2.—Gymnobranch Annelids.

[Polychaets. H. E.]

Arenicola.
Amphinoma.
Aphrodite.
Nereis.
—
Serpula.
Spirorbis.
—
Terebella.
Amphitrite.
Sabellaria.
Siliquaria [mollusc. H. E.].
Dentalium [mollusc. H. E.].

CIRRHIPEDES.

(Class IX. of the Animal Kingdom.)

Oviparous and testaceous animals without a head or eyes, but having a mantle which covers the inside of the shell, jointed arms whose skin is horny, and two pairs of maxillae.

Respiration by gills; a ganglionic longitudinal cord; vessels for circulation.

Observations.

Although only a small number of genera belonging to this class are yet known, the character of the animals contained in these genera is so

[1] In order to perfect these animals' organs of locomotion, nature had to abandon the system of jointed legs, which are independent of a skeleton, and to establish the system of four limbs depending on an internal skeleton, which is characteristic of the most perfect animals; this is what she has done in the annelids and molluscs, where she has paved the way for commencing with the fishes the type of organisation peculiar to vertebrates. Thus in the annelids she has abandoned jointed legs, and in the molluscs she has gone still farther,—she has discarded the use of a ganglionic longitudinal cord.

singular that we have to set them apart as constituting a special class.

Seeing that the cirrhipedes have a shell, a mantle and no head or eyes they cannot be crustaceans; their jointed arms prevent us from placing them among the annelids, and their ganglionic longitudinal cord does not allow us to unite them with the molluscs.

TABLE OF CIRRHIPEDES.

Tubicinella.	Balanus.
Coronula.	Anatifa.

Remark. We see that the cirrhipedes still resemble the annelids by their ganglionic longitudinal cord; but in these animals nature is preparing to form the molluscs, since they have like them a mantle covering the inside of their shell.

MOLLUSCS.

(Class X. of the Animal Kingdom.)

Oviparous animals with soft unjointed bodies, and having a variable mantle.

Respiration by very diversified gills; no spinal cord, nor ganglionic longitudinal cord, but nerves terminating in an imperfect brain.

The majority are enclosed in a shell; others have one that is more or less completely embedded within them, and others again have none at all.

Observations.

The molluscs are the most highly organised of invertebrates; that is to say, their organisation is the most complex and the nearest to that of the fishes.

They constitute a numerous class which terminates the invertebrates, and which is sharply distinguished from the other classes by the fact that the animals composing it are the only ones which, although having a nervous system like many others, have neither a ganglionic longitudinal cord nor a spinal cord.

Nature is here about to begin the formation of the system of organisation of the vertebrates; and appears to be preparing for the change. Hence the molluscs, which have altogether lost the type of articulations, and the support given by a horny skin to animals belonging to this type, are very slow in their movements and appear in this respect even more imperfectly organised than the insects.

Finally, since the molluscs constitute a link between the invertebrates and the vertebrates, their nervous system is intermediate, and exhibits neither the ganglionic longitudinal cord of the invertebrates which have nerves, nor the spinal cord of the vertebrates. This is highly characteristic of them, and clearly distinguishes them from the other invertebrates.

TABLE OF MOLLUSCS.

ORDER 1.—ACEPHALIC MOLLUSCS.

[First group, *Brachiopods*; last group, *Tunicates*; the rest, *Lamellibranchs*. H. E.]

No head; no eyes; no organ of mastication; they reproduce without copulation. The majority have a shell with two valves which articulate at a hinge.

Brachiopods.

Lingula.
Terebratula.
Orbicula.

Ostracians.

Radiolites.
Calceola.
Crania [Brachiopod. H. E.].
Anomia.
Placuna.
Vulsella.
Oyster.
Gryphaea.
Plicatula.
Spondylus.
Pecten.

Byssifera.

Pedum.
Lima.
Pinna.
Mytilus (mussel).
Modiola (?).
Crenatula.
Perna.
Malleus.
Avicula.

Chamaceans.

Etheria.
Chama.
Diceras.
Corbula.
Pandora.
—

Naiads.

Unio (fresh-water mussel).
Anodonta.

Arcaceans.

Nucula.
Petunculus.
Arca.
Cucullaea.
Trigonia.
—

Cardiads.

Tridaena.
Hippopus.
Cardita.
Isocardia.
Cardium (cockle).

Conchs.

Venericardia.
Venus.
Cytherea.
Donax.
Tellina.
Lucinia.
Cyclas.
Galathea.
Capsa.

Mactraceans.

Erycina.
Ungulina.
Crassatella.
Lutraria.
Mactra.
—

Myarians.

Mya.
Panorpa.
Anatina.

Solenaceans.

Glycimeris.
Solen.
Sanguinolaria.
Petricola.
Rupellaria.
Saxicava.

Pholadarians (Boring-mussels).

Pholas.
Teredo.
Fistulana.
Aspergillum.
—

Ascidians.

Ascidia.
Salpa.
Mammaria.

ORDER 2.—CEPHALIC MOLLUSCS.

A distinct head and eyes, and two or four tentacles in the majority, jaws or a proboscis at the mouth; reproduction by copulation.

The shell of those, which have one, never consists of two valves articulated at a hinge.

(1) *Pteropods.*

Two opposite, swimming fins.

Hyalea.
Clio.
Pneumoderma.

(2) *Gastropods.*

(A) *Straight body, united to the foot throughout the whole or nearly the whole of its length.*

Tritonians.

Glaucus.
Aeolis.
Scyllaea.
Tritonia.
Tethys.
Doris.

Phyllidians.

Pleurobranchus. Patella (limpet).
Phyllidia. Fissurella.
Chiton. Emarginula.

Laplysians.

Laplysia (sea-hare). Bullaea.
Dolabella. Sigaretus.

Limacians.

Onchidium. Vitrina (glass-snail).
Limax (slug). Testacella.
Parmacella. —

(B) *Body spiral ; no syphon.*

Colymaceans.

Helix (snail). Amphibulimus.
Helicina. Achatina.
Bulimus. Pupa.

Orbaceans.

Cyclostoma. Planorbis.
Vivipara [Paludina. H. E.]. Ampullaria.

Auriculaceans.

Auricula. Melania.
Melanopsis. Limnaea.

Neritaceans.

Neritina. Nerita.
Navicella. Natica.

Stomataceans.

Haliotis.
Stomatia.
Stomatella.

Turbinaceans.

Phasianella. Scalaria.
Turbo. Turritella.
Monodonta. Vermicularia (?).
Delphinula.

Heteroclites.

Volvaria.
Bulla.
Janthina.

Calyptraceans.

Crepidula. Solarium.
Calyptraea. Trochus.

(C) *Body spiral : a syphon.*

Canalifera.

Cerithium. Pyrula.
Pleurotoma. Fusus.
Turbinella. Murex.
Fasciolaria.

Wing-shells.

Rostellaria.
Pteroceras.
Strombus.

Purpuraceans.

Cassis (helmet-shell).
Harpa (harp-shell)
Dolium (tun).
Terebra (auger-shell).
Eburna.
Buccinum (whelk).
Concholepas.
Monoceros.
Purpura.
Nassa.

Columellarians.

Cancellaria.
Marginella.
Columbella.
Mitra (mitre-shell).
Voluta.

Convolutes.

Ancilla [Ancillaria. H. E.].
Oliva (olive-shell).
Terebellum.
Ovula.
Cypraea (cowry).
Conus (cone-shell).

(3) *Cephalopods.*

(A) *With multilocular test.*

Lenticulaceans. [Foraminifera. H. E.]

Miliola.
Gyrogonita.
Rotalia.
Renulites.
Discorbina.
Lenticulina.
Nummulites.

Lituolaceans.

Lituola, Spirolinites [Foraminifera. H. E.].
Spirula [Cephalopod. H. E.].
Orthoceras [Cephalopod. H. E.].
Hippurites [Lamellibranch. H. E.].
Belemnites [Cephalopod. H. E.].

Nautilaceans.

Baculites.
Turrilites.
Ammonoceras.
Ammonites.
Orbulites.
Nautilus.

(B) *With unilocular test.*

Argonautaceans.

Argonauta.
Carinaria [Gastropod. H. E.].

(C) *Without test.*

Sepiated.

Octopus.
Calamary.
Cuttle-fish.

VERTEBRATE ANIMALS.

They have a vertebral column consisting of a number of short articulated bones following one another in succession. This column serves to support their body, it is the basis of their skeleton, constitutes a sheath for their spinal cord, and terminates anteriorly in a bony case containing the brain.

FIFTH STAGE OF ORGANISATION.

Nerves terminating in a spinal cord and a brain which does not fill up the cavity of the cranium; heart with one ventricle and cold blood.

(*Fishes and Reptiles.*)

FISHES.

(Class XI. of the Animal Kingdom.)

Oviparous vertebrate animals with cold blood; living in water, breathing by gills, covered with a skin either scaly or almost naked and slimy, and having for their locomotive movements only membranous fins supported by a bony or cartilaginous framework.

Observations.

The organisation of the fishes is much more perfect than that of the molluscs and animals of the anterior classes, since they are the first animals to have a vertebral column, the rudiments of a skeleton, a spinal cord and a cranium enclosing the brain. They are also the first in which the muscular system derives its support from internal parts.

Nevertheless their respiratory organs are still analogous to those of the molluscs, cirrhipedes, annelids and crustaceans; and like all the animals of the preceding classes, they are still without a voice and have no eyelids.

The shape of their body is adapted to their necessity for swimming; but they maintain the symmetrical shape of paired parts started in the insects; lastly, among them as among the animals of the three following classes, the type of articulations is altogether internal, and only occurs in the parts of their skeleton.

N.B.—In the preparation of the tables of vertebrate animals I have used M. Duméril's work entitled *Zoologie Analytique*, and I have permitted myself to make but few changes in his arrangement.

TABLE OF FISHES.

ORDER 1.—CARTILAGINOUS FISHES.

Vertebral column soft and cartilaginous; no true ribs in a great number.

(1) *No operculum over the gills, and no membrane.*

Trematopneans. [Hole-breathing. H. E.]
Respiration through round holes.

1. *Cyclostomes.*
Gasterobranchus (hagfish, myxine).
Lamprey.

2. *Plagiostomes.*

Torpedo.
Skate.
Rhinobatus.
Squatina (angel-fish).
Squalus.
Aodon.

(2) *No operculum over the gills, but a membrane.*

Chismopneans. [Cleft-breathing. H. E.]
Gills opening by clefts at the sides of the neck; four paired fins.

3. . . . .

Batrachus [Teleost. H. E.].
Lophius (frog-fish) [Teleost. H. E.].
Balistes [Teleost. H. E.].
Chimaera [Elasmobranch. H. E.].

(3) *An operculum over the gills, but no membrane.*

Eleutheropomes. [Free operculum. H. E.]
Four paired fins; mouth under the snout.

4. . . . .
Polyodon [Ganoid. H. E.].
Pegasus [Teleost. H. E.].
Accipenser (sturgeon) [Ganoid. H. E.].

(4) *An operculum and a membrane over the gills.*

Teleobranchs. [Complete gills. H. E.]
[Teleosts. H. E.]
Gills complete, with an operculum and a membrane.

5. *Aphyostomes.* [Sucking-mouth. H. E.]
Macrorhyncus.
Solenostoma.
Centriscus (snipe-fish).

6. *Pteroptera.* [United fins. H. E.]
Cyclopterus (lump-sucker).
Lepadogaster.

7. *Osteoderms.* [Bony skin. H. E.]

Ostracion (copper-fish).
Tetraodon (globe-fish).
Ovoides.
Diodon (porcupine-fish).
Spherodon.
Syngnathus.

ORDER 2.—BONY FISHES.

Vertebral column with bony vertebrae, that are not flexible.

(1) *An operculum and membrane over the gills.*

Holobranchs. [Complete gills. H. E.]

Apode Holobranchs.

No inferior paired fins. [Eel-shaped. H. E.]

8. *Peropterous holobranchs.* [Finless. H. E.]

Coecilia [Amphibian. H. E.].
Monopterus.
Leptocephalus [immature eel. H. E.].
Gymnotus (Electric eel).
Trichiurus.
Notopterus.
Ophisurus.
Apteronotus.
Regalecus.

9. *Pantopterous Holobranchs.* [With all unpaired fins. H. E.]

Muraena (eel).
Ammodytes (sand-eel).
Ophidium.
Macrognathus.
Xiphias (sword-fish).
Anarrhichas (sea-wolf).
Comephorus.
Stromataeus.
Rhombus.

Jugular Holobranchs.

Inferior paired fins situated under the throat, or in front of thoracic fins.

10. *Auchenopterous Holobranchs.* [Fins on neck. H. E.]

Murenoid.
Calliomorus.
Uranoscopus.
Weever.
Cod.
Batrachoides.
Blenny.
Oligopod.
Kurtus.
Chrysostrome.

Thoracic Holobranchs.

Inferior paired fins situated under the pectorals.

11. *Petalosome Holobranchs.* [Blade shaped. H. E.]

Lepidopus.
Cepola (band-fish).
Taenioid.
Bostrichthys.
Bostrichoid.
Gymnetrus.

12. *Plecopod Holobranchs.* [Inferior fins united. H. E.]

Gobius (goby).
Gobioid.

13. *Eleutheropod Holobranchs.* [Inferior fins free. H. E.]

Gobiomore.
Gobiomoroid.
Echeneis.

14. *Atractosome Holobranchs.* [Spindle-shaped. H. E.]

Scomber (mackerel).
Scomberoid.
Caranx (horse-mackerel).
Trachinote.
Caranxomorus.
Caesio.
Caesiomorus.
Scomberomorus.
Gasterosteus (stickleback).
Centropodus.
Centronotus.
Cephalacanthus.
Istiophorus.
Pomatomus.

15. *Leiopome Holobranchs.* [Smooth-operculed. H. E.]

Hiatula.
Coris (rainbow wrasse).
Gomphosus.
Osphronemus.
Trichopod.
Monodactyl.
Plectorhyncus.
Pogonias.
Labrus (wrasse).
Chilinus.
Cheilodipteron.
Ophiocephalus.
Hologymnosa.
Sparus.
Dipterodon.
Cheilio.
Mullet.

16. *Osteostome Holobranchs.* [Bony-mouthed. H. E.]

Scarus.
Ostorhincus.
Leiognathus.

17. *Lophionotous Holobranchs.* [Crested-back. H. E.]

Coryphaena.
Emipteronota.
Coryphaenoid.
Taenionotus.
Centrolophus.
Eques.

18. *Cephalotous Holobranchs.* [Large-headed. H. E.]

Gobiesox.
Aspidophora.
Aspidophoroides.
Cottus.
Scorpaena (scorpion-fish).

19. *Dactylous Holobranchs.* [Pectorals in distinct rays, like fingers. H. E.]

Dactylopterus.
Prionotus.
Trigla (gurnard).
Peristedion.

20. *Heterosomatous Holobranchs.* [Irregular-shaped. H. E.]

Pleuronectes.
Achirus.

21. *Acanthapome Holobranchs.* [Spiny opercula. H. E.]

Lutjanus.
Centropomus.
Bodianus.
Taenianotus.
Sciaena.
Micropterus (black bass).
Holocentrum.
Perca (perch).

22. *Leptosome Holobranchs.* [Slender-bodied. H. E.]

Chetodon.
Acanthinion.
Pomadasis.
Pomacanthus.
Holacanthus.
Enoplosus.
Glyphisodon.
Acanthurus.
Aspisurus.
Chetodipteron.
Pomacentrus.
Acanthopod.
Selene.
Argyriosus.
Zeus.
Galeoides.
Chrysostose.
Capros (boar-fish).

Abdominal Holobranchs.

Inferior paired fins placed a little in front of the anus.

23. *Siphonostome Holobranchs.* [Tube-like mouths. H. E.]

Fistularia.
Aulostoma.
Solenostoma.

24. *Cylindrosome Holobranchs.* [Cylindrical. H. E.]

Cobitis.
Misgurnus.
Anableps.
Fundulus.
Colubrine.
Amia [Ganoid. H. E.].
Butirinus.
Tripteronotus.
Ompolk.

25. *Oplophore Holobranchs.* [Armed. H. E.] [Catfishes. H. E.]

Silurus.
Macropteronotus.
Malapterurus.
Pimelodus.
Doras.
Pogonathus.
Cataphractus.
Plotosus.
Ageniosus.
Macrorhamphosus (snipe-fish).
Centranodon.
Loricaria.
Hypostome.
Corydoras.
Tachysurus.

26. *Dimerid Holobranchs.* [Two-membered. H. E.]

Cirrhites.
Cheilodactylus.
Polynemus.
Polydactylus.

27. *Lepidome Holobranchs.* [Scaly opercula. H. E.]

Mugil (grey mullet).
Mugiloid.
Chanos.
Mugilomorus.

28. *Gymnopome Holobranchs.* [Naked opercula. H. E.]

Argentina.
Atherina.
Hydrargyrus.
Stolephorus.
Buro.
Clupea (herring).
Mystus.
Clupanodon.
Gasteropleucus.
Mene.
Dorsuaria.
Xystera.
Cyprinus (carp).

29. *Dermopterous Holobranchs.* [Skin-fins. H. E.]

Salmo.
Osmerus (smelt).
Corregonus.
} [Salmonidae. H. E.]

Characinus.
Serrasalmo.
} [Characinidae. H. E.]

30. *Siagonote Holobranchs.* [Long-jawed. H. E.]

Elops.
Megalops (tarpon).
Esox.
Synodon.
Sphyraena.
Lepisosteus.
Polypterus [Ganoid. H. E.].
Scombresox.

(2) *An operculum over the gills, but no membrane.*

Sternoptyges. [Bent sternum. H. E.]

31. ....

Sternoptyx.

(3) *No operculum over the gills, but a membrane.*

Cryptobranchs. [Gills hidden. H. E.]

32. ....

Mormyrus.
Stylophorus.

(4) *No operculum nor membrane over the gills: no inferior paired fins.*

Ophichthians. [Snake-fishes. H. E.]

33. ....

Unibranch aperture.
Sphagebranchus.
Murenophis.
Gymnomuraena.

Remark. Seeing that the formation of a skeleton begins in the fishes, those called cartilaginous are probably the least perfect fishes. Consequently the most imperfect of all should be Gasterobranchus, which Linnæus, under the name of *myxine*, had regarded as a worm.

Thus, in the order that we are following, the genus Gasterobranchus must be the first of the fishes, because it is the least perfect.

REPTILES.

(Class XII. of the Animal Kingdom.)

Oviparous vertebrate animals with cold blood; breathing incompletely by a lung, at all events in later life; and having the skin smooth or else covered either with scales or with a bony shell.

Observations.

Progress in the perfection of organisation is seen to be very remarkable in the reptiles when they are compared with fishes; for it is among them that we find lungs for the first time, and we know that it is the most perfect respiratory organ since it is the same as that of man; but it is still only rudimentary and indeed some reptiles do not have it in early life: as a matter of fact they only breathe incompletely, for it is only a part of the blood that passes through the lung.

It is also among them that for the first time we distinctly see the four limbs, which are included in the plan of the vertebrates and are appendages of the skeleton.

TABLE OF REPTILES.

ORDER 1.—BATRACHIAN REPTILES [AMPHIBIANS. H. E.].

Heart with one auricle: skin naked: two or four legs: gills during immaturity: no copulation.

Urodela.

Siren. Triton.
Proteus. Salamander.

Anura.

Tree-frog. Pipa.
Frog. Toad.

ORDER 2.—OPHIDIAN REPTILES (OR SNAKES).

[Snakes and Apodal Lizards. H. E.]

Heart with one auricle: elongated narrow body without legs or fins: no eyelids.

Homoderms.

Coecilia [Amphibian. H. E.]. Ophisaurus.
Amphisboena. Slow-worm.
Acrochordus. Hydrophis (sea-snake).

Heteroderms.

Crotalus. Erix.
Scytale. Viper.
Boa. Coluber.
Erpeton. Platurus.

ORDER 3.—SAURIAN REPTILES.

[Legged Lizards and Crocodiles. H. E.]

Double-auricled heart; body scaly and having four legs; claws on the digits; teeth in the jaws.

Tereticauds [Round-tailed. H. E.].

Chalcides. Agama.
Scincus. Lacerta.
Gecko. Iguana.
Anolis. Stellio.
Dragon. Chamaeleon.

Planicauds [Flat-tailed. H. E.].

Uroplates.
Tupinambis.
Basiliscus.
Lophura.
Dracaena.
Crocodile [Crocodilia. H. E.].

ORDER 4.—CHELONIAN REPTILES.

Double-auricled heart; body with a carapace and four legs; jaws without teeth.

Chelonia.
Chelys.
Emys.
Tortoise.

SIXTH STAGE OF ORGANISATION.

Nerves terminating in a spinal cord, and in a brain which fills up the cavity of the cranium; heart with two ventricles and warm blood.

(*Birds and Mammals.*)

BIRDS.

(Class XIII. of the Animal Kingdom.)

Oviparous vertebrate animals with warm blood; complete respiration by adherent and pierced lungs; four jointed limbs, two of which are shaped as wings; feathers on the skin.

OBSERVATIONS.

Assuredly birds have a more perfect organisation than reptiles or any other animals of the preceding classes, since they have warm blood, a heart with two ventricles, and their brain fills up the cavity of the cranium,—characters which they have in common only with the most perfect animals composing the final class.

Yet the birds are clearly only the penultimate step of the animal scale; for they are less perfect than the mammals, in that they are still oviparous, have no mammae, are destitute of a diaphragm, a bladder, etc., and have fewer faculties.

In the following table it may be noticed that the first four orders include birds whose young can neither walk nor feed themselves, when they are hatched; and that the last three orders, on the other hand, comprise birds whose young walk and feed themselves as soon as they come out of the egg; finally, the 7th order, that of the palmipeds seems to me to contain those birds which are most closely related to the first animals of the following class.

TABLE OF BIRDS.

ORDER 1.—CLIMBERS.

Two digits in front, and two behind.

Levirostrate Climbers. [Slender-billed. H. E.]

Parrot.
Cockatoo.
Macaw.
Puff-bird.
Touraco.
Trogon.
Musophaga(plantain-eater).
Toucan.

Cuneirostrate Climbers. [Wedge-shaped beaks. H. E.]

Woodpecker.
Wryneck.
Jacamar.
Ani.
Cuckoo.

ORDER 2.—BIRDS OF PREY.

A single digit behind; anterior digits entirely free; beak and claws hooked.

Nocturnal Birds of Prey.

Owl.
Eagle-owl.
Surnia.

Bare-necked Birds of Prey.

Condor.
Vulture.

Feather-necked Birds of Prey.

Griffon.
Secretary-bird.
Eagle.
Buzzard.
Goshawk.
Falcon.

ORDER 3.—PASSERES.

A single digit behind; the two front external ones united; tarsus of medium height.

Crenirostrate Passeres. [Notched beaks. H. E.]

Tanagra.
Shrike.
Flycatcher.
Ampelis (wax-wing, etc.).
Thrush.

Dentirostrate Passeres. [Tooth-beaked. H. E.]

Hornbill.
Motmot.
Phytotoma (plant-cutter).

Plenirostrate Passeres. [Full-beaked. H. E.]

Grackle.
Bird of Paradise.
Roller.
Crow.
Pie.

Conirostrate Passeres. [Conical beaks. H. E.]

Ox-pecker.	Crossbill.
Glaucopis.	Grosbeak.
Oriole.	Colius (mouse-bird).
Cacicus.	Finch.
Starling.	Bunting.

Subulirostrate Passeres. [Subulate beaks. H. E.]

Mannakin.	Lark.
Titmouse.	Wagtail.

Planirostrate Passeres. [Flat-beaked. H. E.]

Martin.
Swallow.
Nightjar.

Tenuirostrate Passeres. [Slender-billed. H. E.]

Kingfisher.	Bee-eater.
Tody.	Humming-bird.
Nuthatch.	Creeper.
Orthorincus.	Hoopoe.

ORDER 4.—COLUMBAE.

Soft, flexible beak, flattened at the base; brood of two eggs.

Pigeon.

ORDER 5.—GALLINACEANS.

Solid, horny beak, rounded at the base; brood of more than two eggs.

Alectride Gallinaceans. [Fowl-like. H. E.]

Bustard.	Guinea-fowl.
Peacock.	Curassow.
Tetras.	Penelope.
Pheasant.	Turkey.

Brachypterous Gallinaceans. [Short-winged. H. E.]

Dodo.	Rhea.
Cassowary.	Ostrich.

ORDER 6.—WADERS.

Tarsus very long, and denuded of feathers as far as the leg; external digits united at their base (waterside birds).

Pressirostrate Waders. [Narrow-beaked. H. E.]

Jacana.	Moorhen.
Rail.	Coot.
Oyster-catcher.	

Cultrirostrate Waders. [Cutting-beaked. H. E.]

Bittern.	Crane.
Heron.	Mycteria.
Stork.	Tantalus.

Teretirostrate Waders. [Round-beaked. H. E.]

Avocet.
Curlew.
Woodcock.
Dunlin.
Plover.

Latirostrate Waders. [Broad-beaked. H. E.]

Boatbill.
Spoonbill.
Phoenicopterus (flamingo).

ORDER 7.—PALMIPEDS.

Digits united by large membranes; tarsus of low height (Aquatic birds, swimmers).

Penniped Palmipeds. [Fin-footed. H. E.]

Anhinga.
Phaëton.
Gannet.
Frigate-bird.
Cormorant.
Pelican.

Serrirostrate Palmipeds. [Serrated beaks. H. E.]

Merganser.
Duck.
Flamingo.

Longipen Palmipeds. [Long-winged. H. E.]

Gull.
Albatross.
Petrel.
Avocet.
Tern.
Scissor-bill.

Brevipen Palmipeds. [Short-winged. H. E.]

Grebe.
Guillemot.
Auk.
Penguin.
King-penguin.

MONOTREMES. (Geoff.)

Animals intermediate between birds and mammals. These animals are quadrupeds without mammae, without any teeth inserted, without lips, and with only one orifice for the genital organs, the excrement and the urine; their body is covered with hair or bristles.

ORNITHORHYNCHUS.
ECHIDNA.

N.B.—I have already spoken of these animals in Chapter VI., page 74, where I showed that they are neither mammals, birds nor reptiles.

MAMMALS.

(Class XIV. of the Animal Kingdom.)

Viviparous animals with mammae; four jointed legs or only two; complete respiration by lungs, not pierced through externally; hair on parts of the body.

Observations.

Nature clearly proceeds from the simplest to the most complex in her operations on living bodies; hence the mammals necessarily constitute the last class of the animal kingdom.

This class undoubtedly comprises the most perfect animals, with the greatest number of faculties, the highest intelligence and, lastly, the most complex organisation.

Since the organisation of these animals approaches most nearly to that of man they display a more perfect combination of senses and faculties than any others. They are the only ones that are really viviparous, and have mammae to suckle their young.

The mammals thus exhibit the highest complexity of animal organisation, and the greatest perfection and number of faculties that nature could confer on living bodies by means of that organisation. They should thus be placed at the end of the immense series of existing animals.

TABLE OF MAMMALS.

Order 1.—Exungulate Mammals.

Two limbs only: they are anterior, short, flattened, suitable for swimming, and have neither claws nor hoofs.

Cetaceans.

Right-whale.
Rorqual.
Physale.
Cachalot.
Sperm-whale.
Narwhal.
Anarnak.
Delphinapterus.
Dolphin.
Hyperodon.

Order 2.—Amphibian Mammals.

Four limbs: the two anterior short, fin-like, with unguiculate digits; the posterior directed backwards, or united with the extremity of the body, which is like a fish's tail.

Seal, Walrus [Pinnipeds. H. E.].
Dugong, Manatee [Sirenia. H. E.].

Observation.

This order is only placed here, on account of the general shape of the animals it contains. (See my observation, p. 74.)

Order 3.—Ungulate Mammals.

Solipeds.

Horse.

Ruminants or Bisulcates.

Ox.
Antelope.
Goat.
Sheep.
Deer.
Giraffe.
Camel.
Musk-deer.

Pachiderms.

Rhinoceros.
Hyrax.
Tapir.
Pig.
Elephant.
Hippopotamus.

Order 4.—Unguiculate Mammals.

Four limbs ; flattened or pointed nails at the extremity of their digits, which do not envelop them.

Tardigrades.

Sloth [Edentate H. E.].

Edentates.

Ant-eater.
Pangolin.
Aardvark.
Armadillo.

Rodents.

Kangaroo [Marsupial. H. E.].
Hare.
Coendu.
Porcupine.
Aye-aye [Lemur. H. E.].
Phascolomys [Marsupial. H. E.].
Hydromys (Australian Water-rat).
Beaver.
Cavy.
Spalax.
Squirrel.
Dormouse.
Hamster.
Marmot.
Vole.
Musk-rat.
Rat.

Pedimana [Marsupials. H. E.].

Opossum.
Bandicoot.
Dasyurus.
Wombat.
Coescoës.
Phalanger.

Plantigrades.

Mole [Insectivore. H. E.].
Shrew [Insectivore. H. E.].
Bear [Carnivore. H. E.].
Kinkajou [Carnivore. H. E.].
Badger [Carnivore. H. E.].
Coati [Carnivore. H. E.].
Hedgehog [Insectivore. H. E.].
Tenrec [Insectivore. H. E.].

Digitigrades. [All carnivores. H. E.].

Otter.
Mongoose.
Skunk.
Weasel.
Cat.
Civet.
Hyaena.
Dog.

Chiroptera.

Galeopithecus [Insectivore. H. E.].	Noctilio.
Rhinolophus.	Bat.
Phyllostome.	Flying Fox.

Quadrumana.

Galago } [Lemurs. H. E.].	Baboon.
Tarsius }	Sapajou.
Loris }	Cebus.
Makis }	African Baboon.
Indris }	Pongo.
Guenon (Old-world Monkey).	Orang.

Remark. According to the order which I have adopted, the quadrumanous family comprises the most perfect of known animals, and especially the later genera of this family; and as a matter of fact the genus orang (*Pithecus*) is at the end of the entire order, just as the monas is at the beginning of it. How great is the difference in organisation and faculties between these two genera!

Naturalists who have considered man exclusively according to the affinities of his organisation, have formed a special genus for him with six known varieties, thus making him a separate family which they have described in the following manner.

BIMANA.

Mammals with differentiated unguiculate limbs; with three kinds of teeth and opposable thumbs on the hands only.

MAN.

Varieties.
- Caucasian.
- Hyperborean.
- Mongolian.
- American.
- Malayan.
- Ethiopian or Negro.

This family has received the name of Bimana, because in man it is only the hands that have a separate thumb opposite to the fingers while in the Quadrumana the hands and feet have the same character as regards the thumb.

Some Observations with regard to Man.

If man was only distinguished from the animals by his organisation, it could easily be shown that his special characters are all due to long-standing changes in his activities and in the habits which he has adopted and which have become peculiar to the individuals of his species.

As a matter of fact, if some race of quadrumanous animals, especially one of the most perfect of them, were to lose, by force of circumstances or some other cause, the habit of climbing trees and grasping the branches with its feet in the same way as with its hands, in order to hold on to them; and if the individuals of this race were forced for a series of generations to use their feet only for walking, and to give up using their hands like feet; there is no doubt, according to the observations detailed in the preceding chapter, that these quadrumanous animals would at length be transformed into bimanous, and that the thumbs on their feet would cease to be separated from the other digits, when they only used their feet for walking.

Furthermore, if the individuals of which I speak were impelled by the desire to command a large and distant view, and hence endeavoured to stand upright, and continually adopted that habit from generation to generation, there is again no doubt that their feet would gradually acquire a shape suitable for supporting them in an erect attitude; that their legs would acquire calves, and that these animals would then not be able to walk on their hands and feet together, except with difficulty.

Lastly, if these same individuals were to give up using their jaws as weapons for biting, tearing or grasping, or as nippers for cutting grass and feeding on it, and if they were to use them only for mastication; there is again no doubt that their facial angle would become larger, that their snout would shorten more and more, and that finally it would be entirely effaced so that their incisor teeth became vertical.

Let us now suppose that a quadrumanous race, say the most perfect, acquired through constant habit among all its individuals the conformation just described, and the faculty of standing and walking upright, and that ultimately it gained the supremacy over the other races of animals, we can then easily conceive:

1. That this race having obtained the mastery over others through the higher perfection of its faculties will take possession of all parts of the earth's surface, that are suitable to it;

2. That it will drive out the other higher races, which might dispute with it the fruits of the earth, and that it would compel them to take refuge in localities which it does not occupy itself;

3. That it will have a bad effect on the multiplication of allied races, and will keep them exiled in woods or other deserted localities, that it will thus arrest the progress of their faculties towards perfection; whereas being able itself to spread everywhere, to multiply without obstacle from other races and to live in large troops, it will

create successively new wants, which will stimulate its skill and gradually perfect its powers and faculties;

4. Finally, that this predominant race, having acquired an absolute supremacy over all the rest, will ultimately establish a difference between itself and the most perfect animals, and indeed will leave them far behind.

The most perfect of the quadrumanous races might thus have become dominant; have changed its habits as a result of the absolute sway exercised over the others, and of its new wants; have progressively acquired modifications in its organisation, and many new faculties; have kept back the most perfect of the other races to the condition that they had reached; and have wrought very striking distinctions between these last and themselves.

The orang of Angola (*Simia troglodytes*, Lin.) is the most perfect of animals: it is much more perfect than the orang of the Indies (*Simia satyrus*, Lin.), called the orang-outang; yet they are both very inferior to man in bodily faculties and intelligence.[1] These animals often stand upright; but as that attitude is not a confirmed habit, their organisation has not been sufficiently modified by it, so that the standing position is very uncomfortable for them.

We know from the stories of travellers, especially as regards the orang of the Indies, that when it has to fly from some pressing danger it immediately falls on to its four feet. Thus, it is said, the true origin of this animal is disclosed, since it is obliged to abandon a deceptive attitude that is alien to it.

No doubt this attitude is alien to it, since it adopts it less when moving about, and its organisation is hence less adapted to it; but does it follow that, because the erect position is easy to man, it is therefore natural to him?

Although a long series of generations has confirmed the habit of moving about in an upright position, yet this attitude is none the less a tiring condition in which man can only remain for a limited period, by means of the contraction of some of his muscles.

If the vertebral column were the axis of the human body, and kept the head and other parts in equilibrium, man would be in a position of rest when standing upright. Now we all know that this is not the case; that the head is out of relation with the centre of gravity; that the weight of the chest and belly, with their contained viscera, falls almost entirely in front of the vertebral column; that the latter has a slanting base, etc. Hence it is necessary as M. Richerand observes, to keep a constant watch when standing, in order to avoid

[1] See in my *Recherches sur les corps vivants*, p. 136, some observations on the orang of Angola.

the falls to which the body is rendered liable by the weight and arrangement of its parts.

After discussing the questions with regard to the erect position of man, this observer expresses himself as follows : " The relative weight of the head, and of the thoracic and abdominal viscera, gives a forward inclination to the axial line of the body, as regards the plane on which it rests ; a line which should be exactly perpendicular to this plane, if standing is to be perfect. The following fact may be cited in support of this assertion : I have observed that children, among whom the head is bulky, the belly protruding and the viscera burdened with fat, find it difficult to get accustomed to standing upright ; it is only at the end of their second year that they venture to trust their own strength ; they continue liable to frequent falls and have a natural tendency to adopt the position of a quadruped " (*Physiologie*, vol. ii., p. 268).

This arrangement of parts, as a result of which the erect position is a tiring one for man, instead of being a state of rest, would disclose further in him an origin analogous to that of the other mammals, if his organisation alone were taken into consideration.

In order to follow out the hypothesis suggested at the beginning of these observations, some further considerations must now be added.

The individuals of the dominant race in question, having seized all the places of habitation which were suitable to them and having largely increased their needs according as the societies which they formed became larger, had to multiply their ideas to an equivalent extent, and thus felt the need for communicating them to their fellows. We may imagine that this will have compelled them to increase and vary in the same degree the signs which they used for communicating these ideas ; hence it is clear that the individuals of this race must have made constant efforts, and turned all their resources towards the creation, multiplication and adequate variation of the signs made necessary by their ideas and numerous wants.

This is not the case with other animals ; for although the most perfect of them such as the Quadrumana mostly live in troops, they have made no further progress in the perfection of their faculties subsequent to the high supremacy of the race named ; for they have been chased away and banished to wild and desert places where they had little room, and lived a wretched, anxious life, incessantly compelled to take refuge in flight and concealment. In this situation these animals contract no new needs and acquire no new ideas ; their ideas are but few and unvaried ; and among them there are very few which they need to communicate to others of their species. Very few different signs therefore are sufficient to make themselves understood by their fellows; all they require are a few movements of the body or parts of it, a few hissings and cries, varied by simple vocal inflections.

Individuals of the dominant race already mentioned, on the other hand, stood in need of making many signs, in order rapidly to communicate their ideas, which were always becoming more numerous and could no longer be satisfied either with pantomimic signs or with the various possible vocal inflections. For supplying the large quantity of signs which had become necessary, they will by various efforts have achieved the formation of articulate sounds. At first they will only have used a small number, in conjunction with inflexions of the voice; gradually they will have increased, varied and perfected them, in correspondence with the growth in their needs and their gain of practice. In fact, habitual exercise of their throat, tongue and lips in the articulation of sounds will have highly developed that faculty in them.

Hence would arise for this special race the marvellous faculty of speaking; and seeing that the remote localities to which the individuals of the race would have become distributed, would favour the corruption of the signs agreed upon for the transmission of each idea, languages would arise and everywhere become diversified.

In this respect, therefore, all will have been achieved by needs alone: they will have given rise to efforts, and the organs adapted to the articulation of sounds will have become developed by habitual use.

Such are the reflections which might be aroused, if man were distinguished from animals only by his organisation, and if his origin were not different from theirs.

ADDITIONS TO THE SUBJECT MATTER OF CHAPS. VII. AND VIII.

During the last few days of June 1809 the menagerie of the Museum of Natural History received a seal known under the name of sea-calf (*Phoca vitulina*) which was sent alive from Boulogne; and I had an opportunity of observing the movements and habits of this animal. Thereupon I acquired a still stronger conviction that this amphibian is much more allied to the unguiculate mammals than to the other mammals, notwithstanding the great differences in general shape between it and them.

Its hind legs, although very short like the fore-legs, are quite free and separate from the tail, which is small but quite distinct, and they can move easily in various ways; they can even grasp objects like true hands.

I noticed that this animal is able to unite its hind feet as we join our hands, and that on then separating the digits between which there are membranes, it forms a fairly large paddle, which it uses for travelling about in the water in the same way as fishes use their tail as a fin.

This seal drags itself about on the ground with some speed by means of an undulatory movement of the body, and without any help from its hind legs, which remain inactive and are stretched out. In thus dragging itself about, it derives help from its fore-legs only by supporting itself on the arms up to the wrists, without making any special use of the hands. It seizes its prey either with its hind feet or with its mouth, and although it sometimes uses its hands to rend the prey that it holds in its mouth, these hands appear to be used principally for swimming or locomotion in the water. Finally, as this animal often remains under water for a longish time and even feeds there in comfort, I have noticed that it easily and completely closes its nostrils just as we close our eyes; this is very useful to it when immersed in the liquid that it inhabits.

As this seal is well known, I shall give no description of it. My purpose here is simply to remark that the amphibians have their hind legs set on in the same direction as the axis of their body, for the simple reason that these animals are compelled to use them habitually as a caudal fin by uniting them and by separating the digits so as to form a large paddle. With this artificial fin they are then able to strike the water either to the right or left, and thus move rapidly in various directions.

The two hind legs of seals are so often united and used as a fin that they would not simply have this backward direction in continuance of the body but would be permanently united as in the walruses, were it not for the fact that the animals in question also use them very frequently for seizing and carrying off their prey. Now the special movements required by these actions prevents the hind legs of seals from becoming permanently united, and only allow them to be joined together momentarily.

Walruses, on the contrary, which are accustomed to feeding on grass, which they come and browse on the shore, only use their hind legs as a caudal fin; so that in most of them these legs are permanently united with one another and with the tail, and cannot be separated.

We find here a new proof of the power of habit over the form and state of the organs, a proof that I may add to all those already set forth in Chapter VII.

I might add still another very striking proof drawn from mammals. The faculty of flight would seem to be quite foreign to them; yet I can show how nature has gradually produced extensions of the animal's skin, starting from those animals which can simply make very long jumps and leading up to those which fly perfectly ; so that ultimately they possess the same faculty of flight as birds, though without having any affinities with them in their organisation.

Flying squirrels (*Sciurus volans*, *aerobates*, *petaurista*, *sagitta*, *volucella*) have more recently acquired this habit of extending their wings when leaping, so as to convert their body into a kind of parachute; they can do no more than make a very long jump by throwing themselves to the bottom of a tree, or leaping from one tree on to another at a moderate distance. Now by frequent repetition of such leaps in the individuals of these races, the skin of their flanks is dilated on each side into a loose membrane, which unites the hind-legs to the fore-legs and embraces a large volume of air ; thus saving them from a sudden fall. These animals still have no membranes between the digits.

The galeopithecus (*Lemur volans*) doubtless acquired this habit earlier than the flying squirrels (*Pteromis*, Geoffr.) ; the skin of their flanks is still larger and more developed ; it unites not only the hind-legs with the fore-legs but also the tail with the hind-legs and the digits with each other. Now these creatures make longer leaps than the preceding, and even perform a sort of flight.

Lastly, the various bats are mammals which probably acquired still earlier than the galeopithecus the habit of extending their limbs and even their digits to embrace a great volume of air, and sustain themselves when they launch forth into the atmosphere.

From these habits, so long acquired and preserved, bats have derived not only lateral membranes but also an extraordinary lengthening of the digits of their four legs (except the thumb) which are united by very large membranes ; so that these membranes of the hands, being continuous with those of the flanks and those which unite the tail to the two hind-legs, constitute in these animals great membranous wings with which they fly perfectly as we all know.

Such then is the power of habit : it has a remarkable influence on the shape of the parts and endows animals, which have long contracted certain habits, with faculties not possessed by those which have adopted different habits.

With regard to the amphibians, of which I spoke above, I should like here to communicate to my readers the following reflections that have been raised in me and ever more strongly confirmed by all the subjects I have dealt with in my studies.

I do not doubt that mammals originally came from the water, nor that water is the true cradle of the entire animal kingdom.

We still see, in fact, that the least perfect animals, and they are the most numerous, live only in the water, as I shall hereafter mention (p. 246); that it is exclusively in water or very moist places that nature achieved and still achieves in favourable conditions those direct or spontaneous generations which bring into existence the most simply organised animalcules, whence all other animals have sprung in turn.

We know that the infusorians, polyps, and radiarians live exclusively in the water; and that some worms even live in it while the rest dwell only in very moist places.

Now the worms appear to form one initial branch of the animal scale, and it is clear that the infusorians form the other branch. We may suppose therefore that such worms as are completely aquatic and do not live in the bodies of other animals, Gordius, for instance, and many others that we are not yet acquainted with, have doubtless become greatly diversified in the water; and that among these aquatic worms, those which afterwards became accustomed to exposure to the air have probably produced the amphibian insects such as gnats, mayflies, etc., etc., while these in turn have given existence to all the insects which live altogether in the air. Several races of these again have changed their habits as a result of their environment and contracted a new habit of living hidden away in solitude: hence the origin of the arachnids, nearly all of which live also in the air.

Finally, those arachnids that frequented water, and gradually became accustomed to live in it until at last they altogether ceased to live in the air, led to the existence of all the crustaceans; this is clearly indicated by the affinities which connect the centipedes with the millipedes, the millipedes with the woodlice, and these again with Asellus, shrimps, etc.

The other aquatic worms, which are never exposed to the air, would have developed in course of time into many different races with a corresponding advance in the complexity of their organisation. They would thus have led to the formation of the annelids, cirrhipedes and molluscs, which form together an unbroken portion of the animal scale.

There seems to us to be a great hiatus between the known molluscs and the fishes; yet the molluscs whose origin I have just named have led to the existence of the fishes through the medium of other molluscs that have yet to be discovered, and it is manifest that the fishes again have given rise to the reptiles.

As we continue to examine the probable origin of the various animals, we cannot doubt that the reptiles, by means of two distinct branches, caused by the environment, have given rise, on the one hand, to the formation of birds and, on the other hand, to the amphibian mammals, which have in their turn given rise to all the other mammals.

After the fishes had led up to the formation of the batrachian reptiles and these to the ophidian reptiles, both of which have only one auricle in their hearts, nature easily succeeded in giving a heart with a double auricle to the other reptiles, which became divided into two separate branches; subsequently she easily achieved the formation of a heart with two ventricles in animals originating from both these branches.

Thus among the reptiles which have a heart with a double auricle, the chelonians appear to have given existence to the birds; for, in addition to their various unmistakable affinities, if I were to place the head of a tortoise on the neck of certain birds, I should find hardly any incongruity in the general appearance of the factitious animals; in the same way the saurians, especially the planicauds, such as crocodiles, seem to have led to the existence of the amphibian mammals.

If the chelonian branch has given rise to the birds, we may suppose that the aquatic palmipeds, and especially the brevipens, such as the penguins and king-penguins, have brought about the formation of the monotremes.

Lastly, if the saurian branch gave rise to the amphibian mammals, it is highly probable that from this branch all the mammals have taken their origin.

I think the belief is justifiable, that the terrestrial mammals originated from those aquatic mammals that we call amphibians. These were divided into three branches by reason of the diversity arising in their habits in the course of time; one of these led to the cetaceans, another to the ungulate mammals, and the third to the various known unguiculate mammals.

Those amphibians indeed which preserved the habit of going on to the beach became divided, owing to their different manner of feeding. Some of them, being accustomed to browsing on grass, as for instance the walruses and manatees, gradually led to the formation of the ungulate mammals such as the pachyderms, ruminants, etc.; the others as, for instance, the seals, having acquired the habit of feeding exclusively on fishes and marine animals, brought about the existence of the unguiculate mammals through the medium of races which as they diversified became altogether terrestrial.

Those aquatic mammals, however, which acquired the habit of never coming out of the water and of only coming to the surface to breathe, probably gave rise to the various cetaceans with which we are acquainted. The cetaceans have been greatly modified in organisation by having dwelt for so long a period exclusively in the sea; hence it is now very difficult to recognise whence they derive their origin.

In consequence of the immense lapse of time during which these animals have lived in the sea without ever using their hind-legs for grasping objects, these unused legs have entirely disappeared, including their bones, and even the pelvis which served for their support and attachment.

The degeneration in the limbs of cetaceans under the influence of the environment and acquired habit is also seen in their fore-feet, which are entirely invested by skin so as not even to show the digits at the end of them ; they thus consist of one fin on each side containing the skeleton of a concealed hand.

Seeing that the cetaceans are mammals, it is assuredly a part of their plan of organisation to have four limbs like all the rest, and consequently a pelvis for the support of their hind-legs. But, here as elsewhere, the loss of these parts is the result of an abortion due to a long disuse of them. When we remember that in seals which still have a pelvis, this pelvis is impoverished, reduced and does not protrude from their haunches, we shall feel that the cause must be the moderate use which these animals make of their hind-legs, and that, if they were to give up using them altogether, the hind-legs and even the pelvis would ultimately disappear.

The arguments which I have just adduced will doubtless seem to be mere guesses, since it is not possible to establish them on direct positive proofs. If we pay attention, however, to the observations set forth in the present work, and if we then closely examine the animals which I have cited and also the effects of their habits and environment, we shall find as a result of this examination that these guesses acquire a high degree of probability.

The table on p. 179 may facilitate the understanding of what I have said. It is there shown that in my opinion the animal scale begins by at least two separate branches, and that as it proceeds it appears to terminate in several twigs in certain places.

This series of animals begins with two branches, where the most imperfect animals are found ; the first animals therefore of each of these branches derive existence only through direct or spontaneous generation.

There is one strong reason that prevents us from recognising the successive changes by which known animals have been diversified and been brought to the condition in which we observe them ; it is this, that we can never witness these changes. Since we see only the finished work and never see it in course of execution, we are naturally prone to believe that things have always been as we see them rather than that they have gradually developed.

Throughout the changes which nature is incessantly producing in every part without exception, she still remains always the same in her totality and her laws ; such changes as do not need a period much

TABLE

SHOWING THE ORIGIN OF THE VARIOUS ANIMALS.

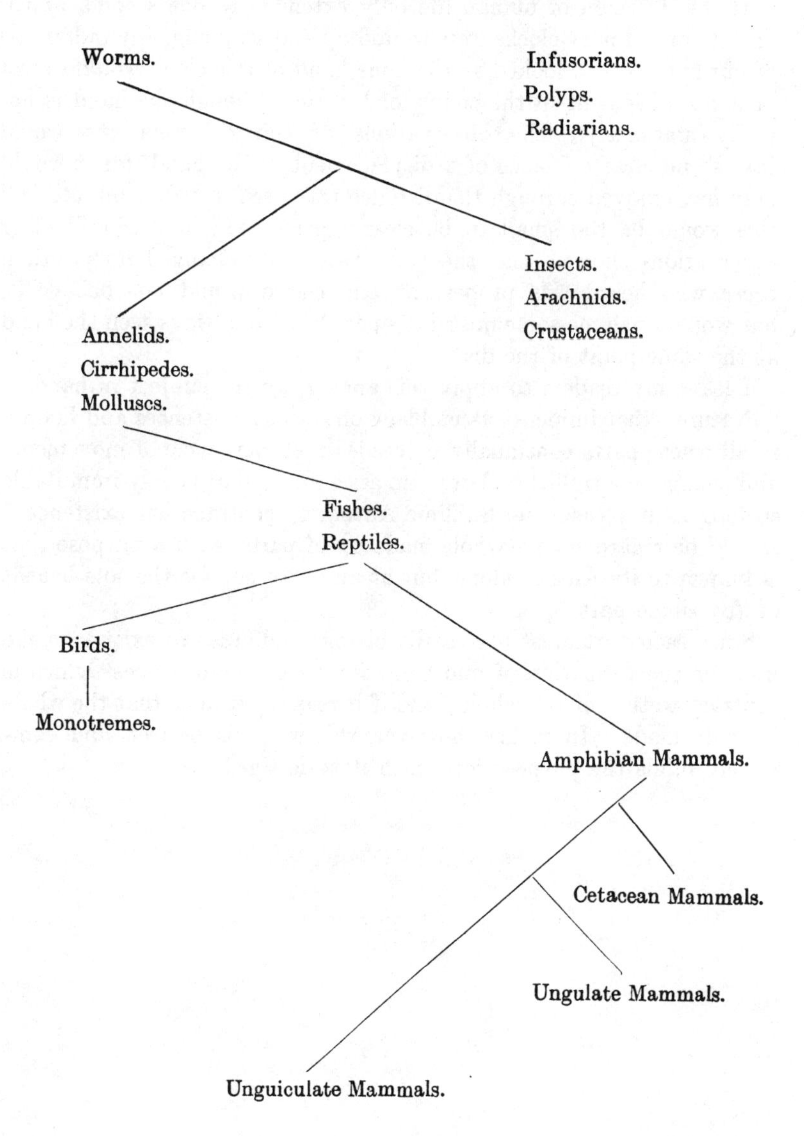

longer than the duration of human life are easily recognised by an observer, but he could not perceive any of those whose occurrence consumes a long period of time.

To explain what I mean let me make the following supposition.

If the duration of human life only extended to one second, and if one of our ordinary clocks were wound up and set going, any individual of our species who looked at the hour hand of this clock would detect in it no movement in the course of his life, although the hand is not really stationary. The observations of thirty generations would furnish no clear evidence of a displacement of the hand, for it would only have moved through the distance traversed in half a minute and this would be too small to be clearly perceived; and if still older observations showed that the hand had really changed its position, those who heard this proposition enunciated would not believe it, but would imagime some mistake, since they had always seen the hand at the same point of the dial.

I leave my readers to apply this analogy to the subject in hand.

Nature—that immense assemblage of various existences and bodies, in all whose parts continually proceeds an eternal cycle of movements and changes controlled by laws—an assemblage that is only immutable so long as it pleases her Sublime Author to continue her existence—should be regarded as a whole made up of parts, with a purpose that is known to its Author alone, but at any rate not for the sole benefit of any single part.

Since each part must necessarily change and cease to exist to make way for the formation of another, each part has an interest which is contrary to that of the whole; and if it reasons, it finds that the whole is badly made. In reality, however, this whole is perfect, and completely fulfils the purpose for which it is destined.

PART II.

AN ENQUIRY INTO THE PHYSICAL CAUSES OF LIFE, THE CONDITIONS REQUIRED FOR ITS EXISTENCE, THE EXCITING FORCE OF ITS MOVEMENTS, THE FACULTIES WHICH IT CONFERS ON BODIES POSSESSING IT, AND THE RESULTS OF ITS PRESENCE IN THOSE BODIES.

INTRODUCTION.

Nature, that word so often spoken as though it referred to a special entity, cannot be for us more than the totality of objects comprising: (1) all existing physical bodies; (2) the general and special laws, which regulate the changes of state and position to which these bodies are liable; (3) lastly, the movement distributed at large among them, which is continually preserved or being renewed, has infinitely varied effects, and gives rise to that wonderful order of things which this totality embodies.

All physical bodies whatever—solid, fluid, liquid or gaseous—are endowed with properties and faculties peculiar to themselves; but as a result of the movement distributed among them, these bodies are liable to different relations and transformations in their state and position. They are liable to contract with one another various kinds of union, combination or aggregation, and then to undergo all kinds of alterations, such as complete or incomplete separation from their other components or from their aggregates, etc.; these bodies thus derive new properties and faculties from the condition in which each of them is placed.

As a further result of the arrangement or position of these same bodies, of their special condition at any period of time, of the faculties possessed by each, of the laws of all the orders which regulate their changes and effects, and, lastly, of the movement which never leaves them in absolute rest, there continually reigns throughout the whole of nature a mighty activity, a succession of movements and transformations of all kinds, which nothing could arrest or annihilate, unless it be the power which has made all things exist.

The idea of nature as eternal, and hence as having existed for all time, is for me an abstract opinion without foundation, finality or probability, and with which my reason could never be satisfied. Since I can have no positive knowledge on this subject, and no power of reasoning about it, I prefer to think that the whole of nature is only an effect: hence, I imagine and like to believe in a First Cause or, in short, a Supreme Power which brought nature into existence and made it such as it is.

As naturalist and physicist, however, I am only concerned in my studies of nature with the bodies that we know or that have been observed, with the qualities and properties of these bodies, with the relations that they may have to one another under different conditions, and finally, with the effects of these relations and of the diverse movements which are distributed and ever preserved among them.

This method, which is the only one open to us, makes it possible to obtain a glimpse of the causes of those multitudinous phenomena which nature exhibits in her various parts, and to arrive at an understanding of the causes of the wonderful phenomena presented by living bodies, in short, the causes of life.

It is no doubt a very important matter to enquire into the nature of what is called life in a body ; what are the conditions of organisation necessary for its existence ; what is the origin of that remarkable force which gives rise to vital movements so long as the state of organisation allows ; lastly, how the various phenomena resulting from the continued presence of life in a body may achieve their result and endow this body with the faculties observed in it ; but of all the problems which man can suggest these are beyond question the most difficult to solve.

It seems to me that it was much easier to determine the course of the stars observed in space, and to ascertain the distance, magnitudes, masses and movements of the planets belonging to our solar system, than to solve the problem of the origin of life in the bodies possessing it, and, consequently, of the origin and production of the various existing living bodies.

However difficult may be this great enquiry, the difficulties are not insuperable ; for in all this we have to deal only with purely physical phenomena. Now it is obvious that the phenomena in question are, on the one hand, only direct effects of the mutual relations of different bodies, and only the result of an order and state of things which give rise to these relations among some of them ; and, on the other hand, it is obvious that these phenomena result from movements set up in the parts of these bodies by a force whose origin it is possible to ascertain.

These early results of our enquiries are unquestionably of very great interest, and give us a hope of obtaining other results no less important. But however well founded they may be, it will perhaps yet still be long before they obtain the attention which they deserve ; because they have to contend with one of the most ancient preconceptions, they have to destroy inveterate prejudices, and present a new field of study very different from any that we are accustomed to.

It is apparently reflections of this kind which caused Condillac to say that "reason has very little force and makes very slow progress, when it has to destroy errors from which no one is exempt" (*Traité des Sensations*, vol. i., p. 1108).

M. Cabanis unquestionably established a very great truth by a series of unexceptionable facts, when he said that the moral and the physical both spring from a common origin; and when he showed that the operations called moral are directly due, like those called physical, to the activity either of certain special organs, or of the living system as a whole; and, finally, that all the phenomena of intelligence and will take their origin from the congenital or fortuitous state of the organisation.

But in order to see more clearly how firmly this great truth is based, we must not confine ourselves to seeking the proofs of it by an examination of the highly complicated organisation of man and the more perfect animals; proof will be obtained still more easily by studying the diverse progress in complexity of organisation from the most imperfect animals up to those whose organisation is the most complex; for this progress will then exhibit in turn the origin of every animal faculty and the causes and developments of these faculties. We shall then acquire a renewed conviction that those two great branches of our existence called the physical and the moral, which exhibit two orders of phenomena apparently so distinct, have a common basis in organisation.

This being so, it is in the simplest of all organisations that we should open our inquiry as to what life actually consists of, what are the conditions necessary for its existence, and from what source it derives the special force which stimulates the movements called vital.

As a matter of fact it is only by a study of the simplest organisations that we can attain a knowledge of the true conditions for the existence of life in a body; for in a complex organisation all the principal internal organs are necessary for the maintenance of life on account of their close connection with other parts of the system, and because the system itself is formed on a plan which requires these organs; but it does not follow that these same organs are essential to the existence of life in all living bodies whatsoever.

This is very important to remember when we are enquiring what are the real conditions for the constitution of life; otherwise we might thoughtlessly attribute to some special organ an existence that is indispensable for the manifestations of life.

The peculiarity of vital movements is to be started and maintained by stimulus and not by transmission. These movements are the only ones of this character in nature, except perhaps for those of fermentation; they differ however from the movements of fermentation in that they can be maintained almost unchanged for a limited period, and in that they lead to a growth subsequently maintained for some time of the body in which they work, whereas those of fermentation irreparably destroy the body subjected to it and increase up to the limit that annihilates them.

Since vital movements are never transmitted but always stimulated, we must enquire what is their exciting cause, that is to say, from what source living bodies derive the peculiar force which animates them.

Whatever may be the state of organisation of a body and of its essential fluids, active life could assuredly not exist in that body without a special cause capable of exciting its vital movements. Whatever hypothesis we may form in this matter, we are always obliged to recognise that some special cause must be present for the active manifestations of life. Now it can no longer be doubted that this cause which animates living bodies is to be found in the environment of those bodies, and thus varies in intensity according to places, seasons, and climates. It is in no way dependent on the bodies which it animates, it exists before they do and remains after they have been destroyed. Lastly, it stimulates in them the movements of life, so long as the state of these bodies allows ; and it ceases to animate them when that state opposes obstacles to the performance of the movements which it stimulates.

In the most perfect animals this exciting cause of life is developed within themselves, and suffices to animate them up to a certain point ; but it still needs the co-operation of that provided by the environment. In the other animals, and in all plants it is altogether external to them ; so that they can only obtain it from their environment.

When these interesting facts have been ascertained and settled, we shall enquire how the first outlines of organisation come to be formed, how spontaneous generation can have occurred and in what part of the two series of living bodies.

If, indeed, bodies which possess life are really productions of nature, she must have had and still have the faculty of producing some of them spontaneously. She must then have endowed them with the faculty of growth, multiplication and increasing complexity of organisation and the power of varying according to time and circumstances. She must have done this if all those that we now observe are really the products of her power and efforts.

After recognising the necessity for these acts of direct creation, we must enquire which are the living bodies that nature may produce spontaneously, and distinguish them from those which only derive their existence indirectly from her. Assuredly the lion, eagle, butterfly, oak, rose, do not derive their existence immediately from nature ;

they derive it as we know from individuals like themselves who transmit it to them by means of reproduction; and we may be sure that if the entire species of the lion or oak chanced to be destroyed in those parts of the earth where they are now distributed, it would be long before the combined powers of nature could restore them.

I propose then to show what is the method apparently used by nature for forming, in favourable places and conditions, the most simply organised living bodies and through them the most perfect animals; how these fragile animals, which are the mere rudiments of animality directly produced by nature, have developed, multiplied and become varied; how at length, after an infinite series of generations, the organisation of these bodies has advanced in complexity and has extended ever more widely the animal faculties of the numerous resulting races.

We shall find that every advance made in complexity of organisation and in the faculties arising from it has been preserved and transmitted to other individuals by means of reproduction, and that by this procedure maintained for very many centuries nature has succeeded in forming successively all the living bodies that exist.

We shall see, moreover, that all the faculties without exception are purely physical, that is, that each of them is essentially due to activities of the organisation; so that it will be easy to show how, from the humblest instinct, the origin of which can be easily ascertained, nature has attained to the creation of the intellectual faculties from the most primitive to the most highly developed.

My readers must not expect to find here a treatise on physiology: the public is already in possession of excellent works of this character in which I have few alterations to suggest; but I must marshal together the general facts and well-established fundamental truths on this subject, because I find that their association leads to new light which has escaped those who have occupied themselves with details, and because this light clearly shows us what the bodies endowed with life really are, why and how they exist, and in what manner they develop and reproduce; lastly, by what methods the faculties observed in them have arisen, and been transmitted and retained in the individuals of each species.

If we wish to grasp the chain of physical causation which brought living bodies into existence, we must pay attention to the principle which I embody in the following proposition:

It is to the influence of the movements of various fluids in the more or less solid substances of our earth that we must attribute the formation, temporary preservation, and reproduction of all living bodies observed on its surface, and of all the transformations incessantly undergone by the remains of these bodies.

If we neglect this important principle, we become involved in an inextricable confusion, and the general cause of observed facts and objects cannot be perceived; our knowledge of this subject then remains without value, coherence or progress, so that instead of comprehensible truths we shall continue to set up those phantoms of our imagination and that love of the marvellous, in which the human mind takes so much delight.

If, on the other hand, we pay to this proposition all the attention to which its importance entitles it, we shall then see that there naturally flow from it a number of subordinate laws which furnish an explanation of all the well-known facts concerning existence, nature, and the various faculties; and, lastly, concerning the transformations of living bodies, and the other more or less complex bodies that exist.

As to the constant but variable movements of the fluids that I am about to discuss, it seems quite clear that they are permanently maintained on our earth by the influence of the sunlight. Sunlight is incessantly causing modifications and displacements of great masses of these fluids in certain regions of the earth, and forcing them to undergo a kind of circulation and various sorts of movements, so that they are able to produce all the observed phenomena.

To establish the accuracy of this statement, I shall merely have to introduce order into my citation of the facts and their relations, and into the application of these principles to observed phenomena.

In the first place it is necessary to distinguish the visible fluids, which are contained in living bodies and there undergo constant change and movement, from certain other subtle fluids which are always invisible but which animate these bodies and are indispensable to the existence of life.

Next, when considering the effects of the activity of the invisible fluids, to which I refer, on the solid, fluid and visible parts of living bodies, we shall easily discern that the organisation of these different bodies and all their movements and modifications are entirely due to the movements of the various fluids occurring in these bodies; that the fluids in question have by their movements organised these bodies, modified them in various ways and modified themselves also, so as gradually to have produced the state of things now observed.

In short, if we give sustained attention to the various phenomena presented by organisation, and especially to those concerned with the development of that organisation mainly in the most imperfect animals, we shall reach the following convictions:

1. That the entire work of nature in her spontaneous creations

consists in organising into cellular tissue the little masses of gelatinous or mucilaginous material which she finds at hand under favourable circumstances; in filling these little cellular masses with fluids and in vivifying them, by setting these contained fluids in motion by means of the stimulating subtle fluids which are incessantly flowing in from the environment;

2. That cellular tissue is the framework in which all organisation has been built, and in the midst of which the various organs have successively developed by means of the movement of the contained fluids which gradually modifies the cellular tissue;

3. That the function of the movement of the fluids in the supple parts of the living bodies which contain them, is to cut out paths and establish depots and exits, to create canals and afterwards various organs; to cause variation in these canals and organs by means of a diversity either in the movements or in the nature of the fluids which produce and modify them; finally to enlarge, elongate, divide and solidify gradually these canals and organs by substances which are formed and incessantly separated off from the essential fluids in movement there; substances of which one part becomes assimilated and united with the organs while the other is thrown out;

4. That, lastly, the function of organic movement is not merely the development of organisation, and the increase and growth of the parts, but also the multiplication of organs and of the function which they fulfil.

After having expounded these great principles which seem to me unquestionable truths although not hitherto recognised, I shall enquire what faculties are common to all living bodies and consequently to all animals; I shall then pass in review the chief of the faculties which are peculiar to certain animals but are not possessed by the rest.

I venture to affirm that grave injury results to the progress of physiological knowledge by the thoughtless supposition that all animals without exception possess the same organs and enjoy the same faculties; as though nature were everywhere forced to employ the same methods to attain her end. Seeing that nothing more than an active imagination is needed for setting up principles if we do not pause to consider facts, it is an easy supposition that all living bodies possess the same organs and hence enjoy the same faculties.

Another subject which I must not neglect in this second part of my work is the question of the immediate results of life in a body. Now I am in a position to show that these results give rise to combinations between principles which, except for this factor, would never have been united together. These combinations accumulate more and more according as the vital energy increases, so that in the most perfect animals there is a high complexity and great intricacy in the combination of principles. Living bodies thus constitute, by their possession of life, nature's principal means for bringing into existence a number of different compounds which would never otherwise have arisen.

It is vain to imagine that living bodies find ready formed in the substances on which they feed all the material required for building up the various parts of their bodies; they only find in these food substances, materials suitable for entering into the combinations which I have mentioned, and not the combinations themselves.

It is no doubt owing to an insufficient study of the power of life in the bodies which possess it, and the failure to perceive the results of this power, that it has been alleged that living bodies find in their ordinary food the material ready prepared which serves for building up their bodies and that these materials have existed in nature for all time.

Such are the subjects which compose the second part of this work: their importance would no doubt justify considerable expansion; but I have confined myself to a concise exposition of what is necessary in order that my observations may be understood.

CHAPTER I.

COMPARISON OF INORGANIC BODIES WITH LIVING BODIES, FOLLOWED BY A PARALLEL BETWEEN ANIMALS AND PLANTS.

I LONG ago conceived the idea of making a comparison between organised living bodies and crude inorganic bodies. I then noticed the extreme difference existing between these two, and I became convinced of the necessity for examining the kind and amount of this difference. It was at that time the general custom to present the three kingdoms of nature arranged in a line, with class distinctions between them; and the enormous difference apparently was not perceived between a living body and a crude lifeless body.

Yet if we wish to arrive at a real knowledge of what constitutes life, what it consists of, what are the causes and laws which control so wonderful a natural phenomenon, and how life itself can originate those numerous and astonishing phenomena exhibited by living bodies, we must above all pay very close attention to the differences existing between inorganic and living bodies; and for this purpose a comparison must be made between the essential characters of these two kinds of bodies.

COMPARISON BETWEEN THE CHARACTERS OF INORGANIC BODIES AND THOSE OF LIVING BODIES.

1. No crude or inorganic body possesses individuality except in its integral molecule; the solid, fluid or gaseous masses that may be formed by a collection of integral molecules have no limits; and the large or small size of these masses neither adds nor subtracts anything that can alter the nature of the body concerned; for this nature is exclusively dependent on that of the integral molecule of the body.

Every living body, on the other hand, possesses an individuality throughout its mass and volume; and this individuality, simple in some and compound in others, is never confined in living bodies to that of their component molecules.

2. An inorganic body may present a truly homogeneous mass or it may constitute a heterogeneous mass ; the aggregation or combination of similar or dissimilar parts can occur without these bodies ceasing to be crude or inorganic. In this respect there is no essential necessity for the masses of this body to be more homogeneous than heterogeneous or *vice versâ* ; it is by chance that they are as we observe them.

All living bodies, on the contrary, even those with the simplest organisations, are necessarily heterogeneous, that is to say, composed of dissimilar parts : they have no integral molecules, but are formed from molecules of a different character.

3. An inorganic body may constitute either a perfectly dry, solid mass or a completely liquid mass or a gaseous fluid.

The contrary holds good in the case of all living bodies ; for no body can possess life unless it is formed from two kinds of necessarily co-existing parts, the one solid, but supple and capable of holding liquids ; the other liquid and contained in the first, but quite independent of the invisible fluids which penetrate the body and develop within it.

The masses which constitute inorganic bodies have no special specific shape ; for whether these masses have a regular shape, as in the case of crystals, or whether they are irregular, their shape does not remain permanently the same ; it is only the integral molecule which has in each kind an invariable shape.[1]

Living bodies, on the contrary, nearly all exhibit a shape peculiar to their species and one which cannot vary without giving rise to a new race.

4. The integral molecules of an inorganic body are entirely independent of one another ; for even when they are combined into a solid, liquid or gaseous mass, each of them continues to exist by itself and to be constituted by the number, proportions, and character of combination of its principles ; its existence is neither conditioned nor increased by the similar or dissimilar molecules in its neighbourhood.

The molecules of a living body, on the other hand, and consequently all the parts of that body, are dependent for their character upon one another ; because they are all subjected to the influence of a factor

[1] The integral molecules which constitute a compound substance all result from combinations of the same number of principles in the same proportions, with exactly the same character of combination : hence they all have the same shape, density and special properties.

But when any causes have produced a variation either in the number of the component principles of these molecules, or in the proportions of the principles, or in the character of their combination, these integral molecules then acquire another shape, density and special properties : they then belong to another species.

which animates them and gives them activity, and because this factor requires their co-operation for a common end both in the separate organs and in the entire individual; because, moreover, the variations in this same factor work similar effects in the state of each molecule and each part.

5. No inorganic body needs any movement in its parts for its preservation; on the contrary, so long as the parts remain at rest the body is preserved without disintegration and might exist in this condition for ever. But as soon as any factor begins to act upon this body and produce movements and changes in its parts, the body at once loses either its shape or its coherence, if the movement and changes produced in its parts merely affect its mass or some part of its mass; and it loses even its fundamental character or is destroyed, if the movements and changes in question penetrate as far as its integral molecules.

Every body possessing life, on the other hand, is permanently or temporarily animated by a special force, which incessantly stimulates movements in its internal parts and uninterruptedly produces changes of state in these parts, at the same time effecting restorations, renewals, developments and a number of phenomena that are entirely peculiar to living bodies; so that in their case the movements stimulated within them produce disintegration and destruction followed by recuperation and renewal. This prolongs the life of the individual so long as the equilibrium between these two opposed elements is not too rudely disturbed.

6. In all inorganic bodies an increase of volume and mass is always accidental and has no necessary limits. This increase only takes place by *juxtaposition*, that is to say, by the addition of new parts to the external surface of the body in question.

The growth of every living body, on the contrary, is always necessary and limited, and only takes place by *intussusception*, that is to say, by internal penetration, or the introduction into the individual of substances which have to be added to it and make part of it after being assimilated. Now this growth is a true development of parts from within outwards, and is exclusively the property of living bodies.

7. No inorganic body has to feed in order to be preserved; for it need never lose any of its parts, and when it does it has no means of restoring them.

All living bodies, on the contrary, necessarily experience in their internal parts successive and constantly renewed movements, changes in the state of the parts, and, lastly, continual losses of substance through the separations and dissipations involved by these changes. Hence no such body can maintain life if it is not constantly feeding, that is to say, if it is not incessantly making good its losses by substances introduced into its interior, in short, if it does not take food whenever it needs it.

8. The masses of inorganic bodies consist of separate parts which are united by accident; these bodies are not born, nor are they ever the produce of a germ or bud whose development gives rise to an individual exactly similar to that from which it springs.

All living bodies, on the contrary, are really born, and are the produce either of a germ which has been vivified and prepared for life by fertilisation, or else simply of an expansible bud. In either of these cases new individuals arise exactly like those which have produced them.

9. Lastly, no inorganic body can die, inasmuch as no such body possesses life. Death is a necessary result of the existence of life in a body, for it is only the complete cessation of organic movements, following upon some disturbance which makes these movements henceforth impossible.

All living bodies, on the contrary, are subject to an inevitable death; for it is a property of life or of the movements constituting life in a body, to bring about after a certain period a condition of the organs which makes it impossible for them to carry on their functions, and which therefore destroys the faculty of performing organic movements.

Hence between crude or inorganic bodies and living bodies there exists an immense difference, a great hiatus, in short, a radical distinction such that no inorganic body whatever can even be approached by the simplest of living bodies. Life and its constituents in a body make the fundamental difference that distinguishes this body from all those that are without it.

How great then is the error of those who try to find a connection or sort of gradation between certain living bodies and inorganic bodies!

Although M. Richerand in his interesting *Physiologie* has dealt with the same subject that I am now treating, I have had to reproduce his views together with modifications of my own; since his studies are very important on the subjects which I still have to set forth.

A comparison between plants and animals does not immediately concern my thesis in this Part II.; nevertheless, as such a comparison assists in the general purpose of this work, I propose here to state a few of its most prominent characteristics. But first let us see what plants and animals actually have in common, in their capacity as living bodies.

The only point in common between animals and plants is the possession of life; hence they both fulfil the conditions of its existence, and possess the general faculties to which it gives rise.

Hence in both cases they are bodies composed essentially of two kinds of parts, the one solid but supple and containing; the other liquid and contained, but independent of the invisible fluids which penetrate and develop within them.

All these bodies possess individuality, either simple or compound; have a shape peculiar to their species; are born at the moment when life begins to exist in them or when they are separated from the body whence they spring; are permanently or temporarily animated by a special force which stimulates their vital movements; are only preserved through nutrition which more or less restores their losses of substance; grow for a limited period by internal development; form for themselves the compound substances of which they are made; reproduce and multiply so as to carry on the species like themselves; lastly, all reach a period when the state of their organisation no longer permits of the maintenance of life within them.

Such are the faculties common to these two kinds of living bodies. Let us now compare the general characters by which they are distinguished from each other.

Comparison between the General Characters of Plants and Animals.

Plants are organised living bodies, not irritable in any of their parts, incapable of performing sudden movements several times in succession, and the vital movements of which are only performed by means of external stimuli, that is to say, by an exciting cause provided by the environment and acting chiefly on the contained and visible fluids of these bodies.

In animals, some or all of the parts are essentially irritable, and have the faculty of performing sudden movements which may be repeated several times in succession. The vital movements are in some performed by means of external stimuli, and in others by a force developing within them. The external stimuli and internal stimulating force affect the irritability of the parts, act upon the visible contained fluids and give rise in all cases to the performance of vital movements.

It is certain that no plant whatever has the faculty of suddenly moving its external parts and repeating such movement several times in succession. The only sudden movements that certain plants display are movements of relaxation or collapse of some part (*v.* p. 52); hygrometric or pyrometric movements also are sometimes performed by certain filaments when suddenly exposed to the air. As to the other movements performed by the parts of plants, such as those which make them bend towards the light, those which cause the opening and closing of flowers, those which give rise to the erection or depression of stamens or peduncles or to the twining of climbing stems and tendrils, finally, those constituting what is called sleeping and waking in plants; none of these movements are ever sudden; they are carried out so slowly as to be altogether imperceptible; and they are only known by their finished results.

Animals, on the contrary, possess the faculty of performing very obvious sudden movements, by means of some of their external parts, and of repeating them several times in succession with or without variation.

Plants, especially those which live partly in the air, grow in a remarkable manner in two opposite directions, in such a way as to exhibit an ascending vegetation and a descending vegetation. These two kinds of vegetation start from a common point which I have elsewhere [1] named the vital knot, because in this point life is specially concentrated when the plant loses its structures, and because the plant only really dies when life ceases to exist in this part. The organisation of this vital knot, otherwise known as the root-collar, is altogether peculiar; from the vital knot the ascending vegetation produces the stem, branches and all the parts of the plant that are in the air; and from the same point the descending vegetation gives birth to roots which are buried in the soil or in water. Finally, in germination, which brings the seeds into life, the early development of the young plant requires ready prepared juices, which the plant cannot yet draw from the soil or from the air; these juices then appear to be furnished by the cotyledons, which are always attached to the vital knot; they suffice for starting the ascending vegetation of the plumule and the descending vegetation of the radicle.

Nothing of the kind is observed in animals. Their development is not limited to two special directions only, but takes place on all sides and in all directions, according to their requirements; finally, their life is never concentrated in an isolated point but is spread throughout the essential special organs, if there are any. In animals in which there are no essential special organs, life is not concentrated in any one part; for when we divide their bodies life is preserved in each separate part.

Plants in general rise perpendicularly, not to the plane of the earth on all occasions, but to that of the horizon; so that according as they grow they shoot upwards towards the sky like a sheaf of rockets in a firework display. Although the twigs and branches which form their tops do not follow the direction of the stem, they always form an acute angle with the stem at their point of insertion. It appears that the stimulating force of the vital movements in these bodies is

[1] *Histoire Naturelle des Végétaux*, édition de Déterville, vol. i., p. 225.

chiefly directed upwards and downwards, and that this is the reason of the peculiar shape and arrangement of these living bodies, in short, of their ascending and descending vegetation. From this it follows that the canals, in which move the essential fluids of these bodies, are parallel to one another and to the longitudinal axis of the plant; for it is always parallel, longitudinal tubes that are formed in their cellular tissue, and these tubes do not diverge except to form the flattened expansions of leaves and petals, or to be distributed in the fruit.

Nothing of all this is found in animals. The longitudinal axis of their bodies is not necessarily directed towards the sky, on the one hand, and the centre of the earth, on the other hand; the force which stimulates their vital movements does not work exclusively in two directions; lastly, the internal canals which contain their visible fluids are turned about in various ways and present no sort of parallelism.

The food of plants consists only of the liquid or fluid substances which they absorb from the environment: this food includes water, atmospheric air, caloric, light, and various gases which they decompose and convert to their own use; hence they never have to carry on digestion, and for this reason they have no digestive organs. Seeing that living bodies themselves elaborate their own substance, it is they which form the first non-fluid combination.

Most animals, on the contrary, feed on substances which are already compound and which they introduce into a tubular cavity suitable to receive them. Hence they have a digestion in order to bring about the complete solution of these substances; they modify existing combinations and load them heavily with new principles; so that it is they which form the most complex combinations.

Lastly, the final residue of destroyed plants is very different from that which emanates from animals, showing that these two kinds of living bodies are indeed of an entirely distinct nature.

In plants, as a matter of fact, solids exist in larger proportion than fluids, mucilage constitutes their softest parts, and carbon predominates among their component principles; whereas in animals fluids are more abundant than solids, gelatine abounds in their soft parts and even in the bones of such as have any, while among their components nitrogen is specially conspicuous.

Moreover, the strata formed out of the residue of plants is chiefly argillaceous and often contains silica, whereas those formed from animals consist either of the carbonate or phosphate of lime.

Some Features of Analogy between Animals and Plants.

Although the nature of plants is very different from that of animals, and although the bodies of the one always possess faculties and even substances that would vainly be sought in the other, the fact remains that they are both living bodies and that nature obviously followed a uniform plan of operations in producing them. In point of fact, nothing is more remarkable than the analogy observed between certain of her operations in these two kinds of living bodies.

In both of them, the most simply organised only reproduce by gemmae or buds. These are reproductive corpuscles which are like eggs or seeds, but require no preliminary fertilisation, and which indeed contain no embryo which has to break through its investments before being able to complete its development. Yet in both animals and plants, when the complexity of organisation was sufficiently advanced to permit of the formation of organs of fertilisation, the reproduction of individuals then became exclusively or chiefly sexual.

Another very remarkable feature of analogy, in the operations of nature, between animals and plants, is the more or less complete suspension of active life, that is to say, of vital movements, which is experienced in certain climates and seasons by a large number of living bodies of both kinds.

In the winter of cold climates, indeed, the woody perennial plants undergo a more or less complete suspension of vegetation, and hence of organic or vital movements; their fluids, which are at these periods less abundant, remain inactive: during these conditions, there occur in the plants no losses or absorptions of food or any alterations or development; in short, their active life is altogether suspended, their bodies become torpid and yet they are not lifeless. Since the truly simple plants can only live for a year, they hurriedly produce their seeds or reproductive corpuscles in cold climates and die on the approach of the bad season.

The phenomena of the more or less complete suspension of active life, that is, of the organic movements composing it, are also witnessed in many animals in very curious forms.

In the winter of cold climates life comes to an end in the most imperfect animals; and among those which retain life, a great many become more or less completely torpid, so that in some all the internal or vital movements are suspended, while in others they still exist but are only performed with extreme slowness. Thus although nearly all the classes contain animals which undergo this more or less complete suspension of active life, it is particularly noticeable in the ants, bees, and many other insects; in the annelids, molluscs, fishes, reptiles (especially snakes), and, lastly, in many mammals such as the bat, marmot, dormouse, etc.

The last feature of analogy that I shall name is no less remarkable; it is this: just as there are simple animals constituting separate individuals, and compound animals adhering together, communicating at their base and sharing a common life such as most of the polyps, so also there are simple plants living as individuals and there are compound plants where several live together, are grafted on to one another and share a common life.

The general rule among plants is to live until they have produced flowers and fruit or reproductive corpuscles. Their lives rarely last for more than a year. Their sexual organs, if they have any, are only of use for a single fertilisation; so that when plants have reached the goal of reproduction (seeds), they die and are completely destroyed.

In the case of a simple plant, death takes place after the production of fruit; and it is difficult, as we know, to propagate it otherwise than by seeds or gemmae.

Annual or biennial plants all appear to be in this position; they are simple plants; and their roots, stems and branches are simply vegetative products; it is by no means every plant however that is in this position, for the greater number of those that are known are in reality compound plants.

Thus, when I see a tree, shrub or perennial, it is not simple plants that I have before me, but a multitude of plants living together upon one another and all sharing a common life.

So true is this that if I were to graft the shoot of a cherry tree on to the branch of a plum tree, and an apricot shoot on to another branch of the same tree, these three species would live together and share a common life while yet remaining distinct.

The roots, trunk, and branches of such a plant consist purely of the vegetable products of this common life, and of separate but adherent plants which live upon it; just as the general substance of a madrepore is the animal product of numerous polyps which live together through successive generations. But every bud in a plant is itself an individual plant, which shares in the common life of all the rest, develops its flower or inflorescence once a year, then produces fruit and may finally give rise to a branch already supplied with other buds, that is, other individual plants. Each of these individual plants either fruits, in which case it does so only once, or produces a branch which gives rise to other similar plants. Any such composite plant is thus a vegetable product, which continues to live after the destruction of all the individuals which have combined to produce it.

By separating off parts of a plant, containing one or more buds or including undeveloped elements, we can form at pleasure a number of new living individuals similar to those from which they are taken, without any necessity for taking the fruit. This in fact is just what horticulturists do when they take slips, layers, etc.

Now just as nature has made compound plants, so too she has made compound animals; and for this purpose she has made no change in the nature of either animals or plants. It is quite as absurd to call compound animals by the name of plant-animals, as it would be to call compound plants by the name of animal-plants.[1]

If the name of zoophyte were given a century ago to compound animals of the class of polyps, the error was excusable; the low state of knowledge then existing about animal nature made this term less obnoxious; but now things have altered, and it cannot be a matter of indifference that a class of animals should receive a name which embodies a false notion of the objects indicated.

Let us now enquire what life is, and what are the conditions for its existence in a body.

[1] When we confine our attention to the substances produced by vegetation or by animals, we often find cases where it is difficult to decide whether they belong to the plant or animal kingdom; chemical analysis of these bodies sometimes decides in favour of animal substances when their shape and organisation are suggestive of true plants. Several genera referred to the family of algae provide examples of this difficulty: it would thus seem to follow that there is an almost imperceptible transition from plants to animals.

I do not think so: on the contrary, I am thoroughly convinced that if it were possible to examine the actual animals which form the membranous or filamentous polyparies so closely resembling plants, the uncertainty as to their true nature would at once be removed.

CHAPTER II.

OF LIFE, WHAT IT CONSISTS OF, AND THE CONDITIONS OF ITS EXISTENCE IN A BODY.

LIFE, said M. Richerand, is a collection of phenomena which succeed one another for a limited period in organised bodies.

He should have said, life is a phenomenon which gives rise to a collection of other phenomena, etc.; for it is not these other phenomena that constitute life, but they are themselves caused by life.

A study of the phenomena resulting from the existence of life in a body provides no definition of life, and shows nothing more than objects that life itself has produced. The line of study which I am about to follow has the advantage of being more exact, more direct and better fitted to illuminate the important subject under consideration; it leads, moreover, to a knowledge of the true definition of life.

Life when studied in living bodies is exclusively due to the relations existing between the three following objects: the parts of the body adapted for containing liquids, the contained liquids moving in them, and the exciting cause of such movements and changes as are carried out.

Whatever efforts we may make by the most profound thought and meditation to decide as to what life consists of, we shall necessarily be compelled to fall back on the principle just enunciated as soon as we pay attention to the teaching of observation on the matter; in fact, life consists of nothing else.

A comparison drawn between life and a watch in active movement is inadequate, to say the least of it; for in the watch there are only two main points to consider: (1) the wheels and machinery of movement; (2) the spring which by its tension and elasticity keeps up the movement so long as that tension continues.

But in a living body, instead of two chief points for study, there are three: (1) the organs or supple containing parts; (2) the essential contained fluids which are always in motion; (3) lastly, the exciting cause of vital movements, from which arises the action of the fluid on the organs and the reaction of the organs on the fluids. It is then purely from the relations between these three objects that the movements, changes, and all the phenomena of life result.

In order to improve the comparison between a watch and a living body we should have to compare the exciting cause of organic movements with the spring of the watch, and regard the supple containing parts, together with the essential contained fluids, as the machinery of the movement in question.

It will then be clear, in the first place, that the spring (exciting cause), is the essential motive power, without which the whole remains inactive, and that its variations of tension must be the cause of the variations of energy and rapidity of movements.

In the second place, it will be obvious that the machinery of movement (the organs and essential fluids) must be in a state and arrangement suitable for the performance of the movements which it has to carry out; hence, when this machinery gets out of order the effective power of the spring is lost.

From this point of view the parallel is complete; a living body may be compared with a watch; and I can easily show the close accuracy of this comparison by reference to known facts and observations.

As to the machinery of movement, its existence and faculties are now well known, as also most of the laws which control its various functions.

But as to the spring, the essential motive power and originator of all movements and activities, it has hitherto escaped the researches of observers: I believe, however, that I shall be able to describe it in the next chapter, in such a way that it cannot in future be neglected.

But first let us continue the enquiry as to what essentially constitutes life.

Seeing that life in a body results exclusively from the relations existing between the containing parts in an appropriate condition, the contained fluids moving in them, and the exciting cause of the movements, activities and reactions which take place, we may include what essentially constitutes life in the following definition.

Life, in the parts of any body which possesses it, is an order and state of things which permit of organic movements; and these movements constituting active life result from the action of a stimulating cause which excites them.

This definition of life, either active or suspended, includes all the positive facts which have to be expressed in it, and covers all special cases. It appears to me impossible to add or subtract a single word without destroying the integrity of the essential ideas contained in it; lastly, it is based on the known facts and observations which have reference to this wonderful natural phenomenon.

To begin with, in this definition active life is kept distinct from that life which, without ceasing to exist, is suspended and appears to be maintained for a limited time without perceptible organic movements; and this, as I shall show, is in accordance with observation.

Then it brings out the fact that no body can possess active life except when the two following conditions are satisfied.

The first is the necessity for a stimulating cause which excites organic movements.

The second is the necessity that a body in order to possess and maintain life should be so ordered in its parts as to possess the property of responding to the action of the stimulating cause and of producing organic movements.

In the animals whose essential fluids are quite simple, such as the polyps and infusorians, if the contained fluids of any of these animals are suddenly removed by a rapid desiccation, such desiccation may be carried out without any disintegration of the organs or containing parts of this animal or any destruction of the order existing in them: in this case life is altogether suspended in the desiccated body; no organic movement occurs in it and it appears no longer to be a living body. Yet it cannot be called dead, for its organs or containing parts have retained their integrity, and if the internal fluids are restored to this body, the stimulating cause, assisted by a gentle warmth, soon excites movements, activities, and reactions in its parts and henceforth it returns to life.

The rotifer of Spallanzani, which was several times reduced to a state of death by rapid desiccation, and afterwards restored to life on being plunged into tepid water, shows that life can be alternately suspended and renewed: it is therefore only an order and state of things in a body, by means of which vital movements can occur when stimulated by a special cause.

In the plant kingdom, the algae and mosses exhibit the same phenomena as the rotifer of Spallanzani; mosses rapidly desiccated and kept in a herbarium even for a century may return to life and fresh vegetation, if they are placed in moisture at a moderate temperature.

Complete suspension of vital movements without degeneration of the parts, and hence with a continued possibility of a return of these movements, may also occur in man himself, though only for a very short time.

We learn from observations made on people that have been drowned, that if anyone falls into the water and is pulled out again after an immersion of three-quarters of an hour or even an hour, he is asphyxiated to the extent that no movement whatever takes place in his organs. Yet it may still be possible to restore him to active life.

If he is left in this condition without any assistance, orgasm and irritability soon become extinct in his internal parts, and thereafter the essential fluids and the softest parts begin to decompose, and this is the sign of death; but if, immediately after his extraction from the water and before the extinction of irritability, the usual aid is administered to him, if, in short, it is possible by means of the usual stimulants to excite in time contractions in his internal parts, and produce movements in his organs of circulation, then all the vital movements quickly resume their course, and active life no longer remains in suspense but is promptly restored.

But when degenerations and disorders of a living body, either in the order or in the state of its parts, are large enough to prevent these parts from yielding to the influence of the exciting cause and producing organic movements, then life is quickly extinguished, and the body henceforth is no longer included among the living.

From what I have just said, it follows that if in a body any disturbance or degeneration affects the order and state of things which endow it with active life, and if this disturbance is of a nature to prevent the performance of organic movements or their restitution after suspension, the body then loses its life, that is to say, it undergoes death.

A disorder resulting in death may be brought about in a living body through various accidental causes; but nature becomes the necessary cause at the end of a certain period; and, in fact, it is a property of life to bring the organs imperceptibly to a condition in which they cannot perform their functions, so that death inevitably ensues; the reason of this I shall explain.

When therefore we affirm that life, in all bodies which possess it, consists only of an order and state of things in the parts of the body, by which these parts are subject to the influence of a stimulating cause and carry out organic movements, we are not expressing a mere conjecture but a fact universally attested, susceptible of many proofs and never liable to be seriously disputed.

This being so, we are only concerned to know what is the order and state of the parts which make a body capable of possessing active life.

But as no precise knowledge of this subject can be directly acquired, let us first investigate the conditions essential to the existence of this order and state of things in the parts of the body, in order that it may possess life.

Conditions essential to the Existence of the Order and Structure of a Body in order that it may possess Life.

First condition. No body can possess life unless it consists essentially of two kinds of parts, viz. supple containing parts and contained fluid substances.

As a matter of fact, no body that is perfectly dry can be alive, nor can any body whose parts are fluid be in possession of life. The first condition essential to life in a body therefore is that it should consist of a mass with two kinds of parts, the one solid and containing, but soft and more or less cohesive, the other fluid and contained.

Second condition. No body can possess life unless its containing parts are cellular tissue or formed out of cellular tissue.

Cellular tissue, as I shall show, is the matrix in which all the organs of living bodies have been successively formed; and the movement of fluids in this tissue is the means adopted by nature for the gradual creation and development of these organs.

Every living body is thus essentially a mass of cellular tissue in which more or less complex fluids move more or less rapidly; so that if the body is very simple, that is, has no special organs, it appears homogeneous and consists only of cellular tissue containing fluids which are slowly moving; but if its organisation is complex, all its organs without exception are invested in cellular tissue down to their smallest parts, and are even essentially formed of it.

Third condition. No body can possess active life except when an exciting cause of its organic movements works within it. Without the impulse of this active stimulus, the solid containing parts of an organised body would be inert, the contained fluids would remain at rest, organic movements would not take place, no vital function would be carried out, and consequently active life would not exist.

Now that we know the three conditions essential to the existence of life in a body it becomes easier for us to ascertain wherein consist the order and state of things necessary to a body for the maintenance of life.

For this purpose, we must not limit our enquiries to living bodies with a highly complex organisation; for we should never learn from them to what cause life is to be attributed, and we might select at hazard factors of no fundamental importance.

But if we fix our attention on that extremity, either of the animal or plant kingdom, in which are found living bodies with the simplest organisations, we shall notice, in the first place, that in each individual the body consists only of a gelatinous or mucilaginous mass of cellular tissue of the feeblest coherence, the cells of which are in communication, and the various fluids of which undergo movements, displacements, dissipations, subsequent renewals, changes of state, and finally deposit parts which become fixed there. We shall then observe that an exciting cause of varying activity, but never entirely absent, incessantly animates the very supple containing parts of these bodies, as well as the essential fluids contained in them, and that this cause keeps up all the movements constituting active life, so long as the parts which have to acquire these movements are in a condition to do so.

Inference.

The order of things necessary for the existence of life in a body is then essentially as follows:

1. A cellular tissue (or organs formed of it) endowed with great suppleness and animated by orgasm, the first result of the exciting cause;

2. Various more or less complex fluids contained in this cellular tissue (or in the organs built up from it), and undergoing as a second result of the exciting cause, movements, displacements, various changes, etc.

In animals the exciting cause of organic movements acts powerfully both on the containing parts and on the contained fluids; it maintains an energetic orgasm in the containing parts, puts them in a condition to react on the contained fluids and hence makes them highly irritable; as to the contained fluids, the exciting cause involves them in a kind of rarefaction and expansion, which facilitate their various movements.

In plants, on the contrary, the exciting cause in question only acts powerfully on the contained fluids, and produces in these fluids such movements and alterations as they are adapted to undergo; but its only effect on the containing parts of these living bodies, even on their most supple parts, is an orgasm or slight erethism which is too feeble to permit of any movement or to cause a reaction on the contained fluids or consequently to endow these parts with irritability. The result of this orgasm has been badly named *latent sensibility*; I shall speak of it in Chapter IV.

In animals, which invariably have parts that are irritable, the vital movements are kept up in some solely by the irritability of the parts, and in others by a combination of irritability with muscular activity of the organs themselves.

In fact, in those animals whose very simple organisation only requires slow movements in the contained fluids, the vital movements are carried out exclusively through the irritability of the containing parts and the agitation produced by the exciting cause in the contained fluids. But as the vital energy increases in proportion to complexity of organisation, there soon arrives a time when irritability and the exciting cause are no longer sufficient by themselves for the acceleration needed in the movements of the fluids ; nature then makes use of the nervous system which increases the effects of the irritability of the parts by adding the activity of certain muscles ; and when this system permits of muscular movement, the heart becomes a powerful motor for accelerating the movement of the fluids ; finally, after the establishment of pulmonary respiration muscular movement is once again necessary to the performance of vital movements on account of the alternate dilatations and contractions occurring in the cavity which contains the respiratory organ and without which there could be no inspirations or expirations.

"Doubtless we are not called upon," says M. Cabanis, "to prove again that physical sensibility is the origin of all the ideas and habits constituting the moral existence of man ; Locke, Bonnet, Condillac and Helvetius have pushed this truth to the last stage of demonstration. Among educated persons who use their reason there is now no one who can throw the smallest doubt upon the matter. From another standpoint, physiologists have proved that *all vital movements are the result of impressions received by sensitive parts*, etc." (*Rapports du Physique et du Moral de l'Homme*, vol. i., pp. 85, 86).

I too recognise that physical sensibility is the source of all ideas, but I am very far from admitting that all vital movements are the result of impressions received by sensitive parts : that at most can only be true with regard to such living bodies as possess a nervous system ; for the vital movements of those which have no such system cannot be the result of impressions received by sensitive parts : this is quite obvious.

If we wish to make a true analysis of life, we must necessarily examine the facts which it presents in all bodies possessing it. Now as soon as we deal with the subject in this way we see that what is really essential to the presence of life in one plan of organisation is by no means essential in another.

No doubt nervous influence is necessary to the maintenance of life in man, and in all animals which have a nervous system ; but this does not prove that vital movements, even in man and in animals provided with nerves, are due to impressions made on sensitive parts : it only proves that their vital movements cannot occur without the help of nervous influence.

It is clear from the above exposition that life in general may exist in a body, although the vital movements are not produced by impressions received by the sensitive parts and although there is no muscular activity ; it may even exist when the body possessing it has no irritable parts whose reaction could assist its movements. It is enough, as we see in plants, that a body possessing it should present internally an order and state of things with regard to its containing parts and contained fluids which permit of the excitation of the characteristic movements and changes, by means of a special force.

But if we consider life in special cases, that is, in various selected bodies, we shall then see that whatever is essential to the plan of organisation of these bodies has also become necessary to the maintenance of life in them.

Thus in man and the most perfect animals, life cannot be maintained without irritability of the reacting parts, without the involuntary muscles to keep up the rapid movement of the fluids, without the nervous influence which by quite a different route from feeling provides for the performance of the functions of the muscles and other internal organs; lastly, without the influence of respiration to restore continually the essential fluids which are so rapidly disintegrated in these systems of organisation.

Now this nervous influence, which is undoubtedly necessary, is exclusively that which sets the muscles in action and not that which produces feeling; for it is not by means of sensations that the muscles act. In fact, no feeling whatever is aroused by the cause which produces the movements of systole and diastole in the heart and arteries; and if we do sometimes perceive the beats of the heart it is when they are stronger and more rapid than usual; this muscle, which is the chief motive power of circulation, then strikes neighbouring sensitive parts. Finally, when we walk or perform any action we never feel the movement of the muscles nor the impulse which drives them.

Hence it is not through the medium of feeling that the muscles carry on their functions, although nervous influence is necessary to them. But since nature was obliged, in order to accelerate the movement of fluids in the most perfect animals, to add the muscular movement of the heart, etc., to the irritability which they possess in common with the rest, nervous influence has become necessary to the maintenance of life in these animals. There can, however, be no justification for the statement that their vital movements are only due to impressions received by sensitive parts, for if their irritability was destroyed they would immediately lose their life; and their feeling, if it still survived, could not alone suffice for its preservation. Moreover, I hope to prove in Chapter IV. of this part that sensibility and irritability are not only quite distinct faculties, but that they have not even a common origin and are due to very different causes.

"Living is feeling," said Cabanis: yes, doubtless for man and the most perfect animals and probably too for a great number of invertebrates. But since the faculty of feeling weakens in proportion to the lower development of the system of organs on which it is based and in proportion to the inferior concentration in the cause which makes this faculty active, we must say that life is rudimentary feeling for those invertebrates that have a nervous system; because this system of organs, especially in the insects, gives them only a very dim feeling.

As to the radiarians, if the nervous system still exists in them, it must be very rudimentary indeed and adapted only to the excitation of muscular movement.

Lastly, since it is impossible that the great majority of polyps or any of the infusorians should possess a nervous system, we must say of them and even of the radiarians and worms, that living is not feeling; as we are obliged also to say in the case of plants.

In dealing with nature, nothing is more dangerous than generalisations, which are nearly always founded on isolated cases: nature varies her methods so greatly that it is difficult to set bounds to them.

According as animal organisation becomes more complex, the order of things essential for life does the same, and life is specialised in each of the principal organs. But all specialised organic life depends on the general life of the individual, just as the latter depends on the specialized life of the principal organs, for there is an intimate connection between each organ and the rest of the organisation. The order of things essential to life in any animal is thus only determinable by a description of that order itself.

In accordance with this principle, it is quite clear that in the most perfect animals, such as mammals, the order of things essential to life includes a system of organs for feeling, consisting of a brain, spinal cord, and nerves, a system of organs for complete pulmonary respiration, a system of organs for circulation with a bilocular heart which has two ventricles, and a muscular system for the movement of internal and external parts, etc.

No doubt each one of these systems of organs has its special life, as Bichat has shown: and on the death of the individual, life becomes extinct in them all. Nevertheless none of these systems of organs could preserve its special life independently, nor could the general life of the individual continue if any of them had lost its own.

From this state of affairs, already generally recognised in the case of mammals, it by no means follows that the order of things essential to life in other bodies, includes a system of organs for feeling, another for respiration, another again for circulation, etc. Nature shows us that these various systems of organs are only essential to life in animals where they form a necessary part of the organisation.

These, I think, are truths against which can be set no known facts and no authoritative observation.

The following is a summary of the principles set forth in this chapter :

1. Life is an organic phenomenon which gives rise to many others; this phenomenon results exclusively from the relations existing between the containing parts of the body, the contained fluids moving in them, and the exciting cause of the movements and changes there occurring;

2. Consequently life in a body is an order and state of things which permits of organic movements, and these movements constituting active life result from the action of a cause which excites them;

3. Without the stimulating and exciting cause of vital movements, life could not exist in any body, whatever the state of its parts;

4. The exciting cause of organic movements acts in vain if the state of things in the parts of the organised body is so disordered that these parts can no longer respond to the action of this cause nor produce the special movements called vital. Life would then be extinct in the body, and could no more be restored;

5. Lastly, in order that the relations between the containing parts of the organised body, the fluids contained in them, and the cause which excites vital movements in them, may produce and maintain the phenomenon of life, the three conditions named in this chapter must be completely fulfilled.

Let us now pass to an examination of the exciting cause of organic movements.

CHAPTER III.

OF THE EXCITING CAUSE OF ORGANIC MOVEMENTS.

We have seen that life is a natural phenomenon which itself produces several others, and that it results from the relations existing between the supple containing parts of an organised body and the contained fluids of that body. We cannot conceive the production of this phenomenon, that is to say, the presence and continuance of the movements constituting active life, unless we imagine a special exciting cause of these movements, a force which animates the organs, controls the activities and all the organic functions,—a spring, in short, of which the permanent though variable tension is the driving energy of all vital movements.

There can be no doubt that the visible fluids of a living body and the solid parts which contain them are irrelevant to the cause that we are here seeking. All these parts together constitute the machinery of movement, if I may revert to the parallel already drawn; and it is not the function of any of them to supply the force in question, that is, the motive power or exciting cause of the movements of life.

We may be certain that if there were no special cause to stimulate and maintain orgasm and irritability in the supple and containing parts of animals, and to produce in plants an obscure orgasm by promoting direct movement of their contained fluids, the blood of animals which have a circulation and the transparent whitish serum of those that have not, would remain at rest and would rapidly decompose together with the solid parts.

In the same way, if there were no exciting cause of vital movements, if there were no force or spring to endow a body with active life, the sap and special fluids of plants would remain motionless, would degenerate and be exhaled, and finally compass the death and desiccation of these living bodies.

The ancient philosophers felt the necessity for a special exciting cause of organic movements; but not having sufficiently studied nature, they sought it beyond her; they imagined a vital principle, a perishable soul for animals, and even attributed the same to plants; thus in place of positive knowledge, which they could not attain from want of observations, they created mere words to which are attached only vague and unreal ideas.

Whenever we abandon nature, and give ourselves up to the fantastic flights of our imagination, we become lost in vagueness, and our efforts culminate only in errors. The only knowledge that it is possible for us to acquire is and always will be confined to what we have derived from a continued study of nature's laws; beyond nature all is bewilderment and delusion: such is my belief.

If it were true that it is really beyond our powers to ascertain the exciting cause of organic movements, it would be none the less obvious that such a cause exists and that it is physical, since we can observe its effects and nature has all the means of producing it. Do we not know that it spreads and maintains movement in all bodies, and that none of the objects submitted to nature's laws really possesses an absolute stability?

I do not wish to go back to the consideration of first causes, nor of all the movements and changes observed in physical bodies of all kinds. We shall confine ourselves to a study of the immediate recognised causes acting on living bodies, and we shall see that they are quite sufficient to maintain in these bodies the movements constituting life, so long as the appropriate order of things is not destroyed.

It would doubtless be impossible to ascertain the exciting cause of organic movement if the subtle, invisible, uncontainable, incessantly moving fluids which constitute it were not disclosed to us in a great variety of circumstances; if we had not proofs that the whole environment in which all living bodies dwell are permanently filled with them; lastly, if we did not know positively that these invisible fluids penetrate more or less easily the masses of all these bodies and stay in them for a longer or shorter time; and that some of them are in a constant state of agitation and expansion, from which they derive the faculty of distending the parts in which they are insinuated, of rarefying the special fluids of the living bodies that they penetrate, and of communicating to the soft parts of these same bodies, an erethism or special tension which they retain so long as their condition is favourable to it.

But it is well known that the question at issue is not insoluble; for no part of the earth inhabited by living beings is destitute of caloric (even in the coldest regions), of electricity, of magnetic fluid, etc. These fluids, some of which are expansive and the others agitated in various ways, are incessantly undergoing more or less regular displacements, renewals or replacements and perhaps in the case of some of them there may actually be a genuine circulation.

We do not yet know how numerous may be these subtle invisible fluids which are distributed in constant agitation throughout the environment. But we do perceive in the clearest manner that these invisible fluids penetrate every organised body and there accumulate with constant agitation, finally escaping in turn after being retained for a longer or shorter period. They thus stimulate movements and life, when they come in contact with an order of things permitting of such results.

With regard to such of these invisible fluids as chiefly constitute the exciting cause under consideration, two of them appear to us to be the essential elements of this cause, viz. caloric and the electric fluid. They are the direct agents which produce orgasm and the internal movements which in organised bodies constitute and maintain life.

Caloric appears to be that of the two exciting fluids in question which causes and maintains the orgasm of the supple parts of living bodies; and the electric fluid is apparently that which provides the cause of the organic movements and activities of animals.

My justification for this division of the faculties assigned to the two fluids in question is based on the following principles.

In inflammations, the orgasm acquires an excessive energy which is at length even destructive of the parts. This is clearly in consequence of the extreme heat developed in inflamed organs: it is, then, especially to caloric that the orgasm must be attributed.

The rapidity of the movements of caloric throughout the bodies which it penetrates is very far from equalling the extraordinary speed of the movements of the electric fluid. Hence this latter fluid must be the cause of the movements and activities of animals; it must be more particularly the genuine exciting fluid.

It is possible, however, that other active invisible fluids combine with the two already named in the composition of the exciting cause; but what appears to me beyond question is that caloric and electricity are the two chief components, and perhaps even the only components of this cause.

In animals with low organisations, the caloric of the environment seems to be sufficient by itself for the orgasm and irritability of their bodies; hence it arises that in extreme reductions of temperature and in the winters of climates in high latitudes, some entirely perish while others become more or less completely torpid. In these same animals the ordinary electric fluid provided by the environment appears to be sufficient for the organic movements and activities.

The case is different with regard to animals of highly complex organisations: in these, the caloric of the environment merely completes or rather aids and favours the power which these living bodies themselves possess of constantly producing caloric within them. It is probable even that this internally produced caloric has undergone modifications in the animal as a result of which it is specialised; and rendered alone suitable for the maintenance of orgasm; for when the state of the organisation has greatly enfeebled the orgasm and irritability, the external caloric arising either from our fires or from a rise of temperature cannot take the place of internal caloric.

The same observation appears to be applicable also to the electric fluid which excites the movements and activities of animals with highly complex organisations. It appears indeed that this electric fluid, which is introduced through the medium of respiration or of food, has undergone some modification in the animal's interior and become transformed into nervous or galvanic fluids.

As to caloric, it is unquestionably one of the principal elements of the exciting cause of life, and is particularly instrumental in producing and maintaining orgasm, without which life could not exist. So true is this that a great reduction of temperature would exterminate all living bodies long before reaching the point of absolute cold. As a matter of fact, the cold of our winters, especially when it is extreme, causes the death of a great many of the animals exposed to it. But we know that on no part of the earth's surface and at no period of the year do we ever find a total absence of caloric.

Let me repeat that without a special exciting cause of orgasm and vital movements—without the force which alone can produce such movements—life could not exist in any body. Now this exciting cause has nothing to do with the visible fluids of living bodies, nor with the solid containing parts of these bodies. This is a fact that can no longer be questioned since it is justified by all observation.

This same exciting cause is also the cause of fermentation, the manifestations of which it alone brings about in all compound non-living matter, whose parts are favourable to it. Thus in great reductions of temperature the activities of life and fermentation are suspended more or less completely, in proportion to the intensity of the cold.

Although life and fermentation are two very different phenomena, they both derive from the same origin the movements by which they are constituted; and in both cases it is necessary that the state of the parts, whether of the organic body capable of life or of the inorganic body capable of fermentation, should be favourable to the performance of these movements. But in bodies possessing life, the existing order and state of things are such that every decomposition of principles is subsequently made good by new and closely similar combinations as a result of continued movements, whereas in the unorganised or disorganised fermenting body, the decompositions which occur cannot be made good by a continuance of fermentation.

As soon as an individual dies, its body, which is then disorganised in reality though often not in appearance, immediately joins the class of bodies liable to fermentation, particularly as regards the more supple of its parts. The exciting cause which gave it life then hastens the decomposition of such of its parts as are capable of fermentation.

We learn then from the principles set forth above that the exciting cause of vital movements must necessarily be sought in the invisible, subtle, penetrating and ever active fluids with which the environment is always supplied; and that the chief element of this cause is the element which maintains an orgasm essential to the existence of life, and, in fact, that it is no other than caloric; as the following observations will further bring out.

I need not cite any special instances on this subject, since the general fact concerned is well known. We are aware that a certain quantity of heat is generally necessary to all living bodies and especially animals. When it is reduced below a certain point, the irritability of animals becomes less intense, their organisation less active and all their functions flag or are performed slowly, especially in those animals where there is no production of internal caloric. When it becomes still further reduced, the most imperfect animals die and a great many of the rest fall into a torpid lethargy and undergo a suspension of life; it cannot be doubted that they would in turn all lose their lives, if this reduction of heat were to be carried much further in the environment.

When the temperature rises, on the contrary—that is, when the heat increases and is distributed everywhere—we constantly notice, if this state of things continues, that life revives and seems to acquire new strength in all living bodies, that the irritability of the internal parts of animals rises proportionally in intensity, that the organic functions are carried on with more energy and promptitude, that the various stages of life succeed each other with greater rapidity, and that life itself comes sooner to an end, but that the new generations are more frequent and abundant.

Although heat is everywhere necessary for the maintenance of life especially in animals, its intensity should not much exceed certain limits; for if it did animals would suffer greatly from it, and in the case of the highly complex would be exposed at the slightest cause to rapid diseases, which would quickly result in death.

We may then be sure that not only is heat necessary to all living bodies, but that when it reaches a certain degree without exceeding the proper limits, it markedly animates all the activities, is favourable to reproduction and appears to expand life everywhere in a wonderful way.

The ease, rapidity, and abundance with which, in tropical countries, nature produces and multiplies the simplest animals are facts in support of this statement. The multiplication of these animals is in fact specially noticeable in favourable times and places, that is to say, in hot climates and in the case of countries of high latitudes in the warm season, especially when there are other conditions favourable to fertility.

Indeed at certain times and in certain climates, the earth (especially at its surface where caloric always accumulates the most) and the body of the waters teem with animated molecules, that is to say, with animalcules of extremely varied genera and species. These animalcules, like many other imperfect animals of different classes, reproduce and multiply with an astonishing fertility—far greater than that of larger animals with a more complex organisation. So rapid are the results of this prodigious fertility, that matter seems to become everywhere animalised. Hence, if it were not for the huge immolation which overtakes the animals of the first orders of the animal kingdom, they would soon overwhelm and perhaps extinguish the more perfect animals of the later classes and orders of this kingdom—so great is the difference between them in the capacity and ease of multiplication!

The above statement as regards the necessity for animals of a caloric distributed throughout the environment, and varying within certain limits, is entirely applicable to plants; but in their case heat only maintains life under certain necessary conditions.

The first and most important condition is that the roots of the plant should have constant access to moisture; for the greater the heat, the more necessary does water become to the plant, to make good the heavy losses of its fluids due to transpiration; and the less the heat, the less it needs moisture, which would then be injurious to its preservation.

The second condition for the elaboration of its products by a plant, is that the plant in addition to heat and water should also have plenty of light.

The third, lastly, makes it dependent upon air, from which it probably appropriates oxygen and the other gases which it finds there, immediately decomposing them and making use of their principles.

From the above statement, it is quite clear that caloric is the first cause of life, in that it produces and maintains orgasm without which no living body could exist. In this it succeeds so long as the state of the parts of the living body does not prevent it.

We find, moreover, that this expansive fluid, especially when its abundance gives it some intensity of action, is the principal factor in the enormous multiplication of living bodies of which I have just spoken. Hence it is universal that in the hot climates of the earth the animal and plant kingdoms exhibit an extremely remarkable wealth and abundance, whereas in the frozen regions of the earth they only exist in a state of the greatest poverty.

A further fact in favour of the principle just established is that there is a great difference between the summer and winter of our own climates as regards the number of animals and plants.

Although caloric is really the first cause of life in the bodies which possess it, yet it could not alone account for its existence nor keep up the movements which constitute it; there is needed in addition, especially for animals, the influence of a fluid to excite their acts of irritability. Now we have seen that electricity possesses all the properties necessary for constituting this exciting fluid, and that it is distributed sufficiently widely notwithstanding its variations, to ensure that living bodies shall always be provided with it.

It may be that some other invisible fluid combines with electricity in making up the cause which is able to excite vital movements and all the organic activities; this is quite possible, but I see no need for supposing it.

It seems to me that caloric and the electric substance together are quite sufficient to constitute the essential cause of life, the one by setting the parts and internal fluids in a proper condition for the existence of life, and the other by arousing in the course of its movements through the body the various stimuli which give rise to the organic activities and the active part of life.

If we were to try to explain how these fluids work, and to determine definitely the number of those that enter into the composition of the exciting cause of all organic movements, we should be abusing the power of our imagination and arbitrarily creating explanations which we have no means of verifying.

It is enough to have shown that the exciting cause of the movements which constitute life does not reside in any of the visible fluids in the interior of living bodies, but that it takes its origin as follows:

1. From caloric, which is an invisible penetrating, expansive, ever active fluid that percolates slowly through the supple parts, distending them and making them irritable; and that is constantly being dissipated and renewed and is never entirely absent from any body that possesses life;

2. From the electric fluid, either the ordinary electric fluid in the case of plants and imperfect animals, or the galvanic fluid for those with very high organisation;—a subtle fluid, moving with extraordinary speed, which instigates sudden local dissipations of the caloric which distends the parts, and thus excites acts of irritability in the non-muscular organs, and movements in the muscles when it extends its action to them.

If the two fluids named above thus combine their special activities, there must ensue in the case of organised bodies submitted to their influence, a powerful cause or force which works effectively, and is controlled by the organisation,—that is, by the regular shape and arrangement of the parts,—and keeps up movements and life so long as there exists in these bodies an order of things which allows of such effects.

This apparently is the mode of action of the exciting cause of life; but it cannot be regarded as established, until it is possible to find proofs of it. The two named fluids may be the only principles contributing to the production of this cause; but that again is a belief of which we cannot be certain. What is quite positive in this respect is that the source, from which nature derives this cause and the resulting force, is to be found in the subtle invisible fluids among which the two just named are unquestionably the chief.

I shall confine myself to the statement that the active and expansive fluids, composing the exciting cause of vital movements, penetrate into or are constantly developing in the bodies which they animate, pass all through them, while harmonising their movements with the nature, order and arrangement of the parts, and are then constantly being exhaled by an imperceptible transpiration. This fact is unquestionable and sheds the brightest light over the causes of life.

Let us now enquire into the special phenomenon that I call orgasm in living bodies, and afterwards into the irritability which this orgasm produces in animals, where the nature of their bodies permits it to be highly developed.

CHAPTER IV.

OF ORGASM AND IRRITABILITY.

It is not the special affection called orgasm that we are now about to discuss, but the general condition known under the same name and characteristic of the supple internal parts of animals during life; a condition which is natural to them, since it is essential to their preservation, and which necessarily comes to an end at or soon after death.

It is certain that, among the solid internal parts of animals, those that are supple are animated throughout life by an orgasm or peculiar kind of erethism, from which they derive the faculty of collapsing and being promptly restored on the receipt of any impression.

An analogous orgasm also exists in the most supple of the solid parts of plants, so long as they are alive; but it is so faint that the parts endowed with it do not derive therefrom any faculty for immediate restoration, after the impressions that they have received.

The orgasm of the supple internal parts of animals contributes to some extent to the production of their organic phenomena; it is maintained by an invisible, expansive, penetrating fluid (possibly several), which slowly passes through the parts affected and produces in them the tension or sort of erethism just mentioned. The orgasm resulting from this state of things in the parts, is maintained throughout life with a strength that is proportional to the favourable disposition of the parts; it is the stronger according as they are more supple and less dried up.

It is this same orgasm, the necessity for which has been recognised for the presence of life in a body, that some modern physiologists have looked upon as a kind of sensibility; hence they have alleged that sensibility is a property of all living bodies, that such bodies are both sensitive and irritable, that all their organs are impregnated with these two necessarily co-existing faculties, in short that they are common to every living thing both animal and plant. Cabanis, who shared this opinion with M. Richerand and apparently others, said indeed that sensibility is the general fact of living nature.

M. Richerand, however, who has in particular developed this opinion in the prolegomena of his *Physiologie*, admits that the sensibility, which gives us the power of receiving sensations and depends on nerves, is not the same thing as that more general kind of sensibility for which no nervous system is necessary. For the former he suggests the name of perceptibility and for the latter that of latent sensibility.

Since these two faculties are different in their origin and results, why should we give a new name to the phenomenon, long known as sensibility, and transfer the name of sensibility to a more recently observed phenomenon of altogether special nature ? Surely it is more convenient to give a particular name to the general phenomenon on which life depends : and this is what I have done by calling it orgasm.

It is probable that without orgasm (latent sensibility), no vital function could go on ; for wherever it exists, there can be no true inertia in the parts ; they are no longer merely passive. This is the element of truth in the idea that all the living parts feel and act in their own way, that they distinguish among the fluids which bathe them whatever is suitable for their nutrition, and that they separate from them those substances which affect their special type of sensibility.

Although we are not definitely aware how each vital function is performed, we should not gratuitously attribute to the parts a knowledge and power of choice among the objects which they have to separate out, and retain or evacuate. We should rather reason thus :

1. The organic movements aroused are simply due to the action and reaction of the parts ;

2. From these actions and reactions it follows that the state and nature of the parts undergo alterations, decompositions, new combinations, etc. ;

3. As a result of these alterations, there occur secretions which are favoured by the width of the secretory canals ; depots are established which are sometimes kept isolated and sometimes attached to these same parts ; lastly, there are various evacuations, absorptions, resorptions, etc.

All these operations are mechanical and subject to physical laws ; they are carried out by means of the exciting cause and of orgasm, which keeps up the movements and activities ; so that by these means and by the shape, arrangement and situation of the organs, the vital functions are varied and controlled, and each works in its special way.

The orgasm dealt with in this chapter is a definite fact, which whatever we call it can no longer be neglected. We shall see that it is very weak and faint in plants where its powers are very limited, but that in animals on the contrary it is very conspicuous ; for it produces in them that remarkable property called irritability by which they are distinguished.

Let us first study it in animals.

Of Animal Orgasm.

By animal orgasm, I mean that curious condition of the supple parts of a living animal, in which there exists at every point a peculiar tension, of such strength that the parts are able immediately to react to any impression that they may experience, and do in fact react on the moving fluids which they contain.

This tension, the degree of which varies in the different parts, constitutes what physiologists call the tone of the parts ; it seems to be due as I have already said to the presence of an expansive fluid which penetrates these parts, remains in them for a certain period and keeps their molecules separated to some distance from one another though without destroying their coherence ; some of the fluid suddenly escapes on any contact, leading to a contraction, but it is promptly restored again.

Thus at the moment of the dissipation of the expansive fluid distending any part, this part subsides on itself as a result ; but it is promptly restored to its previous distention by the arrival of new expansive fluid. Hence it follows that the orgasm of this part gives it the property of reacting on the visible fluids which affect it.

This tension in the soft parts of living animals does not go so far as to break up the cohesion of their molecules, or to destroy their adhesiveness, agglutination and firmness, so long as the intensity of orgasm does not exceed a certain point. But the tension prevents the falling together and collapse of the molecules which would occur if this tension did not exist ; for the soft parts do actually subside in a remarkable way as soon as the tension is removed.

Indeed among animals, and even among plants, the extinction of orgasm, which only occurs on the death of an individual, gives rise to a relaxation and subsidence of the supple parts, making them softer and more limp than in the living state. This has given rise to the belief that these limp parts found in old people after death have escaped the rigidity which gradually comes over the organs during life.

The blood, and especially the arterial blood, of animals whose organisation is highly complex, itself possesses a sort of orgasm ; for it is during life suffused with various gases which develop within it and become modified there. Now these gases may also contribute to the stimulation of the organic acts of irritability, and consequently to the vital movements, when the blood affected flows to the organs.

The excessive tension of orgasm under certain conditions in some or all of the soft parts of an individual, although not great enough to break up the cohesion of those parts, is known under the name of erethism. When it is very strong it produces inflammation, whereas when the orgasm is extremely reduced, though not to the point of disappearance, it is generally designated by the name of atony.

The tension which constitutes orgasm may vary within certain limits, without on the one hand destroying the cohesion of the parts, or on the other hand ceasing to exist. This variation renders possible those sudden contractions and distensions which occur when the cause of orgasm is momentarily removed and then restored. This seems to me to be the chief cause of animal irritability.

The cause which produces orgasm, or that peculiar tension of the supple internal parts of animals, is no doubt an element in what I have called the exciting cause of organic movements. It is to be found principally in caloric, either in that provided by the environment, or in this combined with the caloric that is constantly being produced in the interior of many animals.

Indeed an expansive caloric is continually emanating from the arterial blood of many animals, and constitutes the principal cause of the orgasm in their supple parts. It is especially in the warm-blooded animals that the continual emanation of caloric becomes remarkable. This expansive fluid is constantly being dissipated from the parts which it distends, but it is as constantly being renewed by the emanations always being given off from the animal's arterial blood.

An expansive fluid, similar to that which we are discussing, is distributed throughout the environment and incessantly provides for the orgasm of living animals, either by completing what was wanting to the internal caloric, or achieving it alone.

As a matter of fact, the caloric of the environment assists more or less in the orgasm of the most perfect animals and suffices by itself for that of the rest; it is especially the cause of orgasm in all animals which have no arteries or veins, that is to say, no circulatory system. Hence all organic movement becomes gradually weaker in these animals, according as the temperature of the environment becomes lower; and if this reduction of temperature continues indefinitely their orgasm is extinguished and they die. We have only to recall the torpidity that overtakes bees, ants, snakes and many other animals when the temperature falls below a certain point, and we shall then be in a position to judge whether my statement has not some foundation.

The reduction of temperature which causes many animals to become torpid works this result by weakening their orgasm and hence by slackening their vital movements. If this reduction of temperature goes too far, I have already observed that it extinguishes the orgasm and causes the death of the animal; but I may remark on this subject, with regard to the effects of cooling, not carried far enough to kill the individual, that there is a certain peculiarity in the case of warm-blooded animals and perhaps of all animals that have nerves: it is as follows.

It is known that no very great fall of temperature is required to bring about the torpid state of apparent sleep in certain mammals, such as marmots, bats, etc. If the heat returns, it penetrates, revives, awakes, and restores them to their usual activity; but if on the contrary the cold increases still further after these animals have become torpid, they do not pass imperceptibly from their condition of apparent sleep into death. If the increase of cold is considerable, it causes an irritation in their nerves which awakes and agitates them, and revives their organic movements and hence their internal heat. If the increase of cold then continues, it soon throws them into a state of disease, ending in death unless heat is quickly restored to them.

Hence it follows that, for warm-blooded animals and perhaps for all animals with nerves, a mere weakening of their orgasm may reduce them to a torpid state. In this case the orgasm is not totally destroyed, since if the cold were great enough for that purpose, it would begin by irritating them, giving them pain and end by killing them.

In the case of animals without nerves, it seems that a reduction of temperature sufficient to weaken their orgasm and make them torpid, may if it increases result in their passing through their stage of lethargic sleep to that of death, without any temporary return of activity.

Those who imagine that the first result of a moderate degree of cold is to slacken the respiration, have mistaken the effect for the cause. Thus the torpid state, into which various animals fall when the temperature is lowered, has been attributed to a direct slackening of the respiration of these animals; whereas in point of fact the slackening of respiration is itself due to another effect of the cold, namely, the enfeeblement of their orgasm.

As regards animals which breathe with lungs, those that fall into a torpid state in certain degrees of cold doubtless undergo a considerable slackening of respiration; but here it is clearly only the result of a great enfeeblement in their orgasm. Now this enfeeblement slackens all the organic movements and all the functions. It diminishes also the production of internal caloric and the losses which these animals are subject to, during their customary activity. Their need for restoration during their lethargy is thus very slight or next to nothing.

Animals breathing with lungs undergo alternate swellings and contractions of the cavity containing their respiratory organ. Now these movements are carried out with greater or less facility, according as the orgasm of the supple parts is more or less energetic. Thus several mammals, such as the marmot and dormouse, and many reptiles, as the snakes, fall into a torpid state on certain reductions of temperature, because they then have a greatly weakened orgasm, from which results secondarily a slackening of all their organic functions including that of respiration.

If this decline in the energy of their orgasm did not take place, there would be no reason why any less air should be breathed by these animals when it is cold. In bees and ants which breathe by tracheae and undergo no alternative swellings and contractions, it cannot be said that they breathe less when it is cold; but there are sound reasons for the belief that their orgasm is then greatly enfeebled, and thus accounts for the torpid state which they undergo in these conditions.

Finally, in warm-blooded animals their internal heat is almost entirely produced within their bodies, either as a result of the decomposition of air in respiration as is generally held, or because it is constantly given off from the arterial blood while changing into venous blood, as I myself believe; in either case the orgasm gains or loses energy, according as the internal caloric increases or diminishes in quantity.

As regards the explanation which I am giving of orgasm, it is a matter of indifference whether the caloric produced in the interior of warm-blooded animals, is the result of the decomposition of air in respiration or is an emanation from the arterial blood as it changes into venous blood. If, however, it is desired to consider this question, I should put forward the following suggestions:

If you drink a glass of spirits, the resulting warmth that you feel in your stomach assuredly does not spring from increased respiration. Now if caloric may be given forth from this liquor as it undergoes changes in your organ, so too it may be exhaled from your blood when it undergoes changes in its component parts.

In fever when the internal heat is greatly increased, it is observed that respiration is also faster, and hence it is inferred that the consumption of air is greater. This supports the view that the internal caloric of warm-blooded animals results from the decomposition of the respired air. I know of no experiment to show definitely whether the consumption of air during fever is really greater than in health. I doubt whether it is so; for if respiration were faster in this diseased condition it may be compensated for by each inspiration being less, owing to the constraint of the parts. What I do know is that when

I have some local inflammation like a boil or any other inflamed tumour, caloric issues in extraordinary abundance from the blood of the parts affected. Yet I do not see how any increase of respiration can in this case have given rise to the local concentration of caloric; I should say on the other hand that the blood, being compressed and concentrated in the diseased part, is liable to disturbances and decompositions (as also the supple parts containing it) which involve the productions of the observed caloric.

I cannot admit that atmospheric air includes in its composition a fluid, which when freed is an expansive caloric; I have elsewhere stated my views on this matter. In point of fact, I believe that air is composed of oxygen and nitrogen; and I know that it contains caloric within it, because absolute cold does not exist anywhere on our earth. I am fully convinced that the component fluid which when freed is changed into expansive caloric was previously a constituent part of our blood; that this fluid in combination is always being partially set free and that by its liberation it produces our internal heat. Evidence that this internal heat does not come from our respiration is furnished by the fact that if we were not constantly making good the losses in our blood by means of food, and hence of a chyle always being renewed, our respiration would not restore to our blood the qualities which it must possess for the maintenance of our existence.

The utility of respiration to animals is not in question; by this method their blood derives a restoration which they could not do without; and the belief appears to be justified that it is by appropriating oxygen from the air that the blood derives one of its most necessary restorations. But in all this there is no proof that caloric comes from the air or its oxygen, rather than from the blood itself.

The same thing may be said with regard to combustion: the air in contact with burning substances may be decomposed, and its disengaged oxygen may be fixed in the residue of combustion; but that is no proof that the caloric then produced comes from the oxygen in the air rather than from the combustible substances, with which I hold that it was previously combined. All the known facts are better and more naturally explained on this latter assumption than on any other.

However this may be, the positive fact remains that in a great many animals an expansive caloric is constantly being produced within them, and that it is this invisible penetrating fluid which maintains the orgasm and irritability of their supple parts; while in other animals orgasm and irritability are chiefly the result of the caloric of the environment.

To refuse to recognise the orgasm of which I have spoken, and to regard it as a hypothetical fact, that is to say a product of the imagination, would be to deny to these animals the existence of that *tone* in the parts which they possess throughout life. Now death alone extinguishes this tone, as also the orgasm which constitutes it.

Plant Orgasm.

In plants the exciting cause of organic movements seems to act chiefly on the contained fluids, and sets them alone in motion; while the cellular tissue of the plant, whether simple or modified into vasculiform tubes, only acquires from it an ill-defined orgasm giving rise to a very slow general contractility, which never acts in isolation or suddenly.

If, during the warm season, a plant grown in a pot or box needs watering, we notice that its leaves, the ends of its branches and young shoots hang down as though about to fade: yet life still continues to exist in them; but the orgasm of the supple parts is then greatly enfeebled. If this plant is watered, we see it gradually erect its drooping parts and show an appearance of life and vigour which it did not present when it had no water.

This restoration of the vigour of the plant is doubtless something more than a mere result of the introduction of new fluids into the plant. It is also the result of a revived orgasm, since the expansive fluid causing this orgasm penetrates the parts of the plant with much greater ease when the juices or contained fluids are abundant.

The ill-defined orgasm of living plants thus causes a slow general contractility in their solid parts, especially in the newest,—a sort of tension which various facts justify us in accepting, although there are no sudden movements. This plant-orgasm does not endow the organs with any faculty for instantaneous reaction on contact with objects which might affect them, and hence it has no power of causing irritability in the parts of these living bodies.

It is not true indeed, as has been alleged,[1] that the canals in which move the visible fluids of these living bodies are sensitive to impressions of stimulating fluids, or that they become relaxed or distended by a prompt reaction in order to achieve the transport and elaboration of their visible fluids;—in short they have no true tone.

Finally, it is not true that the peculiar movements observed at certain periods in the reproductive organs of various plants, nor the movements of leaves, petioles or even the small twigs and plants called sensitive, are the product and proof of an irritability existing

[1] Richerand, *Physiologie*, vol. i., p. 32.

in these parts. I have observed and watched these movements and am convinced that they have nothing in common with animal irritability. See what I have already said, pp. 51-53.

Although nature has doubtless only one general plan for the production of living things, she has everywhere varied her means, when diversifying her productions, according to the circumstances and objects on which she worked. But man is always striving to confine her to the same methods ; for the idea that he has formed of nature is still indeed far from that which he ought to entertain.

How great are the efforts that have been made to discover sexual reproduction everywhere throughout the two kingdoms of living bodies ; and in the case of animals to attribute to all of them nerves, muscles, feeling and even will, which is necessarily an act of intelligence! How profoundly different nature would be, if she was really limited in the ways that we imagine!

We have just seen that orgasm has very different degrees of intensity, and consequently has effects that vary according to the nature of the living bodies in which it occurs, and that in animals alone does it give rise to irritability. We now have to enquire into the nature of this remarkable phenomenon.

Irritability.

Irritability is the faculty possessed by the irritable parts of animals, of producing sudden local manifestations which may occur at any point on the surface, and may be repeated as often as the exciting cause acts upon the susceptible regions.

The manifestations consist in a sudden contraction and shrinkage of the irritated point, a shrinkage characterised by the closing in of neighbouring points upon that which is affected, but soon followed by a contrary movement, that is to say by a distension of the irritated point and neighbouring parts ; so that the natural condition of the parts distended by orgasm is promptly re-established.

I said at the beginning of this chapter that orgasm is formed and maintained by caloric, that is a penetrating expansive invisible fluid which passes slowly through the soft parts of animals and produces in them a tension or kind of erethism. Now if some impression is made upon any such part so as to instigate a sudden dissipation of the invisible fluid distending it, the part immediately shrinks and contracts : but if a new supply of expansive fluid is instantly developed and distends it afresh, it then reacts immediately and so produces the phenomenon of irritability.

Lastly, since the parts in the neighbourhood of the point affected themselves suffer a slight dissipation of the expansive fluid distending them, their consecutive shrinkage and restoration throw them momentarily into a condition of quivering.

Thus a sudden contraction of the part affected, followed by an equally sudden distension which restores the part to its original condition, constitutes the local phenomenon of irritability.

The production of this phenomenon does not need any special organ ; for the state of the parts and the instigating cause are sufficient in themselves ; it is in fact observed in the simplest of animal organisations : moreover the impression giving rise to the phenomenon is not conveyed by any special organ to a centre of communication or nucleus of activity ; the whole process is confined to the immediate site of the impression ; every point of the surface of irritable parts is capable of producing it and of repeating it always in the same way. This phenomenon is obviously quite different from that of sensations.

From these principles it will be clear that orgasm is the source from which irritability arises ; but this orgasm exhibits very different degrees of intensity, according to the nature of the bodies in which it is produced.

In plants, where it is ill-defined and devoid of energy, and where it works extremely slowly in causing the shrinkages and distensions of the parts, it has no power to produce irritability.

In animals on the contrary, where orgasm is highly developed on account of the nature of their body-substance, it rapidly produces the contractions and distensions of the parts on the stimulus of the exciting causes ; in them it constitutes a marked irritability.

Cabanis, in his work entitled : *Rapports du physique et du moral de l'homme*, has endeavoured to prove that sensibility and irritability are phenomena of the same nature and have a common origin (*Histoire des sensations*, vol. i., p. 90) ; his intention no doubt was to reconcile what we know of the most imperfect animals with the ancient and universally received belief that all animals without exception possess the faculty of feeling.

The arguments adduced by this savant for showing the identity of nature between feeling and irritability, appear to me neither clear nor convincing ; hence they do not affect the following propositions by which I distinguished these two faculties.

Irritability is a phenomenon peculiar to animal organisation, requiring no special organ, and continuing to subsist some time after the death of the individual. This faculty may exist just the same whether there are or are not any special organs, and it is therefore universal for all animals.

Sensibility on the contrary is a phenomenon peculiar to certain animals ; it can only be manifested in those which have a special organ essentially distinct, and adapted solely for producing it; and it invariably ends with life or even slightly before death.

We may be sure that feeling cannot occur in an animal without the existence of a special organ adapted for producing it,—that is without a nervous system. Now this organ is always quite discernible; for it cannot exist without a centre of communication for the nerves, and hence could not remain unperceived when it is present. This being so, seeing that many animals have no nervous system, it is obvious that sensibility is not a faculty common to all animals.

Finally, feeling as compared with irritability presents in addition this distinctive peculiarity, that it comes to an end with life or even a little before; whereas irritability is still preserved some time after the death of the individual, and even after it has been divided into fragments.

The time during which irritability is preserved in the parts of an individual after death varies no doubt with the system of organisation of that individual; but in all animals it is probably true that irritability continues to be manifested after the cessation of life.

In man the irritability of his parts scarcely lasts more than two or three hours after death, or even less, according to the cause of death: but thirty hours after having removed a frog's heart it is still irritable and capable of producing movements. There are insects, in which movements are manifested still longer after they have been deprived of their internal organs.

From the above account, we see that irritability is a faculty peculiar to animals; that all animals possess it in a high degree in some or all of their parts and that an energetic orgasm is the source of it: we see moreover that this faculty is entirely distinct from that of feeling; that the one is of very different character from the other, and that since feeling can only result from the functions of a nervous system provided as I have shown with its centre of communication, it only occurs in those animals which possess the required system of organs.

Let us now consider the importance of cellular tissue in all kinds of organisation.

CHAPTER V.

OF CELLULAR TISSUE, REGARDED AS THE MATRIX IN WHICH ALL ORGANISATION HAS BEEN CAST.

As we observe the facts presented to us in the various parts of nature, it is curious to remark how the simplest causes of observed facts are often those which remain the longest unperceived.

It is no new discovery that all the organs of animals are invested by cellular tissue, even down to their smallest parts.

It has indeed been long recognised that the membranes which form the investments of the brain, nerves, vessels of all kinds, glands, viscera, muscles and their fibres, and that even the skin itself are all the produce of cellular tissue.

Yet in this multitude of harmonious facts, nothing more appears to have been seen than the mere facts themselves; and no one that I know of has yet perceived that cellular tissue is the universal matrix of all organisation, and that without this tissue no living body could continue to exist.

Thus when I said [1] that cellular tissue is the matrix in which all the organs of living bodies have been successively formed, and that the movement of fluids through it is nature's method of gradually creating and developing these organs out of this tissue, I was not afraid of coming upon any facts which might testify to the contrary; for it is by examining the facts themselves that the conviction is acquired that every organ whatever has been formed in cellular tissue, since it is everywhere invested with it even down to its smallest parts.

Hence we see that in the natural order, both of animals and plants, those living bodies whose organisation is the simplest and which are consequently placed at one extremity of the order, consist only of a mass of cellular tissue in which there are to be seen neither vessels, nor glands nor any viscera; whereas those bodies, whose organisation

[1] Opening address of the course of invertebrate animals delivered in 1806, p. 33. Since the year 1796, I have stated these principles in the early lessons of my course.

is the most complex and which are therefore placed at the other extremity of the order, have their organs so deeply imbedded in cellular tissue that this tissue always forms their investments and constitutes for them a bond of communication. Hence the possibility of those sudden metastases, so well known to those who practice the art of medicine.

Compare the very simple organisation of the infusorians and polyps, presenting nothing more than a gelatinous mass of cellular tissue, with the highly complex organisation of the mammals, which still presents a cellular tissue though enveloping a large number of different organs; you will then be in a position to judge whether the principles, which I have drawn up on this important subject, are merely the results of an imaginary system.

Compare in the same way in plants the very simple organisation of the algae and fungi with the more complex organisation of a big tree or any other dicotyledonous plant, and you will perceive that the general plan of nature is everywhere the same, notwithstanding the infinite variations of her individual operations.

Take for instance the algae which grow under water, such as the widely spread *Fucus* which constitutes a great family with different genera, and such as the *Ulva*, *Conferva*, etc.; their scarcely modified cellular tissue is conspicuous enough to prove that it alone forms the whole substance of these plants. In several of these algae, the movements of the internal fluids in this tissue have not yet given rise to any signs of a special organ, and in the others they have only traced out a few canals through which food is supplied to those reproductive corpuscles, which botanists take for seeds because they often find them invested in a capsular vesicle in the same way as the gemmae of many known examples of *Sertularia*.

We may then convince ourselves by observation that in the most imperfect animals, such as the infusorians and polyps, and in the least perfect plants, such as the algae and fungi, there sometimes exists no trace of any vessels and sometimes only a few rudimentary canals; lastly, we may recognise that the very simple organisation of these living bodies consists only of a cellular tissue, in which slowly move the fluids which animate them; and that these bodies being destitute of special organs only develop, grow, and multiply or reproduce themselves by a faculty of growth and separation of reproductive parts; for they possess this faculty in a very high degree.

In plants indeed, even including those with the most perfect organisation, there are no vessels comparable to those of animals which have a circulatory system.

Thus the internal organisation of plants really consists only of a cellular tissue more or less modified by the movement of fluids, a tissue which is very slightly modified in the algae, fungi and even the mosses, whereas it is much more modified in the other plants and especially in the dicotyledons. But everywhere, even in the most perfect animals, there is really nothing in their interior but a cellular tissue modified into a large number of different tubes, most of which are parallel to one another in consequence of the ascending and descending movement of the fluids. Yet the structure of these tubes is not comparable to that of the vessels of animals which possess a circulatory system. Nowhere do these vegetable tubes intertwine or form those special groups of vessels, folded and interlaced in infinite variety, that we call conglomerate glands in animals which have a circulation. Finally in all plants without exception, there is no special organ within their bodies: there is nothing but more or less modified cellular tissue, longitudinal tubes for the movement of fluids, and harder or softer fibres also longitudinal, for strengthening the stem and branches.

If we admit on the one hand that every living body whatever is a mass of cellular tissue, in which are embedded various organs of a number proportionate to the complexity of organisation; and if on the other hand we also admit that all bodies contain within them fluids that move faster or slower according as the state of organisation allows of a more or less active or energetic life, we are forced to the conclusion that it is to the movement of fluids in the cellular tissue that we have primarily to attribute the formation of every kind of organ in the midst of this tissue, and hence that each organ must be invested by it both in its gross outlines and in its most minute parts, as indeed we actually find.

With regard to animals, I have no need to show that in various of their internal parts the cellular tissue is squeezed aside by the moving fluids, which open a passage through it; and that in these regions it has been forced back upon itself; it has then been compressed and transformed into investing membranes round about these running streams of fluid; while on the outside these living bodies are incessantly compressed by the environing fluids (either water or atmospheric fluids) and modified by external impressions and by deposits upon them. Their cellular tissue has thus come to form that universal investment of every living body, that is called skin in animals and bark in plants.

I was then fully justified when I said "that the function of the movement of fluids in the supple parts of living bodies, and especially in the cellular tissue of the simplest among them, is to carve out routes, places of deposit and exits, to create canals and thereafter diverse organs, to vary these canals and organs in accordance with the diversity of the movements or character of the fluids causing them, finally to enlarge, elongate, divide, and gradually solidify these canals and organs. This is effected by the substances which are incessantly being formed in the fluids, and are then separated from them, and in part assimilated and united to the organs, while the remainder is rejected" (*Recherches sur les corps vivants*, pp. 8 and 9).

I was equally justified when I said "that the state of organisation in every living body has been gradually acquired by the increasing influence of the movement of fluids (firstly in the cellular tissue and afterwards in the organs formed in it), and by the constant changes in the character and state of these fluids owing to the continual wastage and renewals proceeding within them."

Lastly, I was authorised by these principles in saying "that every organisation and every new shape, acquired by this agency and contributing circumstances, were preserved and transmitted by reproduction, until yet further modifications had been acquired by the same method and in new circumstances" (*Recherches sur les corps vivants*, p. 9).

It follows from the above that the function of the movement of fluids in living bodies, and consequently of organic movement, is not merely the development of organisation, for this development continues so long as the movement is not weakened, through the hardness which overtakes the organs during life; but that this movement of fluids has in addition the faculty of gradually increasing the complexity of organisation and of multiplying the organs and their functions according as new modes of life or new habits acquired by individuals stimulate it in various ways, create a necessity for new functions, and consequently for new organs.

Let me add that the faster the movement of fluids in a living body, the more does it complicate the organisation, and the greater the branching of the vascular system.

It is from the uninterrupted co-operation of these factors and of long periods of time, combined with an infinite variety of environment, that all the orders of living bodies have been successively formed.

Vegetable Organisation is also cast in a Cellular Tissue.

Imagine a cellular tissue in which, for various reasons,[1] nature could not establish irritability, and you will have an idea of the matrix in which all vegetable organisation has been cast.

If we then reflect that the movements of fluids in plants are only

[1] Chemical analysis has shown that animal substances abound in nitrogen, while vegetable substances are destitute of this material or contain only very small proportions of it. Hence there is a distinct difference between animal and vegetable substances: now this difference may be the reason why the factors which produce orgasm and irritability in animals cannot establish these faculties in living plants.

excited by external influences, we shall be convinced that in this kind of living bodies life can only have a feeble activity, even in times and climates when vegetation is rapid; and consequently that their complexity of organisation is necessarily confined within very narrow limits.

Infinite pains have been taken to become acquainted with the details of plant organisation: search has been made in them for peculiar or special organs of the same kind as some of those known in animals; and the results of all these researches have done no more than show us that the containing parts consist of a more or less compressed cellular tissue, with elongated cells that communicate with each other by pores and by vascular tubes of various shape and size, mostly having lateral pores or sometimes clefts.

All the details ascertained on this subject furnish little in the way of clear general ideas, and the only ones which we need recognise are as follows:

1. Plants are living bodies with less perfect organisation than animals and with less active organic movements; their fluids move more slowly and the orgasm of the containing parts is very faint;

2. They are essentially composed of cellular tissue; for this tissue is to be found in every part of them, and indeed it is found almost by itself and with very slight modifications in the simplest of them (algae, fungi and probably all the agamous plants);

3. The only change undergone by cellular tissue in monocotyledons and dicotyledons as a result of the fluids moving within them, consists in the transformation of certain parts of this cellular tissue into vascular tubes of varied size and shape open at the extremities, and mostly having lateral pores.

Let me further add that since the movement of fluids in plants is either upwards or downwards their vessels are naturally almost always longitudinal, and approximately parallel to one another and to the directions of the stem and branches.

Lastly, the outer part of that cellular tissue, which constitutes the bulk of every plant and the matrix of its low organisation, is squeezed and compressed by the contact, pressure and collision of the environment, and is thus thickened by deposits and transformed into a general integument [1] called bark which is comparable to the skin of animals. Hence we may understand how the external surface of this bark, being even more disorganised than the bark itself by the causes named,

[1] If the stems of palms and some ferns appear to have no bark, it is because they are only elongated root collars, the exterior of which shows a continuous succession of scars left by old leaves that have fallen; this prevents the possibility of a continuous or uninterrupted bark; but it cannot be denied that each separate part of this exterior has its special bark, although more or less imperceptible on account of the small size of these parts.

comes to constitute that outer pellicle called epidermis both in animals and plants.

If then we study plants from the point of view of their internal organisation, all that we can find is, among the simplest, a cellular tissue without vessels but variously modified and stretched or compressed according to the special shape of the plant; and in the more complex, an assemblage of cells and vascular tubes of various sizes, mostly with lateral pores, and a varying number of fibres, resulting from the compression and hardening that a portion of the vascular tubes has undergone. This is all that the internal organisation of plants presents, as regards their containing parts; even their pith is no exception.

But if we study plants from the point of view of their external organisation, the most general and essential points to observe are as follows:

1. Their various peculiarities of shape, colour and consistency, both in them and their parts;

2. The bark which invests them, and gives communication by pores with the environment;

3. The more or less complex organs which develop on the exterior in the course of the plant's life, and serve for reproduction; they perform their functions once only, and are highly important in the determination of the characters and true affinities of each plant.

It is then in the study of the external parts of plants, and especially in that of their reproductive organs, that the means must be sought for describing the characters of plants and determining their natural affinities.

Since the above exposition is a positive result of knowledge acquired by observation, it is obvious that on the one hand the true affinities among animals can only be determined by their internal organisation, which provides the only features of real importance; and that on the other hand these affinities cannot be similarly determined among plants, as regards any of the divisions which mark their classes, orders, families and genera. In their case affinities can only be determined by a study of their external organisation; for their internal organisation is insufficiently complex, and its various modifications too vague, to provide the means for fulfilling this purpose.

We have now seen that cellular tissue is the matrix, in which all organisation was originally cast; and that it was by means of the movement of the internal fluids of living bodies that all their organs were created in this matrix and out of its substance. We have now to enquire briefly whether we are justified in attributing to nature the power of forming direct generations.

CHAPTER VI.

OF DIRECT OR SPONTANEOUS GENERATION.

LIFE and organisation are products of nature, and at the same time results of the powers conferred upon nature by the Supreme Author of all things and of the laws by which she herself is constituted : this can no longer be called in question. Life and organisation are thus purely natural phenomena, and their destruction in any individual is also a natural phenomenon, necessarily following from the first.

Bodies are incessantly undergoing transformations in their condition, combination and character ; in the course of which some are always passing from the inert or passive condition to that which permits of the presence of life in them, while the rest are passing back from the living state to the crude and lifeless state. These transitions from life to death and from death to life are evidently part of the immense cycle of changes of every kind to which all physical bodies are liable as time goes on.

Nature, as I have already said, herself creates the rudiments of organisation in masses where it did not previously exist ; subsequently, use and the vital movements cause the development and increasing complexity of the organs (*Recherches sur les corps vivants*, p. 92).

However extraordinary this proposition may appear, we shall be obliged to abandon any opinion to the contrary if we take the trouble to examine and reflect seriously upon the principles which I am about to advance.

The ancient philosophers had observed the power of heat, and noted the extreme fertility which it confers on the various parts of the earth's surface in proportion to its abundance ; but they omitted to reflect that the co-operation of moisture is the essential condition for making the heat so fertile and necessary to life. Since however they perceived that life in all living bodies derives its support and activity from heat, and that the privation of heat everywhere results in death, they concluded with justice not only that heat was necessary for the maintenance of life, but that it could even create both life and organisation.

They recognised then that direct generations do occur, that is to say, generations wrought directly by nature and not through the intermediary of individuals of a similar kind: they called them somewhat inappropriately spontaneous generations; and perceiving as they did that the decomposition of animal or vegetable substances provided nature with conditions favourable to the direct creation of new organisms, they wrongly imagined that such organisms were the produce of fermentation.

I am in a position to show that the ancients were not mistaken when they attributed to nature the faculty for direct generation; but that they were very much mistaken in applying this moral truth to a number of living bodies, which neither are nor can be produced by this sort of generation.

As a matter of fact, sufficient observations had not then been collected on this subject; and it was not known that nature, by means of heat and moisture, directly creates only the rudiments of organisation. This direct creation is confined to those living bodies which are at the beginning of the animal and vegetable scales, and perhaps of some of their branches. Thus the ancients, of whom I speak, thought that all the animals with low organisations—which they called for this reason imperfect animals—were the result of these spontaneous generations.

Lastly, since natural history in those times had scarcely advanced at all, and very few facts had been observed as to the productions of nature, the insects and all the animals then designated as worms were generally regarded as imperfect animals, which are born in favourable times and places, as a result of the action of heat on various decaying substances.

It was then believed that putrid flesh directly engendered larvae, which were subsequently metamorphosed into flies; that the extravasated juice of plants, which, as a result of pricks by insects, gives rise to gall nuts, directly produced the larvae which are transformed into *Cinips*, etc., etc.: all of which is without foundation.

Thus the mistake of the ancients in assuming direct generations in cases where they do not occur, was propagated and transmitted from age to age; it was bolstered up by the erroneous beliefs named above, and became the cause of a new mistake for moderns after they had recognised the old one.

When people perceived the necessity for collecting facts and making precise observations as to what actually occurs, the mistake into which the ancients fell was disclosed: men famous for their attainments and powers of observation, such as Rhedi, Leuwenhoek, etc., proved that all insects without exception are oviparous or sometimes apparently viviparous, that worms are never found to appear on putrid meat, except when flies have been able to deposit their eggs on it; and lastly that all animals, however imperfect they may be, themselves have the power of reproducing and multiplying the individuals of their species.

But unfortunately for the progress of knowledge we are nearly always extreme in our opinions, as we are in our actions; and it is only too common for us to compass the destruction of an error, and then throw ourselves into the opposite error. How many examples I might cite in illustration, even in the present state of accredited opinions, if such details were not foreign to my purpose!

It was thus proved that all animals without exception have the power of reproducing themselves; it was recognised that the insects and all the animals of the later classes only do so by the method of sexual generation; bodies resembling eggs had been seen in the worms and radiarians; and lastly the fact had been verified that the polyps reproduce themselves by gemmae or kinds of buds. Hence the inference was drawn that the direct generations attributed to nature never take place, but that every living body springs from a similar individual of its own species, by a generation that is either viviparous, oviparous or even gemmiparous.

This conclusion is erroneous in being too universal: for it excludes the direct generations wrought by nature at the beginning of the animal and vegetable scales, and perhaps also at the beginning of certain branches of those scales. Moreover, supposing that the bodies, in which nature has established life and organisation directly, obtain at the same time the faculty of reproducing themselves, does it necessarily follow that these bodies spring only from individuals like themselves? Unquestionably no; and this is the mistake into which we have fallen, after recognising that of the ancients.

Not only has it been impossible to demonstrate that the animals with the simplest organisation, such as the infusorians and among them especially *Monas*, and also the simplest plants such perhaps as the *Byssus* of the first family of algae, have all sprung from individuals similar to themselves; but moreover there are observations which go to show that these extremely small and transparent animals and plants, of gelatinous or mucilaginous substance, of very slight coherence, curiously ephemeral, and as easily destroyed by environmental changes as brought into existence, are unable to leave behind them any permanent security for new generations. It is on the contrary far more probable that their new individuals are a direct result of the powers and faculties of nature, and that they alone perhaps are in this position. Hence we shall see that nature has played only an indirect part in the existence of all other living bodies, since they are all derived one after the other from the original individuals; meanwhile in the course of long periods, she wrought changes and an increase of complexity in their organisation, and ever preserved by reproduction the modifications acquired and the development attained.

If it is admitted that all natural bodies are really productions of nature, it must be quite clear that in bringing the various living bodies into existence, she must necessarily have begun with the simplest, that is with those which are in truth the veriest rudiments of organisation and which we scarcely venture to look upon as organised living bodies. But when by means of the environment and of her own powers, nature has set going in a body the movements constituting life, the repetition of these movements develops organisation in it and gives rise to nutrition, the earliest of the faculties of life; from the latter soon arises the second of the vital faculties, namely, growth of the body.

Excess of nutrition in causing growth of this body prepares in it the material for a new being with a similar organisation; and thus provides it with the power to reproduce itself. Hence originates the third of the faculties of life.

Finally, the continuance of life in a body gradually increases the hardness of its containing parts and their resistance to the vital movements: it proportionally enfeebles nutrition, sets a limit to growth, and finally compasses the death of the individual.

Thus as soon as nature has endowed a body with life, the mere existence of life in that body, however simple its organisation may be, gives birth to the three faculties named above; and its subsequent stay in this same body slowly works its inevitable destruction.

But we shall see that life, especially in favourable conditions, tends incessantly by its very nature to a higher organisation, to the creation of special organs, to the isolation of these organs and their functions, and to the division and multiplication of its own centres of activity. Now since reproduction permanently preserves all that has been acquired, there have come from this fertile source in course of time the various living bodies that we observe; lastly, from the remains left by each of these bodies after death, have sprung the various minerals known to us. This is how all natural bodies are really productions of nature, although she has directly given existence to the simplest living bodies only.

Nature only establishes life in bodies that are at the time in a gelatinous or mucilaginous state, and that are sufficiently soft to respond easily to the movements which she communicates to them by means of the exciting cause that I have spoken of, or of another stimulus which I shall hereafter endeavour to describe. Thus every germ, at the moment of its fertilisation, that is at the instant when by an organic act it is rendered suitable for the possession of life, and every body which derives immediately from nature the rudiments of organisation and the movements of the most elementary life, are at that time necessarily in a gelatinous or mucilaginous state; although they are yet composed of two kinds of parts, the one containing and the other contained, the latter being essentially fluid.

Comparison between the Organic Act called Fertilisation, and that Act of Nature which gives rise to Direct Generations.

However little we may know of the two phenomena that I am now about to compare, it is quite obvious that they are related, for the results accruing from them are almost identical. Indeed the two acts in question both give origin to life; or give it the power of establishing itself in bodies where it was not previously found, and which could not possess it except through their agency. Thus a careful comparison between them cannot but enlighten us to a certain extent on the real nature of these acts.

I have already said [1] that in the reproduction of mammals, the vital movement in the embryo appears to follow immediately upon fertilisation; whereas in oviparous animals there is an interval between the act of fertilising the embryo and the first vital movement induced by incubation; and we know that this interval may sometimes be very long.

Now during this interval the fertilised embryo cannot yet be reckoned among living bodies; it is ready no doubt for the reception of life, and to that end requires only the stimulus of incubation; but so long as organic movement has not been originated by this stimulus, the fertilised embryo is only a body prepared for the possession of life and not actually possessing it.

If the fertilised egg of a fowl or any other bird is preserved for a certain time without incubation or any increase of temperature, it is not found to contain a living embryo; in the same way, the seed of a plant, which is really a vegetable egg, does not enclose a living embryo unless it has been exposed to germination.

Now if, owing to special circumstances, there occurs no incubation or germination to start the vital movement in the egg or seed, the result is that after a period, dependent on the species and the environment, the parts of this fertilised embryo degenerate; the embryo, since it has never actually had life, will not suffer death; it will

[1] *Recherches sur les corps vivants*, p. 46.

merely cease to be in a condition for receiving life and will ultimately decompose.

I have already shown in my *Mémoires de Physique et d'Histoire naturelle*, p. 250, that life may be suspended for some time and afterwards resumed.

I here wish to observe that preparation for life may be made either by an organic act, or by the direct agency of nature without any such act; so that certain bodies, without possessing life, are yet made ready for its reception by an impression, which does no doubt trace out in these bodies the earliest outlines of organisation.

What indeed is sexual reproduction but an act for achieving fertilisation? What again is fertilisation but an act preparatory to life, an act in short which disposes the parts of a body for the reception and enjoyment of life?

We know that in an unfertilised egg we yet find a gelatinous body which presents a complete external resemblance to a fertilised embryo, and is indeed nothing else than the germ previously existing in the egg although it has not been fertilised.

Yet what is the unfertilised germ of an egg but an almost inorganic body,—a body not prepared internally for the reception of life and incapable of acquiring life even by the most complete incubation?

The fact is generally known that every body which receives life, or which receives the first outlines of organisation preparing it for the possession of life, is at the time necessarily in a gelatinous or mucilaginous state; so that the containing parts of this body have the weakest coherence and the greatest flexibility possible, and are consequently in the highest possible condition of suppleness.

This must necessarily have been the case: the solid parts of the body must have been in a state closely allied to fluids, in order that the disposition, which makes the internal parts of the body ready for life, may be easily achieved.

Now it seems to me certain that sexual fertilisation is nothing else than an act for establishing a special disposition in the internal parts of a gelatinous body; a disposition which consists in a particular arrangement and distension of the parts, without which the body in question could not receive life.

For this purpose it is enough that a subtle penetrating vapour, which escapes from the fertilising material, should be insinuated into the gelatinous corpuscles capable of receiving it; that it should spread throughout its parts and by its expansive movement break up the adhesion between these parts, and so complete the organisation already begun and dispose the corpuscles for the reception of life, that is of the movements constituting life.

It appears that there is this difference between the act of fertilisation which prepares an embryo for the possession of life, and the act of nature which gives rise to direct generations; that the former acts upon a small gelatinous or mucilaginous body, in which the organisation was already outlined, whereas the latter is only carried out upon a small gelatinous or mucilaginous body, in which there was no previous trace of organisation.

In the first case, the fertilising vapour which penetrates the embryo merely breaks asunder by its expansive movement the parts which in the rudiments of organisation ought no longer to adhere together, and arranges them in a particular way.

In the second case, the subtle surrounding fluids, which are introduced into the mass of the small gelatinous or mucilaginous body, enlarge the interstices within it and transform them into cells; henceforth this small body is only a mass of cellular tissue, in which various fluids can be introduced and set in motion.

This small gelatinous or mucilaginous mass, transformed into cellular tissue, may then possess life, although not yet having any organ whatever; since the simplest living bodies, both animal and plant, are really only masses of cellular tissue which have no special organs. On this subject, I may observe that, whereas the indispensable condition for the existence of life in a body is that the body shall be composed of non-fluid containing parts and of contained fluids moving in these parts, the condition is fulfilled by a body consisting of a very supple cellular tissue, the cells of which communicate by pores: the possibility of this is attested by the fact.

If the small mass in question is gelatinous, it will be animal life that is established in it; but if it is simply mucilaginous, then vegetable life only will be able to exist in it.

With regard to the act of organic fertilisation, if you compare the embryo of an animal or plant that has not yet been fertilised with the same embryo after it has undergone this preparatory act of life, you will observe no appreciable difference between them: because the mass and consistency of these embryos are still the same and the two kinds of parts which compose them are extremely vaguely marked out.

You may then conceive that an invisible flame or subtle and expansive vapour (*aura vitalis*) which emanates from the fertilising material, and which penetrates a gelatinous or mucilaginous embryo, that is, enters its mass and spreads throughout its supple parts, does nothing more than establish in these parts a disposition which did not previously exist there, break up the cohesion at the proper places, separate the solids from the fluids in the way required by the organisation, and dispose the two kinds of parts in this embryo for the reception of the organic movement.

Lastly, you may conceive that the vital movement, which follows immediately on fertilisation in mammals but which is on the contrary only set up in oviparous animals and in plants by incubation of various kinds for the former and by germination for the latter, must subsequently develop by slow degrees the organisation of individuals endowed with it.

We cannot penetrate farther into the wonderful mystery of fertilisation; but the principle which I have just set forth is indisputable, and rests on definite facts which I think cannot be called in question.

It is important then to note that, in a different state of affairs, nature in her direct generations imitates her own procedure for fertilisation which she employs in sexual generations; and that for this purpose she does not require the assistance or produce of any pre-existing organisation.

But we must first remember that a subtle penetrating fluid in a more or less expansive condition, and apparently very analogous to the fluid of the fertilising vapours, is distributed everywhere throughout the earth, and that it provides and ever maintains the stimulus which like orgasm is at the base of every vital movement; so that we may rest assured that in places and climates where the intensity of action of the fluid is favourable to organic movement, this movement never ceases until the changes which come over the organs of the living body no longer permit these organs to lend themselves to continuous movement.

Thus in hot climates where this fluid abounds, and especially in places where a considerable dampness acts as a co-operating agency, life seems to be born and multiply everywhere; organisation is formed directly in any appropriate mass where it did not previously exist; and in those where it did exist, it develops rapidly and runs through its various stages in each individual with very remarkable speed.

It is known indeed that in very hot times and climates, the more complex and perfect the organisation of animals may be, the more rapidly does the influence of a high temperature make them traverse the various stages of their existence; this influence accelerates the various stages and the termination of their life. It is well known that in tropical countries, a girl becomes nubile very early, and that she also reaches very early the age of decay or senility. It is, lastly, an admitted fact that intensity of heat increases the danger of the various known diseases, by causing them to run through their stages with astonishing rapidity.

From these principles, we may conclude that any great heat is universally injurious to animals living in the air, because it greatly rarefies their essential fluids. It has thus been noted that in hot countries, especially at the time of day when the sun is most powerful, the animals appear to suffer, and hide themselves so as to avoid too strong a glare.

Aquatic animals, on the other hand, derive from heat, however great it may be, results that are invariably favourable to their movements and organic development. Among them, it is especially the most imperfect such as the infusorians, polyps and radiarians that benefit the most, since the condition is advantageous to their multiplication and reproduction.

Plants, which only possess a faint and imperfect orgasm, are in absolutely the same condition as the aquatic animals of which I have spoken : for, however great the heat may be, so long as these living bodies have enough water at their disposal, they vegetate all the more vigorously.

We have now seen that heat is indispensable to the most simply organised animals ; let us enquire if there are not grounds for the belief that it may itself, with the co-operation of a favourable environment, have fashioned the earliest rudiments of animal life.

Nature, by means of heat, light, electricity and moisture, forms direct or spontaneous generations at that extremity of each kingdom of living bodies, where the simplest of these bodies are found.

This proposition is so remote from the current notion on this matter, that for a long time to come it is likely to be rejected as an error, and even to be regarded as a product of imagination.

But since men who are free even from the most ubiquitous prejudices, and who are observers of nature, will sooner or later perceive the truth contained in this proposition, I wish to contribute towards their perception of it.

I believe I have shown by a collection of comparative facts, that nature under certain circumstances imitates what occurs in sexual fertilisation and herself endows with life isolated portions of matter which are in a condition to receive it.

Why indeed should not heat and electricity, which in certain countries and seasons are so abundantly distributed throughout nature, especially at the surface of the earth, not work the same result on certain substances of a suitable character and in favourable circumstances, that the subtle vapour of the fertilising substances works on the embryos of living bodies by fitting them for the reception of life ?

A famous savant (Lavoisier, *Chimie*, vol. i., p. 202) said with truth that God, when he made light, distributed over the earth the principle of organisation, feeling and thought.

Now light is known to generate heat, and heat has been justly regarded as the mother of all generations. These two distribute over our earth at least, the principle of organisation and feeling; and since feeling in its turn gives rise to thought as a result of the numerous impressions made on its organ by external and internal objects through the medium of the senses, the origin of every animal faculty may be traced to these foundations.

This being the case, can it be doubted that heat, that mother of generations, that material soul of living bodies, has been the chief means employed directly by nature for working in appropriate material the rudiments of organisation, a harmonious arrangement of parts, in short, an act of vitalisation analogous to sexual fertilisation?

Not only has the direct formation of the simplest living bodies actually occurred, as I am about to show, but the following principles proves that such formations must still be constantly carried out and repeated where the conditions are favourable, in order that the existing state of things may continue.

I have already shown that the animals of the earliest classes (infusorians, polyps, and radiarians) do not multiply by sexual reproduction, that they have no special reproductive organ, that fertilisation does not occur in them and that consequently they lay no eggs.

Now if we consider the most imperfect of these animals, such as the infusorians, we shall see that in a hard season they all perish, or at least those of the most primitive orders. Now seeing how ephemeral these animalcules are, and how fragile their existence, from what or in what way do they regenerate in the season when we again see them? Must we not think that these simple organisms, these rudiments of animality, so delicate and fragile, have been newly and directly fashioned by nature rather than have regenerated themselves? This is a question at which we necessarily arrive, with regard to these singular creatures.

It cannot then be doubted that suitable portions of inorganic matter, occurring amidst favourable surroundings, may by the influence of nature's agents, of which heat and moisture are the chief, receive an arrangement of their parts that foreshadows cellular organisation, and thereafter pass to the simplest organic state and manifest the earliest movements of life.

If it is true that unorganised and lifeless substances, whatever they may be, could never by any concurrence of circumstances form directly an insect, fish, bird, etc., or any other animal which has already a complex and developed organisation; such animals certainly can only derive their existence through the medium of reproduction, so that no fact of animalisation can concern them.

But the earliest outlines of animal organisation, the earliest acquisition of a capacity for internal development, namely, by intus-susception, lastly, the earliest rudiments of the order of things and of the internal movement constituting life, are formed every day under our very eyes, although hitherto no attention has been paid to it, and give existence to the simplest living bodies which are placed at one extremity of each organic kingdom.

It is useful to note that one of the conditions, essential to the formation of these earliest outlines of organisation, is the presence of moisture and especially of water in a fluid mass. So true is it that the simplest living bodies could not be formed or perpetually be renewed except in the presence of moisture, that none of the infusorians, polyps or radiarians are ever met with except in water; so that we may regard it as an undoubted fact that the animal kingdom originated exclusively in this fluid.

Let us continue the enquiry into the causes which have created the earliest outlines of organisation in suitable masses, where it did not previously exist.

If, as I have shown, light generates heat, heat in its turn generates the vital orgasm that is produced and maintained in animals, where the cause of it is not within them; thus heat may create the earliest elements of orgasm in suitable masses, which have attained the earliest stages of organisation.

When we remember that the simplest organisation needs no special organ distinct from other parts of the body and adapted to a special function (as is made clear by the simplification of organisation observed in many existing animals), we can conceive that such organisation may be wrought in a small mass of matter which has the following qualification.

The body that is most fitted for the reception of the first outlines of life and organisation is any mass of matter apparently homogeneous, of gelatinous or mucilaginous consistency, and whose parts though cohering together are in a state closely resembling that of fluids, and have only enough firmness to constitute the containing parts.

Now the subtle expansive fluids, distributed and constantly moving throughout the environment, incessantly penetrate and are dispersed in any such mass of matter; in passing through it they regulate the internal arrangement of its parts; they convert it into the cellular state; and they make it fit for continually absorbing and exhaling the other environmental fluids which may penetrate within it, and are capable of being contained there.

We have indeed to distinguish the fluids, which penetrate living bodies, in two categories:

1. *Containable fluids*, such as atmospheric air, various gases, water, etc. The nature of these fluids does not permit them to pass through the walls of the containing parts, but only to go in and escape through the exits;

2. *Uncontainable fluids*, such as caloric, electricity, etc. These subtle fluids are naturally capable of passing through the walls of investing membranes, cells, etc., and hence cannot be retained or preserved by any body, except for a brief period.

From the principles set forth in this chapter, it appears to me certain that nature does herself carry out spontaneous or direct generations, that she has this power, and that she utilises it at the anterior extremity of each organic kingdom, where the most imperfect living bodies are found; and that it is exclusively through their medium that she has given existence to all the rest.

To me then it seems a truth of the highest certainty, that nature forms direct or so-called spontaneous generations at the beginning of the plant and animal scales. But a new question presents itself: is it certain that she does not give rise to similar generations at any other point of these scales? I have hitherto held that this question might be answered in the affirmative, because it seemed to me that, in order to give existence to all living bodies, it was enough for nature to have formed directly the simplest and most imperfect of animals and plants.

Yet there are so many accurate observations, so many facts known, which suggest that nature does form direct generations elsewhere than at the beginning of the animal and vegetable scales, and we know that her resources are so wide and her methods so varied in different circumstances, that it is quite possible that my view, according to which the possibility of direct generations is limited to the most imperfect animals and plants, has no true foundation.

Why indeed should nature not give rise to direct generations at various points in the first half of the animal and plant scales, and even at the origin of certain separate branches of these scales? Why should she not establish, in favourable circumstances, in these diverse rudimentary living bodies, certain special systems of organisation, different from those observed at the points where the animal and vegetable scales appear to begin?

Is it not plausible, as able naturalists have believed, that intestinal worms which are never found except in the body of other animals, are direct generations of nature; that certain vermin, which cause diseases of the skin or pullulate there as a result of such diseases, also have a similar origin? And why should not the same hold good among such plants as moulds, the various fungi, and even the lichens, which are born and multiply so abundantly on the trunks of trees and on rocks favoured by moisture and a mild temperature?

Doubtless as soon as nature has directly created an animal or plant, the existence of life in this body not only endows it with the faculty of growth, but also with that of separating off some of its parts, and in short of forming granular corpuscles suitable for reproducing it. Does it follow that this body which has just obtained the faculty of propagating individuals of its own species must necessarily have sprung itself from corpuscles similar to those that it forms? This is a question which in my opinion is well worthy of examination.

Whether the kind of direct generations, here referred to, do or do not actually take place, as to which at present I have no settled opinion, it seems to me certain at all events that nature actually carries out such generations at the beginning of each kingdom of living bodies, and that she could never, except through this medium, have brought into existence the animals and plants which live on our earth.

Let us now pass to an enquiry as to the immediate results of life in a body.

CHAPTER VII.

OF THE IMMEDIATE RESULTS OF LIFE IN A BODY.

THE laws controlling all the transformations that we observe in nature, although everywhere the same and never in contradiction with one another, produce very different results in living bodies from what they cause in lifeless bodies. The results indeed are quite opposite.

In the former, by virtue of the order and state of things characteristic of living bodies, these laws are constantly striving and succeeding in forming combinations between principles which otherwise would never have been joined together, and in complicating these combinations and adding to them a superfluity of constituent elements; so that the totality of living bodies may be regarded as an immense and ever active laboratory from which all existing compounds were originally derived.

In the latter, on the contrary, that is to say in bodies without life, where there is no force to harmonise their movements and maintain their integrity, these same laws are incessantly tending to decompose existing combinations, to simplify them or reduce the complexity of their composition; so that in course of time they disengage nearly all their constituent principles from their state of combination.

This line of thought leads to developments, which when thoroughly understood and applied to all the known facts, cannot but show more and more the truth of the principle which I have been setting forth.

This course of study however is very different from that which has hitherto occupied the attention of savants; they had observed that the results of the laws of nature in living bodies were quite different from those produced in lifeless bodies, and they attributed the curious facts observed in the former to special laws, although in reality they are only due to the difference of the conditions between these bodies and in bodies that are destitute of life. They did not see that the nature of living bodies, that is, the state and order of things which produce life in them, give to the laws which regulate them a special direction, strength and properties that they cannot have in lifeless bodies; so that, by their omission to reflect that one and the same cause necessarily has varied effects when it acts upon objects of different nature and in different conditions, they have adopted for the explanation of the observed facts a route altogether opposite from what they ought to have followed.

It has indeed been said that living bodies have the power of resisting the laws and forces to which all non-living bodies or inert matter are subject, and that they are controlled by laws peculiar to themselves.

Nothing is more improbable, and nothing moreover is so far from being proved as this alleged property of living bodies for resisting the forces to which all other bodies are submitted.

This doctrine, which is very widely accepted, and is to be found set forth in all modern works on this subject, appears to me to have been invented in the first place to escape from the difficulty of explaining the causes of the various phenomena of life, and in the second place to afford some explanation of the faculty which living bodies possess of forming for themselves their own substance, of making good the wastage undergone by the material composing their parts, and lastly of giving rise to combinations which would never have existed without them. Thus in the absence of any solution, the difficulty has been shelved by the invention of special laws, without any effort being made to ascertain what they are.

In order to prove that bodies possessing life are subject to a different set of laws from that followed by lifeless bodies, and that the former possess in consequence a special force of which the chief property is said to be their release from the sway of chemical affinities, M. Richerand cites the phenomena presented by the living human body, viz.: "the decomposition of food by the digestive organs, the absorption of their nutritive material by the lacteals, the circulation of these nutritive juices in the blood, the changes which they undergo in the lungs and secretory glands, the capacity for receiving impressions from external objects, the power of approaching or flying from them, in short all the functions carried on in the animal economy." In addition to these phenomena, this savant names as more direct proofs, sensibility and contractility, two properties with which are endowed the organs that carry out the functions of the animal economy (*Éléments de Physiologie*, vol. i., p. 81).

Although the organic phenomena just mentioned are not universal to living bodies nor even to animals, they are yet characteristic of a great number of the latter and of the living human body; and they do undoubtedly show the existence of a special force animating living bodies; but this force in nowise results from laws peculiar to these bodies; it finds its origin in the exciting cause of vital movements. Now this cause, which in living bodies may give rise to the force in question, could not produce it in crude or lifeless bodies, nor could it animate them, even though it acted upon them to the same extent.

Moreover, the force in question does not altogether withdraw the various parts of living bodies from the sway of chemical affinity; and M. Richerand himself agrees that there occur, in the living machinery, effects that are quite obviously chemical, physical, and mechanical; but these effects are always influenced, modified, and weakened by the forces of life. To M. Richerand's reflections on this subject I may add the remark that the decompositions and alterations produced in living bodies by chemical affinities which tend to break up the state of things adapted for the maintenance of life, are incessantly being repaired, although more or less completely, by the results of the vital force which acts on these bodies. Now in order to bring this vital force into existence and endow it with its recognised properties, nature has no need of special laws; those which control all other bodies are amply sufficient for the purpose.

Nature never uses more complex methods than necessary: if it was possible for her to produce all the phenomena of organisation by means of the laws and forces to which all bodies are universally subjected, she has doubtless done so; and did not create laws and forces for the control of one section of her productions, opposite to those that she uses for the control of the rest.

It is enough to know that the cause, which produces the vital force in bodies whose organisation and structure permit that force to exist and excite the organic functions, could not give rise to any such power in crude or inorganic bodies, where the state of the parts does not permit of the activities and effects observed in living bodies. This cause, of which I have just spoken, only produces in the case of crude bodies or inorganic substances a force which incessantly works towards their decomposition, and which regularly achieves it by mingling its effects with that of the chemical affinities, when the closeness of their combination does not prevent it.

There is then no difference in the physical laws, by which all living bodies are controlled; but there is a great difference in the circumstances under which these laws act.

The vital force, we are told, keeps up a perpetual struggle against the forces which lifeless bodies obey; and life is only a prolonged combat between these different forces.

For my own part, I see in both cases only one force, which is synthetic in one order of things and analytic in another. Now since the conditions established by these two orders of things are always combined in living bodies, though not in the same parts at the same time, and since they follow each other in turn as a result of the incessant changes wrought by the vital movements, there does exist in these bodies throughout life a perpetual struggle between those conditions which make the vital force synthetic, and those others, always being renewed, which make it analytic.

Before developing this doctrine, let us consider several principles which should not be lost sight of.

If all the activities of life, and all the organic phenomena, without exception, are merely the result of the relations existing between the containing parts in an appropriate state and the contained fluids set in movement by a stimulating cause, the effects named below must necessarily ensue from the existence of this order and state of things in a body.

In point of fact as a result of these relations and of the movements, actions and reactions produced by the stimulating cause, there do incessantly occur the following events in all bodies possessing an active life :

1. Changes in the state of the containing parts of this body (especially the most supple), and in that of its contained fluids ;

2. Real losses in these containing parts and contained fluids, caused by the changes wrought in their state or nature ; losses which give rise to deposits, dissipations, evacuations and secretions of substances, some of which can be no more utilised, while others may be turned to various purposes ;

3. The constant need for making good the losses undergone ; a need which perpetually requires the introduction of new and suitable substances into the body, and which is actually assuaged by food in animals, and by absorptions in plants ;

4. Lastly, various kinds of combinations, which the conditions and results of the various activities of life are alone able to bring about ; combinations which, but for these results and conditions, would never have occurred.

Thus throughout life in a body, combinations are incessantly formed which are as heavily loaded with principles as the organisation of the body is adapted for ; and among them also, decompositions are always taking place, and ultimately destructions which perpetually give rise to the losses experienced.

This is the main positive fact, that is always confirmed by a close observation of vital phenomena.

Let us now return to the study of the two important principles of which I spoke above, and which furnish us to some extent with the key to all the phenomena connected with complex bodies. These principles are as follows:

The first deals with a universal and ever active factor, which more or less rapidly destroys all existing compounds.

The second concerns a power which is incessantly forming combinations, increasing their complexity and adding new principles to them, according as the circumstances are favourable.

Now although these two powers are in opposition, they both derive their origin from laws and forces which are certainly not opposed, but which work out very different effects on account of the very different circumstances.

I have already established the fact in several of my works,[1] that, by means of nature's laws and forces, every combination and every compound substance tends to be destroyed; and that this tendency is greater or less, faster or slower in its realisation, in proportion to the nature, number, proportions, and closeness of combination between the principles composing it. The reason of this is that some of these principles in combination have been forced into that condition by an external force, which modifies them while fixing them; so that these principles have a constant tendency to liberate themselves; a tendency to which they give effect, on the advent of any favouring factor.

Hence, but little attention is necessary to convince us that nature (the activity of movement established in all parts of our earth) works unceasingly towards the destruction of all existing compounds, the liberation of their principles from the combined state by constantly bringing forward factors which make for such liberation, and the restoration of these principles to that state of freedom, in which they recover their special faculties and which they tend to preserve for ever; this is the first of the two doctrines enunciated above.

But I have shown at the same time that there also exists in nature a peculiar, powerful, and ever active cause, which has the faculty of forming combinations, of increasing and varying them, and which incessantly tends to add to them new principles. Now this powerful cause, which is comprised by the second of the two doctrines cited, resides in the organic activity of living bodies, where it is always forming combinations that would never have existed without it.

This special cause is not found in any laws adapted to living bodies, and opposite to those which regulate other bodies; but it takes its origin in an order of things essential to the existence of life, and especially in a force which results from the exciting cause of organic movements. Hence the special cause, which builds up the complex

[1] *Mémoires de Physique et d'Histoire naturelle*, p. 88; *Hydrogéologie*, p. 98 *et seq.*

substances of living bodies, is due to the sole condition capable of giving it existence.

In order to be understood, I should mention that two hypotheses have been tried with a view of explaining all the facts bearing on existing compounds and their transformations, and on the elementary combinations that we can ourselves form, break up and then re-establish.

The one generally received is the hypothesis of affinities: it is well known.

The other, which is my own special theory, rests on the assumption that no simple substance whatever can have a tendency of its own to combine with any other, that the affinities between certain substances should not be regarded as forces but as harmonies which allow of the combination of these substances, and lastly, that they can never combine except when constrained by a force external to themselves, and then only when their affinities or harmonies permit of it.

According to the received hypothesis of affinities, to which chemists attribute active special forces, the whole environment of living bodies tends to their destruction; so that unless these bodies possessed within them a principle of reaction, they would soon succumb to the action of surrounding substances. For this reason men have been unwilling to admit the fact that there exists an exciting force of movements in the environment of every body, living or inanimate; and that among the former it succeeds in setting up the phenomena which they present, whereas among the latter it brings about a series of changes permitted by the affinities and finally destroys all existing combinations. The supposition is preferred that life only maintains and develops that series of phenomena found in living bodies, because these bodies were subjected to laws that were altogether peculiar to them.

It will no doubt be recognised some day that affinities are not forces, but that they are harmonies or kinds of relationships between certain substances, which enable them to enter into a more or less close combination through the agency of a general force outside themselves which constrains them to it. Now since the affinities vary between different substances, those substances which displace others from their combinations only do so because they have a greater affinity with certain of the principles in those combinations; they are assisted in this act by that general exciting force of movements and by that which works for the approximation and union of all bodies.

As to life, all that ensues from it during its residence in a body results, on the one hand, from a tendency of the constituent elements of compounds to free themselves from their state of combination, especially those which have been forced into it, and, on the other hand, from the results of the exciting force of movements. It is indeed easily perceived that in an organised body, this force regulates its activity in all the organs of the body, that it preserves harmony in its activities through the connection of these organs, that so long as they maintain their integrity it everywhere makes good the wastage wrought by the first cause, that it profits by the changes taking place in the compound moving fluids to appropriate from these fluids the assimilated substances which they carry and to fix them in their right positions, lastly, that by this order of things it always conduces to the preservation of life. This same force also conduces to growth of the parts in a living body; but this growth soon comes to an end in almost every part, for a special reason which I shall state in its proper place; and it then endows the body with the faculty of reproducing itself.

Let me repeat then that this singular force, which is derived from the exciting cause of organic movements, and which in organised bodies brings about the existence of life and produces so many wonderful phenomena, is not the result of any special laws but of certain conditions and of a certain order of things and acts which give it the power of producing such effects. Now among the effects to which this force gives rise in living bodies, we must include that of building up diverse combinations, of making them more complex, of loading them with such principles as can be forced into combination, and of incessantly creating substances which, but for it and but for the combination of circumstances in which it works, would never have existed in nature.

It is true that the trend of arguments, generally received by the physiologists, physicists and chemists of our century, is very different from that of the principles which I have set forth and developed elsewhere.[1] It is however not my purpose to endeavour to change this tendency of thought, and thus convert my contemporaries; but I was obliged to state here the two doctrines concerned, because they complete the explanation that I have given of the phenomena of life, and because I am convinced of their accuracy, and know that without them we shall always have to imagine for living bodies laws contrary to those which regulate the phenomena of other bodies.

It appears to me beyond question, that if we enquire sufficiently as to what happens in the objects concerned, we shall soon be convinced:

That the organic functions of all living beings confer on them the faculty, in some cases (plants) of forming direct combinations, that is, of uniting free elements after modification, and of immediately pro-

[1] *Hydrogéologie*, p. 105.

ducing compounds; in other cases (animals) of modifying these compounds and altering their character by the addition of new principles to a remarkable extent.

I must then again impress the fact that living bodies form for themselves, by the activity of their organs, the substance of their bodies and the various secretions of their organs; and that they neither find this substance ready formed in nature, nor the secretions, which come purely from them alone.

It is by means of food, which animals and plants are obliged to use for the preservation of their life, that the organs of these living bodies work their effects. These effects consist in a modification of the food resulting in the formation of special substances, which would never have existed without this cause, and in building up by perpetual alterations and renewals of these substances, the entire body which they constitute, as also its products.

Whereas all animal and vegetable substances are composed of principles in very complex combinations, and many of which have been forced into these combinations, man has no power to do the like; all that he can do is to decompose, alter, or destroy them, or to convert them into various special combinations, always less and less complex. It is only the movements of life that can produce these substances.

Thus plants, which have no intestinal canal nor any other organ for digestion, and which consequently use for food only fluid substances or substances whose molecules are not aggregated (such as water, atmospheric air, caloric, light, and the gases that they absorb), yet form out of such material, by means of their organic activity, all the juices that are proper to them, and all the substances of which their body is composed; that is, they form for themselves the mucilages, gums, resins, sugar, essential salts, fixed and volatile oils, feculae, gluten, extractive and woody matter; all of them substances arising direct from immediate combinations, and none of which can ever be formed by art.

Plants certainly cannot take from the soil by means of their roots the substances which I have just named: they are not there, and those which are there are in a more or less advanced condition of degradation or decomposition; lastly, if there were any in a state of complete integrity, plants would not be able to make use of them without having previously decomposed them.

Plants then have formed directly the substances to which I refer; but when they are outside plants, these substances can only be useful as manure; that is to say, only after being altered in nature, broken up, and having undergone the necessary degradation to fit them for manure, the essential function of which is to keep up a favourable moisture round about the roots of plants.

Animals cannot build up direct combinations like plants: hence they use compound substances for food; they have to carry out digestion (at least nearly all of them), and they consequently have organs for this purpose.

But they also form for themselves their own substance and secretions: now for this purpose they are not obliged to use as food either these secretions, or a substance like their own: out of grass or hay the horse forms by the action of its organs its blood and other humours, its flesh and its muscles, the substances of its cellular tissue, vessels and glands, its tendons, cartilages and bones, and lastly the horny matter of its hoofs, and the hair of its body, tail and mane.

It is then in forming their own substance and secretions, that animals build up to a high degree the combinations that they produce, and give to these combinations the astonishing number of principles that enter into animal substances.

Let us now remark that the substance of living bodies, as also the secretions which they produce by their organic activity, vary in quality according to the following circumstances:

1. The actual nature of the living being which forms them: thus vegetable productions are in general different from animal productions; and among the latter the productions of vertebrates are in general different from those of invertebrates.

2. The nature of the organ which separates them from other substances after their formation: the secretions of the liver are not the same as those of the kidneys, etc.

3. The vigour or debility of the organs of the living being and of their action: the secretions of a young plant are not the same as those of the same plant when it is very old; nor are those of a child the same as those of a grown man.

4. The integrity of the organic functions: the secretions of a healthy man cannot be the same as those of a diseased man.

5. The abundance of caloric which is continually formed on the surface of the earth although in quantities varying in different climates, and which favours the organic activity of the living bodies which it penetrates; or the rarity of caloric, as a result of which this organic activity is greatly enfeebled: as a matter of fact in hot climates the secretions formed by living bodies are different from those that they produce in cold climates; and in cold climates again the secretions of these bodies differ among themselves, according as they are formed in the hot season or during the rigours of winter.

I shall not here further emphasise the fact that the organic action of living bodies incessantly builds up combinations, which would not have arisen without it: but I shall again repeat that if it is true, as can hardly be doubted, that all compound mineral substances such as earths and rocks, and all metallic, sulphurous, bituminous, saline substances, etc., arise from the remains of living bodies,—remains which have undergone successive decompositions on and under the surface of the earth and waters; it is equally true to say that living bodies are the original source from which all known compound substances have arisen. (See my *Hydrogéologie*, p. 91 *et seq.*)

It would thus be a vain task to try to make a rich and varied collection of minerals in certain regions of the earth, such as the vast deserts of Africa, where for many centuries there have been no plants and only a few stray animals.

Now that I have shown that living bodies form their own substance for themselves as well as the various matters that they secrete, I must say a word about the faculty of feeding and growing which all these bodies possess within certain limits, since these faculties again are the result of vital activities.

CHAPTER VIII.

OF THE FACULTIES COMMON TO ALL LIVING BODIES.

It is a well-known and established fact that living bodies have faculties which are common to all of them and belong to them as a consequence of life itself.

But I think that little attention has been paid to the fact that those faculties common to all living bodies do not need any special organs as a basis, whereas those faculties which are peculiar to certain bodies only, are necessarily based on some special organ capable of producing them.

Doubtless no vital faculty can exist in a body without organisation; and organisation is itself simply a collection of organs in combination. But those organs, whose combination is necessary for the existence of life, are not peculiar to any one portion of the body they compose; they are, on the contrary, distributed throughout this body, and they bring life to every part of it, as also the essential faculties which spring from life. Hence the faculties common to all living bodies are exclusively due to the same causes which lead to the existence of life.

The case is different with the special organs that give rise to the faculties belonging only to certain living bodies: life can exist without them; but when nature achieved their creation, the chief of them have so close a connection with the order of things existing in the body, that they then become necessary for the maintenance of life in that body.

Thus it is only in the simplest organisations that life can exist without special organs; these organisations are then incapable of producing any other faculty than those common to all living bodies.

On starting an investigation as to the essential properties of life, we must distinguish the phenomena belonging to all bodies which possess life, from those which are peculiar to some of those bodies: and since the phenomena presented by living bodies are a measure of their faculties, we may usefully adopt this method of distinguishing the faculties common to all from those that are peculiar to some.

The faculties common to all living bodies,—that is, the only faculties that they have in common, are as follows :

1. Feeding, by means of incorporating food substances ; the continual assimilation of a part of these substances ; lastly, the fixation of the assimilated substances, which repair at first plentifully and afterwards less completely the loss of substance which these bodies undergo at all periods of their active life.

2. Building up their bodies ; that is to say, forming for themselves the substances of which they are made by means of materials which only contain the principles of these substances, and which are mainly supplied in the form of food.

3. Developing and growing, up to a certain limit which varies according to the species ; this growth being more than a mere aggregation of matter added externally.

4. Lastly, reproducing themselves, that is producing other bodies which are exactly like them.

Whether a living body, animal or plant, has a very simple or very complex organisation, whatever may be its class, order, etc., it necessarily possesses the four faculties enumerated above. Now since these faculties are the only ones common to all living bodies, they may be regarded as constituting the essential phenomena presented by these bodies.

Let us now enquire how much we can ascertain with regard to nature's methods for the production of these phenomena.

If nature only creates life directly in bodies which did not previously possess it ; if she only creates the simplest type of organisation (Chapter VI.) ; lastly, if she only maintains organic movements by means of an exciting cause of these movements (Chapter III.) ; we may ask how the movements kept up in an organised body, can give rise to the nutrition, growth and reproduction of that body, and at the same time confer on it the faculty of forming its own substance for itself.

I have no desire to provide an explanation of all the details of this wonderful work of nature ; for such an attempt would expose us to the probability of error and might discredit the main truths yielded by observation. I believe that the question propounded above is sufficiently answered by the following observations and reflections :

The activities of life, or the organic movements, necessarily produce alterations of state both in the containing parts and in the contained fluids of a living body, as a result of affinities and of the decomposition of principles previously in combination : such decomposition being due to these organic movements and the penetration of subtle fluids. Now from these alterations which give rise to various new combinations, there result different kinds of substances; some of which are dissipated or evacuated as the vital movement continues, while others are merely separated from parts, which do not thereby suffer any fundamental change. Among these separated substances some are deposited in particular parts of the body or are reabsorbed through canals and serve certain purposes; such are lymph, bile, saliva, the generative substance, etc.; but the rest acquire a special character, and are carried off by the general force which animates all the organs and drives all the functions, and are then fixed in similar or corresponding parts, either solid or supple and containing. They make good the wastage of these and enlarge their size, in proportion to their abundance and the possibilities of the case.

It is therefore by means of these assimilated substances, which have become adapted to particular regions, that nutrition is carried out. Nutrition, the first of the faculties of life, is thus essentially a mere restoration of the losses undergone; it is merely a means for reversing the tendency towards decomposition, which all compound substances are liable to. Now this reformation is achieved by means of a force which conveys the newly assimilated substances to their destined positions, and not by any special law, as I have already endeavoured to show. In fact, each kind of part in the animal body appropriates and stows away, by a true affinity, the assimilated molecules capable of being incorporated with it.

But nutrition is more or less abundant according to the state of organisation of the individual.

During youth, nutrition is exceedingly abundant in all organised living bodies; and it then does more than repair losses, for it adds to the size of the parts.

Indeed in a living body all the newly formed containing parts are extremely supple and of weak consistency, as a result of the causes of their formation. Nutrition under these conditions is carried on so easily as to be excessive. Not only does it completely make good the losses; but by an internal fixation of assimilated particles, it adds successively to the size of the parts and gives rise to growth of the young individual.

But after a certain period, varying with the organisation in each race, the parts, including even the most supple parts of this individual, lose much of their suppleness and vital orgasm; and their faculty of nutrition is then proportionally diminished.

Nutrition in this case is limited to the restoration of losses; the body maintains a stationary condition for some time; it is indeed in the height of its vigour, but it grows no more. Now the surplus of the parts prepared, being of no further use for nutrition or growth, is destined by nature to another purpose, and becomes the source through which she arranges for the reproduction of new individuals like the others.

Hence reproduction, the third of the vital faculties, derives its origin like growth from nutrition, or rather from the materials prepared by nutrition. But this faculty of reproduction only acquires intensity when the faculty of growth begins to decline : this fact is confirmed by common observation, since the reproductive organs (sexual parts) both in plants and animals only begin to develop when the growth of the individual is nearing an end.

I should add that, since the materials prepared for nutrition are assimilated particles of as many different kinds as there are parts in the body, the union of these diverse particles left over from nutrition and growth constitute the elements of a very small organised body, exactly similar to that from which it sprang.

In a very simple living body with no special organs, when nutrition has attained the limit of growth for the individual, the excess is then diverted to the formation and development of a part which thereupon separates from the organism and continues to live and grow, constituting a new individual like the old one. Such indeed is the method of reproduction by fission and by gemmae or buds, which occurs without any need for a special organ.

Ultimately after a still longer period—a period that varies even in the individuals of one race according to their habits and climate—the most supple parts of the living body acquire so great a rigidity and suffer so great a diminution of orgasm, that nutrition thereafter repairs the losses only incompletely. The body then gradually wastes away ; and if some slight accident or some internal disorder, that the diminished vital forces cannot cope with, do not put an end to the individual, its increasing old age is necessarily terminated by a natural death which supervenes when the existing state of things no longer permits of the performance of organic movements.

This rigidity of the soft parts, which increases during life, has been denied on the ground that after death the heart and other soft parts of an old man shrink more and become more flaccid than in a child or young man who has just died. But the fact has been overlooked that orgasm and irritability which still continue sometime after death, lasts longer and is more intense in young individuals than in the old, among whom these faculties are greatly weakened and are extinguished almost simultaneously with life. This cause alone gives rise to the observed effects.

This is the place to show that nutrition cannot be carried out, without slowly increasing the consistency of the parts restored.

All living bodies, and especially those in which internal heat is developed and maintained throughout life, continually have a part of their humours and even of their bodily tissue in a real state of decomposition; hence they are incessantly undergoing real losses, and it cannot be doubted that it is to the effects of degradations of the solids and fluids of living bodies, that the various substances formed in them are due. Of these, some are secreted and deposited or retained, while others are evacuated by various routes.

These losses would soon lead to degeneration of the organs and fluids of the individual, if nature had not given to living bodies a faculty essential to their preservation: that of making good the losses. Now as a result of these continuous losses and repairs, it follows that after a certain period the body cannot have in its parts any of the molecules which originally composed it.

It is known that the repairs are effected by means of nutrition; but they are more or less complete according to the age and state of the organs of the individual, as I remarked above.

Besides this inequality in the relation of losses to restorations according to the ages of the individuals, there exists another which is very important, and which yet appears to have received no attention. It concerns the constant inequality between the substances assimilated and fixed by nutrition, and those which are liberated as a result of the continual degradation above mentioned.

I have shown in my *Recherches* (vol. ii., p. 202) that the cause of this inequality is as follows:

Assimilation (the nutrition resulting from it) always provides more solid principles or substances, than are removed or dissipated by the losses.

The successive losses and repairs, which never cease in living bodies, have long been recognised; and yet it is only during the last few years that the conviction has grown that these losses are due to degradations continually being undergone by the fluids and even the solids of the body. Some people still have a difficulty in believing that the formation of the various secretions is the result of these degradations and changes or combinations always going on in the essential fluids of living bodies: but this fact I have already established.[1]

Now if it is true, on the one hand, that the losses of the body consist less of solid, earthy and concrete substances than of fluid substances and especially volatile substances; and if, on the other hand, it is also

[1] *Mémoire de Physique et d'Histoire naturelle*, pp. 260-263; and *Hydrogéologie*, pp. 112-115.

true that nutrition gradually provides the parts with more solid substances than fluid and volatile substances ; it will follow that the organs will gradually acquire increasing rigidity, making them less fitted for carrying out their functions, as is actually the case.

It is far from being true that the whole environment of living bodies tends to their destruction, as is repeated in all modern physiological works. I am convinced that, on the contrary, they only maintain their existence by means of external influences, and that the cause leading to the death of the individuals is within them and not without them.

Indeed I see clearly that this cause is due to the difference between the substances assimilated and fixed by nutrition, and those thrown out or dissipated by the continual wastage to which living bodies are subject, since volatile substances are always the first and the easiest to be freed from their state of combination.

I see, in short, that this cause, which brings about old age, decrepitude, and finally death, resides in the progressive hardening of the organs ; a hardening which gradually produces rigidity, and which in animals reduces to a corresponding extent the intensity of orgasm and irritability, stiffens and narrows the vessels, and imperceptibly destroys the action of the fluids on the solids, and *vice versâ*. Lastly, it disturbs the order and state of things necessary to life, which ultimately is entirely extinguished.

I believe I have proved that the faculties common to all living bodies are those of feeding ; of building up for themselves the various substances of which their bodies are composed ; of developing and growing up to a certain limit that varies in each case ; of propagating, that is, of reproducing other individuals like themselves ; lastly, of losing their life by a cause that is within themselves.

I shall now examine the faculties that are peculiar to some living bodies ; and shall confine myself, as I have just done, to an exposition of the general facts without any attempt to enter into the details that may be found in works on physiology.

CHAPTER IX.

OF THE FACULTIES PECULIAR TO CERTAIN LIVING BODIES.

JUST as there are faculties common to all bodies that enjoy life, as I have shown in the preceding chapter, so too we find in certain living bodies faculties peculiar to themselves and not shared by the rest.

We are now confronted with a circumstance of capital importance, to which the utmost attention should be paid if further progress is to be made in natural science; it is this.

It is quite clear that both animal and vegetable organisation have, as a result of the power of life, worked out their own advancing complexity, beginning from that which was the simplest and going on to that which presents the highest complexity, the greatest number of organs, and the most numerous faculties; it is also quite clear that every special organ and the faculty based on it, once obtained, must continue to exist in all living bodies which come after those which possess it in the natural order, unless some abortion causes its disappearance. But before the animal or plant which was the first to obtain this organ, it would be vain to seek either the organ or its faculty among simpler and less perfect living bodies; for neither the organ nor its faculty would be found. If this were otherwise, then all known faculties would be common to all living bodies; every organ would be present in each one of these bodies, and there would be no progress in complexity of organisation.

It is, on the contrary, well established that organisation exhibits an obvious progress in complexity, and that all living bodies do not possess the same organs. Now I propose to show that, from want of sufficient study of nature's order in her productions and of the remarkable progress that occurs in complexity of organisation, naturalists have made altogether fruitless attempts to trace in certain classes, both of animals and plants, organs and faculties which could not possibly be there.

We must then first determine the point in the natural order, say of animals, at which some organ began to exist, in order to save ourselves from seeking that organ in much earlier points of the order. Otherwise science would be retarded by our hypothetically referring to parts with which we are little acquainted, faculties which they could not have.

Thus several botanists have made useless attempts to find sexual reproduction in agamous plants (the cryptogams of Linnæus), and others have thought that they had found, in what are called the tracheae of plants, a special organ for respiration. In the same way several zoologists have wanted to prove the existence of lungs in certain molluscs, a skeleton in star-fishes, gills in jelly-fishes: lastly, a learned society has this year set, as a prize subject, the question whether there exists a circulation in radiarians.

Such attempts prove indeed how little we are yet impressed by the natural order of animals, by the progress in the complexity of their organisation, and by the general principles which result from the knowledge of that order. In a matter of organisation, moreover, when the objects dealt with are very small and unknown, people think they actually see what they want to see, and they thus find whatever they want: as, for instance, already happens in the arbitrary reference of faculties to parts of whose nature and function we are ignorant.

Let us now enquire what are the chief faculties peculiar to certain living bodies and let us see at what point in the natural order of animals and plants each of these faculties, with its attached organ, began to exist.

The chief of the faculties peculiar to certain living bodies, and consequently not shared by the rest, are as follows:

(1) The digestion of food;
(2) Respiration by a special organ;
(3) The performance of acts and movements by muscular organs;
(4) Feeling, or the capacity for experiencing sensations;
(5) Multiplication by sexual reproduction;
(6) A circulation of their essential fluids;
(7) The possession of a certain degree of intelligence.

There are many other special faculties, of which examples are found among living bodies and especially among animals; but I shall confine myself to the consideration of these few, because they are the most important, and because what I have to say about them is sufficient for my purpose.

The faculties which are not common to all living bodies are based in every case without exception on special organs which cause them, and hence on organs that are not possessed by all living bodies; and the acts which make up these faculties are functions of those organs.

I shall consequently not enquire whether the functions of such organs are being performed uninterruptedly or only intermittently, nor shall I consider whether these functions subserve the preservation of the individual or of the species, nor whether they act as links between the individual and surrounding bodies that are foreign to it. I shall merely state briefly my views on the organic functions which give rise to the seven faculties named above. I shall prove that each of them is limited to particular animals, and cannot be common to the entire animal kingdom.

Digestion. This is the first of the special faculties, and is possessed by the greater number of animals. It is at the same time an organic function carried on in a central cavity of the individual; a cavity which, although varying in shape in different races, is generally like a tube or canal, which is sometimes open at one of its extremities only, and sometimes at both.

This function, which acts only on compound substances, called alimentary substances and not a part of the individual, consists firstly in destroying the aggregation of the component molecules of the alimentary substances, introduced into the digestive cavity; and then of changing the state and properties of these molecules, in such a way that part of them become fitted for the formation of chyle, and for renewing or restoring the essential fluid of the individual.

Various liquids, delivered into the digestive organ by the excretory ducts of various glands in the neighbourhood, liquids which are chiefly poured forth when digestion has to be performed, facilitate in the first place the dissolution, that is to say, the destruction of the aggregation of the molecules of the food substances; and then contribute to bringing about the changes which these molecules have to undergo. Thereafter, such of the molecules as have been adequately altered and prepared, are suspended in the digestive and other liquids, and penetrate through the absorbent pores of the walls of the alimentary or intestinal tube into the lacteals or subordinate canals, and there constitute that precious fluid which is destined to restore the essential fluid of the individual.

All the molecules or coarser parts, which are of no use for the formation of chyle, are afterwards rejected from the alimentary cavity.

Thus the special organ of digestion is the alimentary cavity, whose anterior opening by which food is introduced bears the name of mouth, while that of the posterior extremity, when there is one, is called the anus.

It follows from the foregoing that no living body, which lacks an alimentary cavity, ever has any digestion to perform; and since all digestion works on compound substances and breaks down the aggregation of the food molecules into solid masses, it results that such living bodies as have no digestion, can only feed on fluid, liquid or gaseous material.

This applies to all plants; they have no digestive organ, nor as a matter of fact do they have any digestion to perform.

Most animals, on the contrary, have a special organ for digestion; but this faculty is not, as has been alleged, common to all animals, and cannot be cited as one of the characters of animality. The infusorians indeed do not possess it; and we should vainly seek an alimentary cavity in a monas, volvox, proteus, etc.; there is none to be found.

The faculty of digesting is then only common to the greater number of animals.

Respiration. This is the second of the faculties peculiar to certain animals, for it is less general than digestion; its function is carried on in a distinct special organ, which varies greatly in different races and different requirements.

This function consists in a restoration of the essential fluid, which in these individuals becomes too rapidly degraded; a restoration for which the slower alternative of food is not sufficient. The restoration in question is effected in the respiratory organ by means of the contact of a special fluid that is breathed in, and decomposes and communicates restorative principles to the individual's essential fluid.

In those animals whose essential fluid is quite simple and only moves slowly, the degradation of this fluid is also slow, and then the method of food alone suffices for the restorations; the fluids capable of providing certain necessary restorative principles penetrate into the individual by this route and also by absorption; and their influence is sufficient without any need for a special organ. Hence the faculty of breathing by a special organ is not necessary to these living bodies. This is the case with all plants and also with a considerable number of animals, such as those that compose the class of infusorians and that of polyps.

The faculty of breathing then should only be attributed to those living bodies that possess a special organ for the purpose; for if those which have no such organ require for their essential fluid any influence analogous to respiration (which is very doubtful), they apparently derive it through some slow general route like that of food or of absorption through external pores, and not by a special organ. Hence these living bodies do not breathe.

The most important of the restorative principles furnished by the fluid breathed to the animal's essential fluid, appears to be oxygen. It is liberated from the respired fluid, combines with the essential fluid of the animal, and restores to the latter qualities which it had lost.

There are, as we know, two different respiratory fluids which provide oxygen for breathing. These fluids are air and water; in general they are the media in which living bodies are immersed, or by which they are surrounded.

Water indeed is the respiratory fluid of many animals which live permanently in its depths. It is believed that this fluid does not decompose when giving up oxygen; but that it always has a certain amount of air mixed up with it, and that it is this air which is decomposed in the act of breathing, and thus provides oxygen for the essential fluid of the animal. This is the way in which fishes and many aquatic animals breathe; but this respiration is less active, and yields its restorative principles more slowly, than that which takes place in free air.

Free atmospheric air is the second respiratory fluid, and that which is breathed by a large number of animals which live permanently in it or within reach of it: it is promptly decomposed in the act of breathing, and thereupon yields up its oxygen to the essential fluid of the animal. This kind of breathing, which is characteristic of the most perfect animals and many others, is the most active: and its activity is proportional to the development of the organ in which it is carried out.

It is not enough to discuss the existence of a special organ for breathing; we must pay attention to the character of this organ, in order to judge of the height of the animal's development, by means of the faster or slower recurrence of the necessity for restoring its essential fluid.

In proportion as the essential fluid of animals becomes more complex and animalised, the degradations which it suffers during life are greater and more rapid, and the restorations required gradually develop in proportion to the changes experienced.

In the simplest and most imperfect animals, such as the infusorians and polyps, the essential fluid is so elementary, so little animalised, and becomes so slowly degraded, that the restorations of the food are sufficient. But soon afterwards, nature begins to require a new method for preserving the essential fluid of animals in a proper condition. It is then that she creates respiration; but at first she only sets up a very weak and inactive respiratory system,—that namely furnished by water, which has itself to convey its influence to every part of the animal.

Nature subsequently varies the type of respiration in accordance with the progressive increase of the requirement. She makes this function ever more active, and ultimately endows it with the highest energy.

Since water-born respiration is the least active, let us examine it first. We shall find that water-breathing organs are of two kinds, which again differ as regards activity; we shall afterwards note the same thing in the case of air-breathing organs.

Water-breathing organs are divided into water-bearing tracheae and gills, just as air-breathing organs are divided into air-breathing tracheae and lungs. It is indeed quite obvious that water-bearing tracheae are to gills what air-breathing tracheae are to lungs. (*Systéme des Animaux sans Vertebrés*, p. 47.)

Water-bearing tracheae consist of a certain number of vessels which ramify and spread in the animal's interior, and open on the outside by a number of small tubes which absorb the water: by this means water continually enters by these tubes, undergoes a kind of circulation all through the animal's interior, carries the respiratory influence there, and appears to issue forth again through the alimentary cavity.

These water-bearing tracheae constitute the most imperfect, the least active, and the earliest respiratory organ created by nature; that moreover which appertains to animals whose organisation is so low that their essential fluid still has no circulation. Striking examples are found in the radiarians, such as the sea-urchins, star-fishes, jelly-fishes, etc.

Gills are also a water-bearing organ, which may moreover become accustomed to breathing free air; but this respiratory organ is always isolated either within or without the animal, and only occurs in animals whose organisation is sufficiently advanced to have a nervous and a circulatory system.

Trying to find gills in radiarians and worms merely because they breathe water, is like trying to find lungs in insects because they breathe air. The air-breathing tracheae of insects constitute therefore the most imperfect of the air-breathing organs; they extend throughout all parts of the animal, carrying with them the valuable influence of respiration; whereas lungs, like gills, are isolated respiratory organs which at their highest development are more active than any other.

For the thorough appreciation of the foregoing doctrine, some attention must be given to the two following principles.

Respiration, in animals which have no circulation of their essential fluid, is carried out slowly without any perceptible movement, and in a system of organs which is distributed to almost every part of the animal's body. In this type of respiration, the respired fluid itself conveys its influence to the parts; the animal's essential fluid goes nowhere in advance of it. Such is the respiration of the radiarians and worms, in which water is the respired fluid; and such again is the respiration of the insects and arachnids, in which the respired fluid is atmospheric air.

But the respiration of animals, which have a general circulation, is of a very different type; it is effected more rapidly, it gives rise to special movements which in the highest animals become regular, and it is carried out in a simple, double or compound organ that is isolated and does not spread throughout the body. The essential fluid or blood of the animal then goes beyond the respired fluid, which only penetrates as far as the respiratory organ: the blood therefore has to undergo in addition to the general circulation a special circulation that I may call respiratory. Now since it is sometimes only a part of the blood that travels to the organ of respiration before being despatched throughout the animal's body, and since in other cases the whole of the blood passes through this organ before its journey in the body, the respiratory circulation is accordingly said to be either complete or incomplete.

Now that I have shown that there are two quite different types of respiration in those animals which have a distinct respiratory organ, I think that the name of general respiration may be given to the first type, such as that of the radiarians, worms, and insects; and that the name of local respiration should be applied to the second type, which belongs to animals more perfect than insects, including perhaps the limited respiration of arachnids.

The faculty of breathing is thus peculiar to certain animals; and the nature of the organ by which they breathe is so well adapted to their needs and to the stage of development of their organisation, that it would be very unreasonable to expect to find in imperfect animals the respiratory organ of more perfect animals.

The Muscular System. This confers upon the animals which possess it, the faculty of performing actions and movements, and of controlling these activities either by the inclination due to habit, or by the inner feeling, or, lastly, by the operations of the intellect.

Since it is admitted that no muscular activity can occur without nervous influence, it follows that the muscular system must have been formed after the rise of the nervous system, at all events in its first outlines. Now if it is true that that function of the nervous system, of which the purpose is to dispatch the subtle fluid of the nerves to the muscular fibres or bundles and set them in action, is much simpler than that other function of producing feeling (as I hope to prove), it must follow that as soon as the nervous system had reached the stage of a medullary mass in which terminate the various nerves, or as soon as it was provided with separate ganglia sending out nervous threads to various parts, it was henceforth capable of giving rise to muscular excitation without however being able to produce the phenomenon of feeling.

From these principles I believe I am justified in drawing the conclusion that the formation of the muscular system is subsequent to that of the earliest stages of the nervous system, but that the faculty of carrying out actions and movements by means of muscular organs is in animals prior to that of experiencing sensations.

Now since the origin of the nervous system is anterior to that of the muscular system, and since its functional existence only dates from the time when it was composed of a main medullary mass from which issue nervous threads, and since no such system of organs can exist in animals with organisations as simple as the infusorians or most polyps, it clearly follows that the muscular system is peculiar to certain animals, that it is not possessed by all, and yet that the faculty of acting and moving by muscular organs exists in a greater number of animals than does the faculty of feeling.

For deciding as to the presence of a muscular system in animals in doubtful cases, it is important to consider whether there are in these animals any points of attachment for muscular fibres, of a certain strength or firmness ; for, being constantly under stress, these points of attachment become gradually stronger.

It is certain that the muscular system exists in insects and all animals of subsequent classes ; but has nature established this system in animals that are more imperfect than insects ? If she has, it can hardly be (as far as the radiarians are concerned) anywhere but in the echinoderms and fistulides : it cannot be in the soft radiarians : perhaps there are rudiments of it in the sea-anemones ; the coriaceous substance of their bodies makes this belief plausible, but its presence cannot be supposed in the hydra nor in most other polyps, and still less in the infusorians.

It is possible that, when nature set out to establish some special system of organs, she selected conditions favourable to their creation ; and that consequently there are several interruptions in our scale of animals near the point at which the system is established, and due to the existence of cases in which its formation was impracticable.

Attentive observation of the operations of nature in the light of these principles will doubtless teach us many things that we do not yet know on these interesting subjects, and may perhaps disclose the fact that although nature was able to begin the muscular system with the radiarians, yet the worms which follow them are still devoid of it.

If this principle is well-founded, it will confirm what I have already urged with regard to worms, viz. : that they appear to constitute a special branch of the animal chain that has started afresh by spontaneous generation (Chapter VI., p. 247).

The plainly marked and well-known muscular system in insects is everywhere found afterwards in animals of the following classes.

Feeling is a faculty which must take the fourth rank among those that are not common to all living bodies; for the faculty of feeling appears to be still less general than those of muscular movement, respiration and digestion.

We shall see farther on that feeling is only an effect; that is to say, the result of an organic act and not a faculty inherent in any of the substances, which enter into the composition of a body that can experience it.

None of our humours and none of our organs, not even our nerves, have the faculty of feeling. It is only by an illusion that we attribute the singular effect, that we call sensation or feeling, to a definite part of our body; none of the substances composing this part does or can really feel. But the very remarkable effect called sensation or, when more intense, pain, is the product of the function of a very special system of organs, the activity of which is dependent on the circumstances which provoke it.

I hope to prove that this effect, constituting feeling or sensation, is an undoubted result of an affective cause which excites action in any part of the special system of organs adapted to it; this action by a repercussion, that is swifter than light and affects every part of the system, delivers its general effect in the common nucleus of sensation and the sensation is then propagated to the point of the body that was affected.

I shall endeavour to describe in the third part of this work, the wonderful mechanism of the effect which we call feeling: I shall here merely remark that the special system of organs for producing such an effect, is known under the name of the nervous system; and I may add that this system only acquires the faculty of giving rise to feeling, when it is so far developed as to have numerous nerves meeting in a common nucleus or centre of communication.

It follows from these principles that no animal, which does not possess a nervous system of the kind named, can experience the remarkable effect in question, nor consequently can it have the faculty of feeling. *A fortiori*, any animal which does not have nerves, terminating in a main medullary mass, must be destitute of feeling.

The faculty of feeling therefore cannot be common to all living bodies, since it is universally admitted that plants have no nerves and can therefore have no feeling. It has however been held that this faculty is common to all animals; this is clearly a mistake, for all animals neither have nor can have nerves; moreover, those in which nerves are just arising, do not yet possess a nervous system that fulfils the conditions for the production of feeling. It is probable indeed that, in its origin or primitive imperfection, this system has no other faculty than that of exciting muscular movement. The faculty of feeling therefore cannot be common to all animals.

If it is true that every faculty that is limited to certain living bodies is based upon a special organ, as is everywhere found to be the case, it must also be true that the faculty of feeling, which is clearly limited to certain animals, is exclusively the product of a special organ or system of organs, whose activities produce it.

According to this principle, the nervous system constitutes the special organ of feeling when it is composed of a single centre of communication and of nerves terminating it. Now it seems probable that it is only in the insects that the nervous system attains a development sufficient for the production of feeling, although still of a vague kind. The faculty recurs in all animals of later classes in a regular progress towards perfection.

But in animals less perfect than insects, such as worms and radiarians, if we do find traces of nerves and separated ganglia, there are strong reasons for the presumption that these organs are only adapted to the excitation of muscular movement, the simplest faculty of the nervous system.

Finally, in animals still more imperfect, such as the majority of polyps and all the infusorians, it is quite certain that they cannot possess a nervous system capable of giving them the faculty of feeling, nor even that of moving by muscles: for them, irritability alone takes its place.

Thus feeling is not a faculty common to all animals, as has been generally held.

Sexual Reproduction. This is a special faculty which is in animals nearly as general as feeling; it results from an organic function, not essential to life, the purpose of which is to attain the fertilisation of an embryo which then becomes fitted for the possession of life, and for constituting after development an individual like that or those from which it sprang.

This function is performed at particular periods, sometimes regular and sometimes not, by the co-operation of two systems of organs called sexual, one being the male organs and the other female.

Sexual reproduction is observed in animals and plants, but it is limited to particular animals and plants and is not a faculty common to all these living bodies; nature could not have made it so, as we shall see.

In the production of living bodies, both animal and plant, nature was originally obliged to create the simplest organisation in the most fragile bodies, where it was impossible to establish any special organs. She soon had to endow these bodies with the faculty of multiplying, for otherwise she would everywhere have been occupied with creations, and this is beyond her power. Now since she could not give her earliest productions the faculty of multiplying by any special system of organs, she hit upon the plan of giving it through the medium of growth, which is common to all living bodies. She conferred the faculty of undergoing divisions, at first of the entire body, and afterwards of certain projecting portions of the body ; in this way were produced gemmae and the various reproductive bodies, which are only parts that grow out, become separated, and continue to live after their separation, and which need no fertilisation, form no embryo, develop without the rupture of any membrane, and yet after growth resemble the individuals from which they spring.

Such is the method employed by nature for the multiplication of those animals and plants, to which she could not give the complicated apparatus of sexual reproduction ; it would be in vain to seek any such apparatus in the algae and fungi, or in the infusorians and polyps.

When the male and female organs are united in the same individual, that individual is said to be hermaphrodite.

In this case a distinction must be drawn between perfect hermaphroditism, which is sufficient to itself, and that which is imperfect and not sufficient to itself. Indeed many plants are hermaphrodites, in which the individual suffices to itself for fertilisation ; but in animals, which combine the two sexes, it is not yet proved by observation that the individuals are sufficient to themselves ; and it is known that many truly hermaphrodite molluscs none the less fertilise one another. It is true that, among hermaphrodite molluscs, those which have a bi-valve shell and are fixed, like oysters, must apparently fertilise themselves : it is however possible that they may fertilise one another mutually through the medium in which they are immersed. If this is so, there are among animals only imperfect hermaphrodites ; and it is known that, among vertebrates, there are not even any true hermaphrodites at all. Perfect hermaphrodites will thus be confined to plants.

The character of hermaphroditism consists in the combination of the two sexes in one individual, but it seems that the monoecious plants constitute an exception ; for although a monoecious shrub or tree carries both sexes, its individual flowers are none the less unisexual.

I may remark in this connection that it is wrong to give the name of *individual* to a tree or shrub or even to herbaceous perennials; for a tree, shrub, etc., is in reality a collection of individuals which live on one another, communicate together, and share a common life, in the same way as the compound polyps of madrepores, millipores, etc.; as I have already proved in the first chapter of this second part.

Fertilisation, the essential result of an act of sexual reproduction, must be divided into two different kinds, one of which is higher or more eminent than the other, since it belongs to the most perfect animals (mammals). This comprises the fertilisation of viviparous animals, while the other, which is inferior and less perfect, includes that of oviparous animals.

The fertilisation of viviparous animals immediately vivifies the embryo exposed to it, and this embryo forthwith continues to live, and feeds and develops at the expense of its mother, with which it remains in communication up to birth. No interval is known between the act which prepares it for the possession of life and the reception of life itself; moreover, this fertilised embryo is enclosed in a membrane which contains no stores of food within it.

The fertilisation of oviparous animals, on the other hand, only prepares the embryo for the reception of life, but does not actually confer life. Now this fertilised embryo of oviparous animals is enclosed with a store of food in investments, which cease to communicate with the mother before being separated from her; and it only receives life when a special factor, which may come sooner or later according to circumstances, or may not come at all, communicates to it the vital movement.

This special factor, which confers life on the embryo of an oviparous animal after it has been fertilised, consists as regards animals' eggs in a mere rise of temperature, and as regards the seeds of plants in the co-operation of moisture with a gentle penetrating warmth. In birds' eggs, for instance, incubation causes this rise of temperature, and in many other eggs a gentle warmth of the atmosphere is enough; lastly, circumstances that favour germination vitalise the seeds of plants.

But eggs and seeds adapted for giving existence to animals and plants must of necessity contain a fertilised embryo enclosed in investments, whence it can only emerge after breaking through them: such eggs and seeds are therefore products of sexual reproduction, since reproductive bodies otherwise originating do not have any embryo enclosed in investments which have to be broken through at the outset of development. Gemmae and the more or less oviform reproductive bodies of many animals and plants cannot assuredly be compared with them: it would be a waste of time to search for sexual generation where nature has had no means for establishing it.

Sexual reproduction is thus peculiar to certain animals and plants: consequently the simplest and most imperfect living bodies cannot possess any such faculty.

Circulation. This is a faculty which only exists in certain animals, and which is much less general in the animal kingdom than the five others of which I have already spoken. This faculty springs from an organic function whose purpose is the acceleration of the movements of the essential fluid of certain animals,—a function which is performed by a special system of organs adapted to it.

This system of organs is essentially composed of two kinds of vessels, viz. arteries and veins, and almost always in addition a thick and hollow muscle, which occupies about the centre of the system, which soon becomes the principal motive power of it, and which is called the heart.

The function carried out by this system of organs consists in driving the animal's essential fluid, which is here known by the name of blood, from an almost central point occupied by the heart (when there is one), through the arteries into every part of the body; whence it returns to the same point by the veins, and is then dispatched anew throughout the body.

It is this movement of the blood, always being driven into every part and always returning to its starting point throughout the duration of life, that has received the name of *Circulation.* It should be qualified as *general* in order to distinguish it from the respiratory circulation, which is undertaken by a special system likewise composed of arteries and veins.

Nature, when initiating organisation in the simplest and most imperfect animals, was only able to give their essential fluid an extremely slow movement. This no doubt is the case in the very simple and scarcely animalised essential fluid that moves in the cellular tissue of infusorians. But afterwards she gradually animalised and developed the essential fluid of animals in proportion as their organisation became more complex and perfect; and she accelerated its movement by various methods.

In the polyps, the essential fluid is nearly as simple and has scarcely more movement than that of the infusorians. The regular shape of the polyps, however, and especially their alimentary cavity begin to furnish means to nature for somewhat increasing the activity of their essential fluid.

She probably took advantage of this in the radiarians, to establish in their alimentary cavity the centre of activity of their essential fluid. The expansive surrounding subtle fluids, in fact, which constitute the exciting cause of these animals' movements, penetrate chiefly into their alimentary cavity ; and by their incessant expansions have greatly developed this cavity, have induced the radiating form of these animals both internally and externally, and moreover cause the isochronous movements observed in the soft radiarians.

When nature had established muscular movement, as she has in the insects and perhaps even a little before, she had a new means for increasing the movement of their serum or essential fluid ; but on reaching the organisation of the crustaceans, this means no longer sufficed, and a special system of organs had to be created for accelerating the essential fluid or blood of these animals. It is indeed in the crustaceans that we find for the first time a complete general circulation ; for this function is only rudimentary in the arachnids.

Every new system of organs acquired is permanently preserved in subsequent organisations ; but nature continues to work towards its gradual perfection.

The general circulation is thus at first provided only with a heart with one ventricle, and indeed in the annelids even a heart is unknown : it is at first accompanied only by an incomplete respiratory circulation, viz. one in which all the blood does not pass through the organ of respiration before being despatched to the parts. This is the case with animals which have imperfect gills ; but in fishes, where the branchial respiration is perfect, the general circulation is accompanied by a complete respiratory circulation.

When nature subsequently created lungs for breathing, as she did in the reptiles, the general circulation was of necessity accompanied only by an incomplete respiratory circulation ; because the new respiratory organ was still too imperfect, and because the general circulation still had a heart with only one ventricle and also because the new fluid breathed is by itself more effectively restorative than water, so that a complete respiration was not needed. But when nature reached that perfection of pulmonary respiration seen in the birds and mammals, the general circulation came to be accompanied by a complete respiratory circulation ; the heart necessarily had two ventricles and two auricles ; and the blood gained its highest velocity ; the high animalisation became capable of raising the animal's internal temperature above that of the environment, and, lastly, the blood became subject to rapid decomposition requiring corresponding restoration.

The circulation of the essential fluid of a living body is then an organic function peculiar to certain animals : it first becomes complete and general in the crustaceans, and is afterwards found gradually becoming more perfect in animals of the following classes ; but it would be vain to seek it in the less perfect animals of the anterior classes.

Intelligence. Of all the faculties peculiar to certain animals, this is the one that is the most limited as regards the numbers which possess it, even in a very imperfect form; but it is also the most wonderful, especially when highly developed; and it may then be regarded as the high-water-mark of what nature can achieve by means of organisation.

This faculty arises from the activities of a special organ which can alone produce it, and which is itself highly complex when it has acquired all the development of which it is capable.

As this organ is actually distinct from that which produces feeling although unable to exist without it, it follows that the faculty of performing acts of intelligence is not only not common to all animals but is not even common to all those that can feel; for feeling may exist without intelligence.

The special organ in which are produced the acts of the understanding appears to be only an accessory of the nervous system; that is, a part added on to the brain, and containing the nucleus or centre of communication of the nerves. The special organ in question is thus adjacent to the nucleus; the nature of the substance of which it is composed appears moreover to differ in no way from that of the nervous system; in it alone, however, acts of intelligence are performed; and it is a special organ, for the nervous system may exist without it.

In the third part I shall take a general survey of the probable mechanism of the functions of this organ. In vertebrates it is confused with the medullary mass under the name of brain, although it only consists of the two wrinkled hemispheres which cover it over. It is sufficient here to note that of those animals which have a nervous system, it is only the most perfect that actually possess the two cerebral hemispheres; probably all invertebrates, except perhaps some of the last order of molluscs, are destitute of it, although a great many of them have a brain to which run directly the nerves of one or more special senses, and although this brain is generally divided into two lobes separated by a furrow.

In accordance with this view the faculty of performing acts of intelligence has only just begun in the fishes or, at the earliest, in the cephalopod molluscs. It is in these animals in a state of extreme imperfection; some development has been achieved in the reptiles, especially in the later orders; much more has been made in the birds, and the faculty reaches its highest point in the latest orders of mammals.

Intelligence is then a faculty limited to certain animals which are able to feel; but the faculty is not common to all those that possess feeling: indeed, as we shall see, among the latter, those that have no special organ for performing acts of intelligence can only have simple perceptions of the objects which affect them, but can form no idea of them, do not make comparisons or judgments, and are guided in all their actions by their habitual needs and inclinations.

Summary of Part II.

By confining myself in the nine preceding chapters solely to the observations with which I was concerned, I have avoided entering into a quantity of details which are doubtless very interesting, but may be found in the good works on physiology already accessible to the public: the principles which I have advanced appear to me sufficient to prove:

1. That life in every body which possesses it consists only of an order and state of things, by which the internal parts can be influenced by an exciting cause and perform movements called organic or vital, from which are produced according to the species the recognised phenomena of organisation;

2. That the exciting cause of vital movements is external to the organs of all living bodies; that the elements of this cause are always found, although in varying abundance, wherever there is life; that it is provided to living bodies by the environment either in whole or part; and that without this same cause no such body could possess life;

3. That every living body whatever is necessarily composed of two kinds of parts, viz.: containing parts consisting of a very supple cellular tissue, in which and out of which every kind of organ has been formed; and visible contained fluids capable of moving about and of undergoing various changes in their condition and nature;

4. That animal nature does not differ essentially from vegetable nature as regards the special organs of these two kinds of living bodies, but chiefly as regards the nature of the substances of which they are composed: for the substance of every animal body is such that the exciting cause can establish in it an energetic orgasm and irritability; whereas the substance of all vegetable bodies merely gives the exciting cause the power of setting in motion the visible contained fluids, while only permitting in the containing parts of a faint orgasm, not enough to produce irritability or to cause any sudden movements by the parts;

5. That nature herself produces direct or so-called spontaneous generations by creating organisation and life in bodies which did not previously possess them; that she must of necessity have this faculty in the case of the most imperfect animals and plants at the beginning of the animal and vegetable scales, and also perhaps of some of their branches; and that she only performs this strange phenomenon in

tiny portions of matter, gelatinous in the case of animals and mucilaginous in the case of plants, transforming these portions of matter into cellular tissue, filling them with visible fluids which develop within them, and setting up in them various movements, dissipations, restorations and alterations by means of the exciting cause provided by the environment;

6. That the laws which control the various transformations in bodies, of whatever nature they may be, are everywhere the same; but that these laws in living bodies work results altogether opposite to those achieved in crude or inorganic bodies, because in the former they find an order and state of things which give them the power to produce all the phenomena of life, while in the latter they find a very different state of things and produce very different effects: so that it is not true that nature has special laws for living bodies, opposite to those which control the transformations observed in lifeless bodies;

7. That all living bodies of both kingdoms and all classes have certain faculties in common; these are the property of the general organisation of such bodies, and of the life which they contain; hence these faculties, common to all living things, need no special organ for their existence;

8. That in addition to the faculties common to all living bodies, some of these bodies, especially among animals, have faculties peculiar to themselves and not found among the rest; but these special faculties are in every case the product of a special organ or system of organs, so that no animal without that organ or system of organs can possibly possess the faculty which it confers on those that have it; [1]

9. Lastly, that the death of every living body is a natural phenomenon, which necessarily results from the presence of life and is brought about by natural causes, unless some accidental cause intervenes first; this phenomenon is nothing else than the complete cessation of vital movements, resulting from some disturbance in the order and state of things necessary for the performance of these movements; in animals with highly complex organisations, the principal systems of organs possess to some extent a life of their own, although closely bound up with the general life of the individual. The death of an animal thus takes place gradually in the separate parts, so that life becomes succes-

[1] In this connection I may observe that plants in general have no special organs within them for particular functions. Every part of a plant contains the organs essential to life, and may therefore either live and vegetate separately, or as a result of grafting share with another plant a life common to both; lastly, from this order of things in plants it follows that several individuals of the same species, or even only of the same genus, may live on one another in the enjoyment of a common life.

I may add that the latent buds found on the branches and even the trunk of woody plants are not special organs, but the rudiments of new individuals, awaiting favourable conditions for their development.

sively extinct in the principal organs in a regular and constant order; and the moment at which life ceases in the last organ is that which completes the death of the individual.

On such difficult subjects as those of which I have been treating, we are closely confined within the limits of knowledge and to the sphere of what we can learn from observation. Everything has reference to the conditions essential to life in a body; conditions established in compliance with facts which prove their necessity.

If things are not really as I have described, or if it is held that the conditions named and the admitted facts which testify to the true foundation of these matters, are not adequate proofs to justify us in admitting them, we shall then have to abandon altogether the enquiry into the physical causes which give rise to the phenomena of life and organisation.

PART III.

AN ENQUIRY INTO THE PHYSICAL CAUSES OF FEELING, INTO THE FORCE WHICH PRODUCES ACTIONS, AND, LASTLY, INTO THE ORIGIN OF THE ACTS OF INTELLIGENCE OBSERVED IN VARIOUS ANIMALS.

INTRODUCTION.

In the second part of this work, I have endeavoured to throw light on the physical causes of life, on the conditions necessary for its existence, and on the origin of that exciting force of vital movements, without which no body could actually possess life.

I now propose to enquire what feeling may be, how the special organ giving rise to it (the nervous system) produces the wonderful phenomenon of sensations, how sensations themselves produce ideas through the medium of the brain, and how ideas cause in that organ the formation of thoughts, judgments and reasoning; in short, of acts of intelligence that are still more wonderful than sensations.

"But," it is said, "the functions of the brain are of a different order from those of the other viscera. In the latter causes and effects are of the same nature (physical nature). . . .

"The functions of the brain are of quite a different order: they consist in receiving sense impressions through the nerves, in transmitting them immediately to the mind, in preserving the traces of these impressions, and in reproducing them with varying rapidity, clearness and fulness whenever the mind needs them in its operations or the laws of association of ideas recall them; lastly, in transmitting to the muscles again through the medium of nerves the commands of the will.

"Now these three functions involve a mutual influence, which has always remained incomprehensible, between divisible matter and the indivisible ego. This has always constituted an impassable hiatus in the system of our ideas, and the stumbling block of all philosophies; they involve us moreover in a further difficulty that has no necessary connection with the first: not only do we not understand nor ever shall understand how impressions on the brain can be perceived by the mind and produce images in it; but however refined our means of investigation, these traces cannot be made visible in any way; and we are entirely ignorant of their nature, although the effect of age and diseases on the memory leave us in no doubt either as to their existence or their seat." (*Rapport à l'Institut sur un Mémoire de MM. Gall et Spurzheim*, p. 5.)

It is, I think, a little rash to fix limits to the conceptions which the human intellect may reach, or to specify the boundaries and the powers of that intellect. How indeed can we know that man will never obtain such knowledge, nor penetrate these secrets of nature? Do we not know that he has already discovered many important truths, some of which seemed to be entirely beyond him?

It is more rash, I repeat, to try to determine positively what man may know and what he never can know, than to study the facts, examine the relations existing between various physical bodies, draw all possible inferences, and then make continuous efforts to discover the causes of natural phenomena; even when the coarseness of our senses does not allow us to reach anything more than moral certainties.

If we were concerned with objects outside nature, with phenomena that are neither physical nor the result of physical causes, the subject would doubtless be beyond the human intellect; for it can never obtain a grasp of anything external to nature.

Now, since in this work we are dealing mainly with animals, and since observation teaches us that there are among them some which possess the faculty of feeling, which form ideas and judgments and carry out intelligent acts, which, in short, have memory, I wish to ask what is the peculiar entity called mind in the passage cited above; a remarkable entity which is alleged to be in relation with the acts of the brain, so that the functions of this organ are of a different order from those of the other organs of the individual.

In this factitious entity, which is not like anything else in nature, I see a mere invention for the purpose of resolving the difficulties that follow from inadequate knowledge of the laws of nature: it is much the same thing as those universal catastrophes, to which recourse is had for giving answers to certain geological questions. These questions puzzle us because the procedure of nature, and the different kinds of transformations that she is always producing, are not yet ascertained.

With regard to the traces impressed on the brain by ideas and thoughts, what matters it that these traces cannot be perceived by our senses, if, as is agreed, observations exist which leave us in no doubt as to their presence and their seat: do we see any more clearly the way in which other organs perform their functions, and, to take a single example, do we see any more clearly how the nerves set the muscles in action? Yet we cannot doubt that nervous influence is indispensable for the performance of muscular movements.

In the sphere of nature, knowledge is extremely important for us and yet very difficult to obtain in any better form than moral certainties; such knowledge can, I believe, only be attained by the following method.

Do not let us be imposed upon by dogmatic utterances which are nearly always ventured with little thought; let us carefully collect such facts as we can observe, let us make experiments wherever we can, and when experiment is impossible let us marshal all the inferences that we can draw from analogy, and let us nowhere make a dogmatic pronouncement: by this method we shall be able gradually to attain a knowledge of the causes of many natural phenomena, including perhaps even those that now appear to us the most incomprehensible.

Since, then, the limits of our knowledge as to what occurs in nature neither are nor can be fixed, I shall endeavour, by the use of such facts as have been collected, to determine in this third part what are the physical causes which confer on certain animals the faculty of feeling, of producing for themselves the movements which constitute their actions, and, lastly, of forming ideas and of comparing these ideas, so as to obtain judgments: in short, of performing various intelligent acts.

The principles which I shall set forth on this matter will as a rule be such as to fill us with an inward moral conviction, although it is impossible to prove positively their accuracy. It seems that, with regard to many natural phenomena, this order of knowledge is alone possible for us; and yet its importance cannot be called in question in innumerable cases where we are called upon to form judgments.

If the physical and the moral have a common origin, if ideas, thought and even imagination are only natural phenomena, and therefore really dependent on organisation, then it must be chiefly the province of the zoologist, who makes a special study of organic phenomena, to investigate what ideas are, and how they are produced and preserved, in short, how memory renews them, recalls them and makes them perceptible once more; from this it is only a short way to perceiving what are thoughts themselves, for thoughts can only be invoked by ideas; lastly, by following the same method and building up from original perceptions, it may be possible to discover how thoughts give rise to reasoning, analysis, judgments and the will to act, and how again numerous acts of thought and judgments may give birth to imagination, a faculty so fertile in the creation of ideas that it even seems to produce some which have no model in nature, although in reality they must be derived from this source.

If all the acts of the intellect, into the causes of which I am now enquiring, are only phenomena of nature, that is to say, acts of the organisation, may I not hope, by acquiring a thorough knowledge of the only means by which the organs perform their functions, to discover how the intellect may give rise to the formation of ideas and preserve their traces or impressions for a longer or shorter period, and finally, by means of these ideas, carry out thought, etc., etc.?

It cannot now be doubted that the acts of the intellect are exclusively dependent on organisation, since it is known that even in man disturbances in the organs which produce these acts involve others in the acts themselves.

An investigation of the causes of which I spoke above appears therefore to me to be obviously possible: I have given attention to the subject; I have devoted myself to an investigation of the only method by which nature can have brought about the phenomena in question; and it is the result of my meditations on this subject that I am now about to present.

The essential point is that, in every system of animal organisation, nature has but one method for making the various organs perform their appropriate functions.

These functions indeed are everywhere the result of the relation between fluids moving in the animal, and the parts of its body which contain these fluids.

There are everywhere moving fluids (some containable, others uncontainable) which act upon the organs; and there are also everywhere supple parts, which are sometimes in erethism and react on the fluids which affect them, and which are sometimes incapable of reacting; but in either case they modify the movement of the fluids taking place among them.

Thus, when the supple parts of organs are capable of being animated by orgasm and of reacting on the contained fluids which affect them, the various resulting movements and changes, both in the fluids and the organs, produce phenomena of organisation which have nothing to do with feeling or intelligence; but when the containing parts are so soft as to make them passive and incapable of reacting, the subtle fluid moving in these parts, and modified by them in its movements, gives rise to the phenomena of feeling and intelligence as I shall endeavour to prove in this Part.

We have therefore to deal only with the relations existing between the concrete supple and containing parts of an animal, and the moving fluids (containable or uncontainable) which act on these parts.

This well-known fact has been for me as a beam of light; it guided me in the research that I have sketched out, and I soon perceived that the intelligent acts of animals are, like their other acts, phenomena of animal organisation, and that they take their origin from the relations existing between certain moving fluids and the organs which produce these wonderful acts.

What matters it that these fluids, whose extreme tenuity prevents us from seeing them or even keeping them in a vessel for making experiments with, only manifest their existence by their effects? These effects constitute a cogent proof that no other cause could have produced them. It is, moreover, easily ascertained that the visible fluids, which penetrate the medullary substance of the brain and nerves, are only nutritive and adapted for secretion; but that they are too slow in their movements to give rise to the phenomena either of muscular movement, feeling, or thought.

In the light of these principles, which restrain the imagination within its proper limits, I shall first show how nature originally succeeded in creating the organ of feeling, and by its means the force productive of actions: I shall afterwards proceed to consider how (by means of a special organ for intellect), ideas, thoughts, judgments, memory, etc., may have arisen in the animals which possess such an organ.

CHAPTER I.

OF THE NERVOUS SYSTEM, ITS FORMATION, AND THE VARIOUS SORTS OF FUNCTIONS THAT IT CAN FULFIL.

THE nervous system, in man and the most perfect animals, consists of various quite distinct special organs and even systems of organs, which are closely connected and form a very complicated whole. It has been supposed that the composition of this system is everywhere the same, except for its greater or lesser development, and the differences of size, form, and situation involved by the various types of organisation. On this theory the various sorts of functions, to which it gives rise in the most perfect animals, were all regarded as being characteristic of it throughout all animal organisation.

This manner of regarding the nervous system throws no light on the nature of the organs in question, on the mode of their origin, on the growing complexity of their parts in proportion to the complication and perfection of animal organisation, nor, lastly, on the new faculties which it confers on animals in proportion to its development. On the contrary, instead of enlightening physiologists on these matters, it leads them to attribute everywhere to the nervous system in various degrees of concentration the faculties which that system confers on the most perfect animals, and this is entirely without foundation.

I shall therefore endeavour to prove: (1) That this system of organs cannot be a property of all animals; (2) that at its origin, that is, at its greatest simplicity, it only confers on the animals which possess it the one faculty of muscular movement; (3) that afterwards, when more highly developed, it endows animals not only with muscular movement but also with feeling; (4) that lastly, on reaching completion, it confers on the animals which possess it the faculties of muscular movement, of experiencing sensations, and of forming ideas, comparing them together, and producing judgments; in short, of having an intellect whose development is proportional to the perfection of organisation.

Before setting forth the proofs of these theories, let us see what general idea we can form of the nature and arrangement of the various parts of the nervous system.

This system, wherever it occurs in animals, presents a main medullary mass, either divided into separate parts or concentrated into a single whole of varying shape, and also nervous threads which run into this mass.

All these organs are composed of three kinds of substances of very different character, viz. :

1. A very soft medullary pulp of peculiar character.

2. An aponeurotic investment, which surrounds the medullary pulp and provides sheaths to its prolongations and threads, including even the finest. The nature and properties of this investment are different from those of the pulp which it encloses.

3. A very subtle invisible fluid, which moves in the pulp without requiring any visible cavity, and which is kept in at the sides by the sheath, through which it cannot pass.

Such are the three kinds of substances which compose the nervous system, and which produce the most astonishing of all organic phenomena as a result of their arrangement, relations and the movements of the subtle fluid contained within the system.

It is known that the pulp of these organs is a very soft medullary substance, white on the inside, greyish on its outer layer, not sensitive, and apparently albumino-gelatinous in character. It forms, by means of its aponeurotic sheaths, threads and cords which proceed to the larger masses of this medullary substance containing the nucleus (simple or divided), or centre of communication of the system.

Both for the performance of muscular movement and for sensations, it is necessary that this system of organs should have a nucleus or centre of communication for the nerves. As a matter of fact, in the first case the subtle fluid which acts upon the muscles issues from a common nucleus and travels towards the parts which it has to actuate ; and, in the second case, the same fluid, being set in motion by the affective cause, starts from the extremity of the affected nerve and travels towards the centre of communication, there producing the disturbance which gives rise to sensations.

A nucleus or centre of communication, in which the nerves terminate, is therefore absolutely necessary in order that the system may carry on any of its functions ; and indeed we shall see that without it the individual would not become cognisant of the acts of the organ of intellect. Now this centre of communication is situated in some part of the main medullary mass, which always constitutes the basis of the nervous system.

The threads and cords of which I spoke above are nerves; and the main medullary mass, which contains the centre of communication of the system, consists in some invertebrates either of separate ganglia or of a ganglionic longitudinal cord; in the vertebrates, it forms the spinal cord and the medulla oblongata which is united to the brain.

Wherever a nervous system exists, however simple or imperfect it may be, there is always a main medullary mass in some form or other; for it constitutes the basis of the system and is essential to it.

It is in vain to deny this truth by such arguments as the following:

1. That it is possible to remove entirely the brain of a tortoise or a frog, which nevertheless continue to exhibit movements showing that they still have sensations and a will: I reply, that this operation only destroys a part of the main medullary mass, and not that part which contains the centre of communication or *sensorium commune*; for this is not contained in the two hemispheres which form the bulk of what is called the brain;

2. "That there are insects and worms, which when cut into two or more pieces, promptly form so many new individuals, each having its own system of sensation and its own will"; I reply again, that as regards insects the alleged fact is untrue, that no experiment has shown that when an insect is cut in two there may result two individuals both capable of life; and even if it were so, each half of the insect would still possess a main medullary mass in its share of the ganglionic longitudinal cord;

3. "That the more evenly the nervous substance is distributed, the less essential is the rôle of the central parts."[1] I reply, for the last time, that this assertion is erroneous; that it has no facts to support it; and that it is only made through ignorance of the functions of the nervous system. Sensibility is neither the property of nervous substance nor any other substance, and the nervous system can only enter upon its functional existence when it is composed of a main medullary mass from which nervous threads take rise.

Not only can the nervous system have no functional existence unless it is composed of a main medullary mass which contains one or more nuclei for starting muscular excitement and from which various nerves proceed to the parts, but we shall also see in Chapter III. that the faculty of feeling in any animal can only arise when the medullary mass contains a single nucleus or centre of communication, to which the nerves of the sensitive system travel from all parts of the body.

It is true that the extreme difficulty of following these nerves to their centre of communication, has led some anatomists to deny the existence

[1] See *L'Anatomie comparée* of M. Cuvier, vol. ii., p. 94; and the *Recherches sur le Système nerveux* of MM. Gall and Spurzheim, p. 22.

of any common nucleus that is essential to the production of feeling; they consider feeling to be an attribute of all the nerves, including even their smallest parts; and, to strengthen their view as to the absence of any centre of communication in the sensitive system, they allege that the need for finding a definite situation for the soul has caused the invention of this common nucleus or circumscribed locality to which all sensations are conducted.

It is quite enough to believe that man possesses an immortal soul; there is no occasion for us to study the seat and limits of this soul in the individual body, nor its connection with the phenomena of organisation: all that we can ever say on this subject is baseless and purely imaginary.

If we are studying nature she alone should occupy our attention; and we should confine ourselves exclusively to the examination of the facts which she presents, in our endeavour to discover the physical laws which control the production of these facts; lastly, we ought never to introduce into our theories any subjects that are outside nature, and about which we shall never be able to know anything positive.

For my own part, I only study organisation in order to arrive at an understanding of the various faculties of animals. I am convinced that many animals possess feeling, and that some of them also have ideas and perform intelligent acts; and I hold that the causes for these phenomena should be sought in purely physical laws. I always make a rule of this in my own researches, and I may add that I am not only convinced that no kind of matter can in itself possess the faculty of feeling, but I am also convinced that this faculty in such living bodies as possess it consists only in a general effect which is set up in an appropriate system of organs, and that this effect cannot occur unless the system possesses a single nucleus or centre of communication, in which terminate all the sensitive nerves.

In the case of vertebrates, it is at the anterior extremity of the spinal cord, in the medulla oblongata itself or perhaps its annular protuberance, that the *sensorium commune* is lodged; that is to say, the centre of communication of the nerves which give rise to the phenomenon of sensibility; for it is towards some point at the base of the brain that these nerves appear to converge. If the centre of communication were farther forward in the interior of the brain, acephalic animals, whose brain had been destroyed, would be devoid of feeling and unable even to live.

But this is not the case: in animals which possess any faculty of intelligence, the nucleus for feeling is confined to some part of the base of what is called their brain; for this name is given to the entire medullary mass contained within the cranial cavity. The two hemispheres, however, which are confused with the brain, should be distinguished from it; because they form together a special organ added on to the brain, have special functions of their own and do not contain the centre of communication of the sensitive system.

Although the true brain, that is to say, the medullary part which contains the nucleus of sensations and to which the nerves lead from the special senses, is difficult to identify and define in man and intelligent animals, on account of the contiguity or union between this brain and the two hemispheres which cover it, it is none the less true that these hemispheres constitute an organ specially related to the functions that it performs.

Indeed it is not in the brain properly so-called that ideas, judgments, thoughts, etc., are formed; but it is in the organ superimposed on it, consisting of the two hemispheres.

Nor is it in the hemispheres that sensations are produced; they have no share in it, and the sensitive system exists satisfactorily without them; these organs may therefore undergo great degeneration without any injury to feeling or life.

I now revert to the general principles concerning the composition of the various parts of the nervous system.

The nervous threads and cords, the ganglionic longitudinal cord, the spinal cord, the medulla oblongata, the cerebellum, the cerebrum and its hemispheres;—all these parts have, as I have already observed, a membranous and aponeurotic investment which serves as a sheath, and which by its peculiar nature retains within the medullary substance the special fluid that moves about there; but at the extremities of the nerves where they terminate in the parts of the body, these sheaths are open and allow the nervous fluid to communicate with the parts.

Details about the number, shape and situation of the parts I have referred to, belong to the sphere of anatomy; an exact description may be found in works which deal with this sphere of our knowledge. Now since my purpose here is simply to investigate the general principles and faculties of the nervous system, and to enquire how nature first conferred it on such animals as possess it, I need not enter into any of the details that are known about the parts of this system.

Formation of the Nervous System.

We certainly cannot positively determine the manner in which nature brought the nervous system into existence; but it is quite possible to ascertain the conditions which were necessary for this purpose. When once we have ascertained and studied these conditions, we may be able to conceive how the parts of the system were formed, and how they were filled with the subtle fluid which moves within them and enables them to carry out their functions.

We may suppose that when nature had advanced so far with animal organisation that the essential fluid of animals had become highly animalised, and the albumino-gelatinous substance been formed, this substance would be secreted from the animal's chief fluid (blood, or its substitute) and deposited in some part of the body: now observation shows that this first occurs in the shape of several small separate masses, and afterwards as a larger mass which becomes lengthened into a ganglionic cord and occupies nearly the whole length of the body.

The cellular tissue is modified by this mass of albumino-gelatinous substance, and so provides it with its investing sheath, and that of its various prolongations or threads.

Now on examining the visible fluids which move or circulate in the bodies of animals, I find that, in the animals with the simplest organisations, these fluids are much less complex and contain much fewer principles than is the case in the more perfect animals. The blood of a mammal is a more complex and animalised fluid than the whitish serum of insects; and this serum again is a more complex fluid than that watery matter which moves in the bodies of polyps and infusorians.

This being so, I am justified in the belief that those invisible and uncontainable fluids which keep up irritability and vital movements in the most imperfect animals, are the same as those existing in animals with a highly complex and perfected organisation. In the latter, however, they undergo so great a modification as to be changed into containable fluids, though still invisible.

It appears indeed that an invisible and very subtle special fluid, which is modified during its presence in the blood of animals, is continually separating out to spread through the nervous medullary masses, and incessantly makes good the wastage due to the various activities of this system of organs.

The medullary pulp of the nervous system, and the subtle fluid moving within it, will thus only be formed when the complexity of animal organisation has reached a sufficient development for the manufacture of these substances.

Just as the internal fluids of animals are progressively modified, animalised and compounded in correspondence with the progress in the complexity and perfection of organisation; so too the organs and solid or containing parts of the body are gradually compounded and diversified in the same way and by the same cause. Now the nervous fluid, which becomes containable after its secretion by the blood, is distributed in the albumino-gelatinous substance of the nerve tissue, for this substance is a natural conductor of it and is adapted for holding it and letting it move freely about; the fluid is kept in by the aponeurotic sheaths which invest this nervous tissue, since these sheaths do not permit of the passage of the fluid.

Thereafter, when the nervous fluid is distributed throughout that medullary substance which was originally arranged in separate ganglia and afterwards in a cord, its movements probably thrust out portions which become elongated into threads and it is these threads which constitute the nerves. It is known that they spring from their centre of communication, and issue in pairs either from a ganglionic longitudinal cord or a spinal cord at the base of the brain, and that they then proceed to their termination in the various parts of the body.

This no doubt was the method employed by nature for the formation of the nervous system: she started by producing several small masses of medullary substance when the animal organisation had advanced sufficiently to enable her to do so: she then collected them into one chief mass; through this mass immediately spread the nervous fluid, which had become containable and was kept in by the nervous sheaths: it was then that its movements gave rise to the medullary mass in question, and to the nervous threads and cords which issue from it to the various parts of the body.

In accordance with this theory, nerves cannot exist in any animal unless there is a medullary mass containing their nucleus or centre of communication; hence those isolated whitish threads which do not lead to a medullary mass are not to be regarded as nerves.

I may add to these reflections on the formation of the nervous system, that if the medullary substance has been secreted by the chief fluid of the animal, it is through the agency of the capillary extremities of certain arterial vessels in red-blooded animals; and since the extremities of these arterial vessels must be accompanied by the extremities of venous vessels, all these vascular extremities, containing coloured blood, are buried in the medullary substance which they have produced, and give rise to the greyish colour, which this medullary substance presents in its external layer: sometimes, indeed, as a result of certain evolutions taking place in the encephalon as it develops, the nutritive organs have penetrated so deeply that the greyish medullary substance is central in some localities, and surrounded in great part by that which is white.

I may add further, that if the extremities of certain arteries have secreted and then maintained the medullary substance of the nervous system, these same vascular extremities may likewise have deposited the nervous fluid which separates off from the blood and is continually poured into that medullary substance which is so well adapted to contain it.

Lastly, I shall conclude these reflections by some remarks on the development of the main medullary mass and of the swellings and expansions which are found in certain parts of that mass. These expansions are proportional to the formation and development of the special systems which compose the common and perfected nervous system.

In the main medullary mass of every nervous system, the particular part which to some extent gave origin to the rest, need not necessarily be larger than the other parts which have grown from it; for the thickness and size of these other parts are always dependent on the use which the animal makes of the nerves that issue from them. I have given sufficient proof of this in the case of all the other organs: the more they are exercised the more they become developed, strengthened, and enlarged. It is because this law of animal organisation has not been recognised, or because no attention had been paid to it, that it is believed that the part of the medullary mass which produces the other parts must of necessity be larger than them.

In vertebrates, the main medullary mass consists of the brain and its accessories, the medulla oblongata and the spinal cord. Now it appears that the part of this mass which produced the rest is really the medulla oblongata; for it is from this part that issue the medullary appendages (the peduncles and crura) of the cerebellum and cerebrum, the spinal cord, and, lastly, the nerves of the special senses. Yet the medulla oblongata is in general smaller or less thick than the brain which it has produced, or the spinal cord which proceeds from it.

Whereas, on the one hand, the brain and its hemispheres are employed in acts of feeling and intellect, while the spinal cord only serves for the excitation of muscular movements [1] and the performance of organic functions; and whereas, on the other hand, the continued use or exercise of the organs causes in them a remarkable development; it must follow that in man, who is continually exerting his senses and intellect, the brain and hemispheres should become much enlarged while the spinal cord, which in general is little used, can only acquire moderate dimensions. Finally, since the chief muscular movements of man are those of the arms and legs, we should expect to find a conspicuous swelling in his spinal cord at the exits of the crural and brachial nerves, and this is confirmed by observation.

In those vertebrates, on the contrary, which make but little use of

[1] With regard to the function of the spinal cord in providing nervous influence to the organs of movement, recent experiments have shown that poisons which act on this cord do actually cause convulsions and attacks of tetanus before producing death.

their senses and particularly of their intellect, and which are chiefly given up to muscular movement, the brain and especially the hemispheres should have undergone slight development, whereas the spinal cord is likely to acquire considerable dimensions. Thus fishes, which are largely confined to muscular movement, have a very large spinal cord and a correspondingly small brain.

Among the invertebrates some have a ganglionic longitudinal cord, instead of a spinal cord, throughout their length, such as the insects, arachnids, crustaceans, etc.; because these animals carry out much movement, and the cord is thereby strengthened and swollen where each pair of nerves issues.

Lastly the molluscs, which have only feeble supports for their muscles and generally only carry out slow movements, have no spinal cord nor longitudinal cord, and exhibit nothing more than a few scattered ganglia from which issue nervous threads.

In accordance with this theory, we may conclude that in the vertebrates the nerves and main medullary mass cannot have been developed from above downwards, that is, from the superior terminal part of the brain; any more than the brain itself can be a production of the spinal cord, that is, of the inferior or posterior part of the nervous system; but that these various parts spring originally from one which produced the rest. Probably this one is the medulla oblongata. Some point in the neighbourhood of its annular protuberance must have given origin to the cerebral hemispheres, the cerebellar peduncles, the spinal cord and the special senses.

It matters not that the medullary bases of the hemispheres are narrowed and much less bulky than the hemispheres themselves; and that the same applies to the peduncles of the cerebellum, etc. It is plain to all that the gradual development of these organs, in proportion to their more frequent use, may have caused in them an expansion which makes them much larger than their roots!

These reflections on the formation of the nervous system are doubtless somewhat indefinite; but they suffice for my purpose, and seem to me interesting, because they are accurate and in accordance with the observed facts.

Functions of the Nervous System.

The nervous system of the most perfect animals is, as we know, highly complicated, and may consequently fulfil various kinds of functions, which confer on the animals possessing them as many special faculties. Now before proving that this system is limited to certain animals and not common to all; and before stating what are the faculties conferred by it in the various degrees of complexity of animal organisation, we must say a word about its functions and the faculties resulting from them. They are of four different kinds, viz.:

1. That of instigating muscular activity;
2. That of giving rise to feeling or to the sensations which constitute it;
3. That of producing the emotions of the inner feeling;
4. That, lastly, of forming ideas, judgments, thoughts, imagination, memory, etc.

I shall endeavour to show that the functions of the nervous system which give rise to these four kinds of faculties are very different in character, and that they are not all performed by the animals which possess this system.

The activities of the nervous system which give rise to muscular movement are altogether distinct from and even independent of those which produce sensations: thus we may experience one or more sensations without any muscular movement ensuing, and we may set in action various muscles without any resulting sensation. These facts are worthy of note and they are unquestionably well-founded.

Muscular movement cannot be executed without nervous influence; and although we do not know how this influence works, we are justified by many facts in the belief that it may be by an emission of nervous fluid which starts from a centre or reservoir and travels down the nerves to the muscles which have to be actuated. In this function of the nervous system then, the movements of the subtle fluid which works the muscles take place from some centre or nucleus towards the parts that have to carry out some action.

It is not only to set the muscles in action that the nervous fluid travels from its nucleus or reservoir towards the parts which have to carry out movements; this emission also takes place apparently in order to assist various organs in the performance of functions, where no distinct muscular movement is involved.

Since these facts are well known I shall not dwell further upon them; but shall adopt the conclusion that the nervous influence, which gives rise to muscular activity and which aids various organs in the performance of their functions, works by an emission of nervous fluid which travels from some centre or reservoir to the parts requiring to be actuated.

On this subject I may record a well-known fact that is relevant to the matter now in hand. It is as follows:

With regard to the nervous fluid which leaves its reservoir on its way to the parts of the body, one portion of this fluid is subject to the will of the individual who starts it moving, by means of the emotions of his inner feeling, when stimulated by some requirement; whereas the other portion is regularly distributed, independently of the individual's will, to those parts of the body which have to be kept incessantly in action for the preservation of life.

It would be highly inconvenient if the movements of our heart or arteries, or the functions of our viscera or secretory or excretory organs, were dependent on our will; but it is equally important for the satisfaction of all our requirements that we should have at our disposal some portion of our nervous fluid, for despatching to the regions that we wish to actuate.

It appears that the nerves which continually convey the nervous influence to the vital organs and to the muscles that are independent of the individual, have a firmer or denser medullary substance than the other nerves; or have some other distinguishing peculiarity, as a result of which not only does the nervous fluid move less rapidly and less freely, but is also to a great extent protected from those general agitations caused by the emotions of the inner feeling. If it were otherwise, every emotion would interfere with the nervous influence necessary to the essential organs and vital movements, and would endanger the life of the individual.

Those nerves, on the contrary, which convey the nervous influence to the muscles dependent on the individual, allow to the subtle fluid which they contain every liberty and rapidity of movement, so that the emotions of the inner feeling easily set these muscles in action.

Observation justifies us in the belief that the nerves which serve for the excitation of muscular movement issue from the spinal cord in vertebrates, from the ganglionic longitudinal cord in such invertebrates as have one, and from the separate ganglia in those which have neither a spinal cord nor a ganglionic longitudinal cord. Now these nerves, destined for muscular movement, have no close connection with the sensitive system, in animals which have feeling, and when they are injured they produce spasmodic contractions and do not interfere with the system of sensations.

Hence there are grounds for the belief that, of the various special systems which compose the nervous system at its highest perfection, that which is engaged in muscular excitation is distinct from that which serves for the production of feeling.

Thus the function of the nervous system, which consists in producing muscular activity and the performance of the various vital functions, can only be fulfilled by the dispatch of the subtle fluid of the nerves from the reservoir to the various regions.

But that other function of the nervous system which induces feeling is of quite a different character; for the production of a sensation cannot occur without nervous influence, and it requires that the subtle fluid of the nerves should always travel from the point of the body that is affected towards the nucleus or centre of communication of the system and there start an agitation which affects all the nerves serving for feeling; their fluid then reacts and sensation is produced.

Not only do these two sorts of functions of the nervous system differ, in that there is no sensation produced by any muscular movement and that there is not necessarily any muscular movement for the production of a sensation; but these functions differ also, as we have just seen, by the fact that in one the nervous fluid is driven from its reservoir to the parts, whereas in the other it is driven from the parts to the nucleus or centre of communication of the system of sensations. These facts are manifest, although we cannot witness the movements which cause them.

The function of the nervous system which consists in bringing about emotions of the inner feeling, and which works by means of a general disturbance of the free mass of nervous fluid—a disturbance which is followed by no reaction and therefore produces no distinct sensation—is yet quite peculiar and very different from the two that I have named; in the account that I shall give of it (Chapter IV.) we shall find that its study is very curious and interesting.

Whereas the function, by which the nervous system sets the muscles in action and assists the performance of organic functions, is different from the function by which this system produces feeeling as also from that which constitutes the emotions of the inner feeling, I have now to remark that when the system is sufficiently developed to have obtained that special accessory organ constituted by the wrinkled cerebral hemispheres, it then has the faculty of performing a fourth kind of function, very different from the three others.

Indeed, by means of the accessory organ that I have mentioned, the nervous system gives rise to the formation of ideas, judgments, thoughts, will, etc.; phenomena which assuredly could not be produced by the first three kinds of functions. Now the accessory organ, in which are carried out functions capable of giving rise to these phenomena, is only a passive organ, on account of its extreme softness; and it receives no excitation because none of its parts would be capable of reacting; but it preserves the impressions received, and these impressions modify the movements of the subtle fluids in its numerous parts.

An ingenious idea, though destitute of proof or any adequate basis, has been expressed by Cabanis, who said that the brain acts on the impressions which the nerves conveyed to it as the stomach acts on the food poured in from the oesophagus; that it digests them in its own way, and that when agitated by movement transmitted to it, it reacts and that this reaction gives birth to a perception which thereafter becomes an idea.

This does not appear to me to be based on a study of the faculties of the cerebral pulp; and I cannot convince myself that so soft a substance is really active, or that it can truly be said to react and give rise to perception when agitated by the movements transmitted to it.

This mistake arises, in the first place, from the fact that Cabanis took no note of the nervous fluid, and was obliged in his mind to attribute the functions of that fluid to the nervous tissue in which it moves; and in the second place, from the fact that he confused sensations with intelligence, whereas the nature of these two organic phenomena is essentially different, and demands in each case an individual system of organs for its production.

Thus there are four very different kinds of functions carried on by the perfected nervous system, that is, when it is completely developed and provided with its accessory organ; but seeing that the organs which give rise to these various functions are not the same, and seeing that they have only come into existence successively, nature formed those which are adapted to muscular movement before those which give rise to sensation, and these latter before setting up the means for producing emotions of the inner feeling; she at length completed the perfection of the nervous system by making it capable of producing the phenomena of intelligence.

We shall now see that all animals neither have nor can have a nervous system, and, moreover, that those which possess this system do not necessarily derive from it the four kinds of faculties named.

The Nervous System is limited to certain Animals.

Doubtless it is only in animals that the nervous system can exist; but does it follow thence that they must all possess it? There are certainly many animals whose organisation is such that they could not possibly have this system of organs; for the system consists necessarily of two kinds of parts, viz.: a main medullary mass, and various nervous threads which unite with it. Now this cannot exist in the elementary organisations of a great many known animals. It is obvious, moreover, that the nervous system is not essential to life, since all living bodies do not possess it, and it would be vainly sought among plants. This system, then, can only have become necessary to those animals in which nature was able to establish it.

In Chapter IX. of Part II., p. 273, I have already shown that the nervous system is peculiar to certain animals: I shall now give a further proof of it by showing the impossibility that all animals should possess this system of organs; whence it follows that those which have not got it can enjoy none of the faculties which it produces.

When people have said that, in animals without nervous threads (such as the polyps and infusorians), the medullary substance which yields sensations was distributed and dissolved in every part of the body, instead of being collected into threads; and that from this it followed that each fragment of these animals became an individual endowed with its particular ego; they have probably paid no attention to the invariable characteristic organic function, which is always due to relations between the containing parts and contained fluids, and to the movements resulting from these relations. There was no adequate knowledge of the essential facts with regard to the functions of the nervous system; it was not known that these functions only worked by causing the movement or transport of a subtle fluid, either from a nucleus towards the parts or from the parts towards the nucleus.

The nervous system cannot then exist, nor fulfil the least of its functions, unless it consists of a medullary mass with a nucleus for the nerves, and also of nervous threads which run into this nucleus. Moreover the medullary matter, or any other animal substance, cannot possess in itself the faculty of producing sensations, as I hope to prove in the third chapter of this part; hence this medullary substance when dissolved as alleged in every part of an animal's body would not give rise to feeling.

If the nervous system at its greatest simplicity is necessarily composed of two kinds of parts, viz. a main medullary mass and nervous threads running into it; we may feel how great was the progress required in the complexity of animal organisation, starting from the *Monas*, which is the simplest and most imperfect of known animals, before nature could have attained the formation of such a system of organs even in its greatest imperfection. Yet when this system begins, its complexity and perfection are still very far from what we find in the most perfect animals; and before it could begin, animal organisation had already made much progress in development and complexity.

To convince ourselves of this truth, let us examine the products of the nervous system at its chief stages of development.

The Nervous System in its Simplest State produces nothing but Muscular Movement.

I have, it is true, nothing more than a mere opinion to offer on this subject; but it is based on considerations of such importance and weight that it may at least be regarded as a moral truth.

If the procedure of nature is attentively examined, it will be seen that in the creating or giving existence to her productions, she has never acted suddenly or by a single leap, but has always worked by degrees towards a gradual and imperceptible development: consequently all her products and transformations are everywhere clearly subject to this law of progress.

If we follow the operations of nature, we shall indeed see that she created by successive stages all the tissues and organs of animals, that she gradually brought them to completion and perfection, and that in the same way by slow degrees she modified, animalised, and compounded all the internal fluids of the animals she had brought into existence; so that in course of time they were brought to the condition in which we now see them.

The nervous system at its origin is assuredly in its greatest simplicity and least perfection. This kind of origin is common to it, as to all the other special organs, which also began in their most extreme state of imperfection. Now it cannot be doubted that, in its greatest simplicity, the nervous system gives to the animals possessing it less numerous and lofty faculties than it bestows on the more perfect animals, where it has reached its highest complexity and acquired its accessories. We only have to observe the facts to recognise the truth of this statement.

I have already proved that when the nervous system is in its greatest simplicity, it necessarily has two kinds of parts, viz.: a main medullary mass and nervous threads which run into this mass; but this same medullary mass may at first exist without giving rise to any special sense, and it may be divided into separate parts, to each of which run nervous threads.

Such appears to be the case in animals of the class of radiarians, or at least in those of the division of echinoderms in which a nervous system is supposed to have been discovered; the system would be reduced to separate ganglia, communicating together by threads and sending out others to the parts.

If the observations, which affirm this state of the nervous system, are well-founded, we have here the system in its greatest simplicity. It possesses several centres of communication for the nerves, that is to say, as many nuclei as there are separate ganglia; lastly, it does not give rise to any of the special senses, not even to sight, which is certainly the first to show itself unequivocally.

By special senses I mean those which result from special organs such as sight, hearing, smell, and taste; as to touch, it is a general sense, a type no doubt of all the rest, but needing no special organ and incapable of being yielded by the nerves until they are competent to produce sensations.

When I come to describe in Chapter III. the mechanism of sensations, we shall see that none of them can be produced except when the whole animal shares in the general effect, by reason of the complexity of its nervous system and of the single common nucleus for the nerves. If this is the case, it follows that in animals with the most elementary nervous systems, where there are different nuclei for the nerves, no effect or agitation can become generalised through the individuals, no sensation can be produced, nor can the separate medullary masses give rise to any special sense. If these separate medullary masses communicate together by threads, it is in order to secure the free distribution of the nervous fluid within them.

Yet as soon as the nervous system exists, however simple it may be, it must be capable of performing some function; we may therefore hold that it has an effective action, even when it cannot yet give rise to feeling.

If we reflect that, for the excitation of muscular movement, which is the least of the faculties of the nervous system, a lower degree of complexity and less extent of its parts are required than for the production of feeling, and moreover that separate centres of communication are no bar to the nervous fluid conveying its influence to the muscles from the individual nuclei, it will then appear very probable that the animals with the simplest nervous systems derive from it the faculty of muscular movement, while yet being destitute of feeling.

Thus on starting the nervous system, nature appears to have formed at first only separate ganglia, communicating together by threads and dispatching other threads to the muscular organs. These ganglia are the main medullary masses; and although they communicate by threads, the separation of the nuclei prohibits the general effect necessary for constituting sensation, though it is not opposed to the excitation of muscular movement: hence the animals which possess such a nervous system are devoid of any special sense.

Having now seen that the nervous system in its extreme simplicity can only produce muscular movement, we shall go on to show that when nature has developed, compounded, and further perfected this system, she proceeded to endow it not only with the faculty of exciting muscular action but also with that of producing feeling.

The Nervous System on reaching a Higher Complexity produces both Muscular Movement and Feeling.

Of all the systems of organs the nervous system is doubtless that which confers upon animals the most lofty and the most marvellous faculties; but unquestionably it only reaches this point after having acquired its highest possible complexity and development. Prior to this stage, the animals which have nerves and a main medullary mass present all degrees of perfection in the faculties derived from them.

I have already said that in its greatest simplicity the nervous system appears to have its main medullary mass divided into several separate parts, each of which contains an individual nucleus for the nerves running into it. In this condition the system is not adapted for producing sensations, though it has the faculty of setting the muscles in action: now, does this very imperfect nervous system, which is alleged to have been identified in the radiarians, also exist in the worms? I do not know; and yet there are grounds for the belief, unless the worms are a branch of the animal scale started afresh by spontaneous generation. All I know is that in animals of the class which follow the worms, the nervous system has reached a much higher stage of development, and is quite easy to see and possesses a very definite form.

Indeed, as we follow the animal scale from the most imperfect to the most perfect animals, the first appearance of the nervous system has hitherto seemed to be in the insects; because in all the animals of this class it is very clearly defined, and presents a ganglionic longitudinal cord, which as a rule extends throughout the animal's length and is greatly diversified in shape according to the species of insect and to the state of larva or perfect insect. This longitudinal cord, which ends anteriorly in a subbilobate ganglion, constitutes the main medullary mass of the system, and from its ganglia, which vary in size and proximity, nervous threads proceed to the various parts of the body.

The subbilobate ganglion at the anterior extremity of the ganglionic longitudinal cord of insects has to be distinguished from the other ganglia of the cord, since it gives rise directly to a special sense—that of sight. This terminal ganglion is, then, really a small and very imperfect brain, and doubtless contains the centre of communication of the sensitive nerves, since the optic nerve runs into it. Perhaps the other ganglia of the longitudinal cord are in the same way special nuclei, which provide for the action of the animal's muscles: if these nuclei exist, they would not prevent the general effect which alone, as I have proved, can produce feeling, since they are united by the nervous cord.

Thus in the insects, the nervous system begins to present a brain and single centre of communication for the production of feeling. These animals, by the complexity of their nervous system, possess then two distinct faculties, viz.: that of muscular movement and that of experiencing sensations. These sensations are probably still only simple and fugitive perceptions of the objects which affect them, but they suffice at least to constitute feeling, although incapable of producing ideas.

This state of the nervous system, which gives rise in insects only to these two faculties, is almost the same in the animals of the five following classes, viz.: arachnids, crustaceans, annelids, cirrhipedes, and molluscs; there are apparently no other differences than those involved by a higher development of the two faculties named.

I have not a sufficient number of observations to be able to indicate which of the animals possessing a nervous system, capable of supporting sensations, are liable to experience emotions of their inner feeling. It may be that as soon as the faculty of feeling exists, that which produces emotions arises also; but the origin of the latter is so vague and imperfect that I believe it can only be recognised in vertebrates. Let us then pass on to a determination of the point in the animal scale at which begins the fourth kind of faculty of the nervous system.

When nature had supplied the nervous system with a true brain, that is, with an anterior medullary swelling, capable of giving rise immediately to at least one special sense such as sight, and of containing in a single nucleus the centre of communication of the nerves, she had not yet completed the development of the system. Indeed she long continued to be concerned with the gradual development of the brain, and started the rudiments of the senses of hearing in the crustaceans and molluscs. But it still continues to be a very simple brain, appearing to be the basis of the organ of feeling, since the sensitive nerves and those of the existing special senses proceed to unite with it.

Indeed the terminal ganglion, which constitutes the brain of insects and of the animals of the following classes up to and including the molluscs, although as a general rule divided by a furrow and to some extent bilobate, still shows no trace of the two wrinkled hemispheres, so susceptible of development, which in the most perfect animals cover over the true brain, viz. that part of the encephalon which contains the nucleus of the sensitive system; hence the functions, for which these new accessory organs are adapted, cannot be performed in any of the invertebrates.

The Nervous System in its Complete State gives rise to Muscular Movements, Feeling, the Inner Emotions and Intelligence.

It is only among the vertebrates that nature has arrived at the completion of every part of the nervous system; and it is probably in the most imperfect of these animals (viz. the fishes) that she started the rudiments of the accessory organ of that brain which consists of two wrinkled hemispheres, situated opposite one another but united at their base. The hemispheres are commonly confused with the brain properly so called, but this name should be confined to that region which contains the sensitive centre.

The accessory organ which, when highly developed, confers marvellous faculties on animals, rests upon the brain and covers it over entirely, so as commonly to be confused with it; for as a rule the name of brain is given to the entire medullary mass enclosed in the cavity of the cranium. We should however distinguish between the brain properly so called and its accessory organ, however difficult the distinction may be; for the accessory organ fulfils altogether special functions, and is neither essential to the brain nor even to the maintenance of life. It therefore deserves a special name, and I propose to call it the *hypocephalon.*

Now this hypocephalon is the special organ in which ideas and all acts of intelligence are carried out; no such phenomena could take place in the true brain, viz. that part of the main medullary mass which contains the centre of communication of the nerves, and where also the nerves of the special senses meet.

If we regard the brain as that medullary mass which serves as the meeting place for the various nerves, contains their centre of communication, and, in short, comprises the nucleus from which nervous fluid is dispatched to the various parts of the body and to which it is returned when it evokes any sensation, it will then be true to say that the brain even of the most perfect animals is always very small. But when the brain is provided with two hemispheres which cover it over and are more or less confounded with it, and when these wrinkled hemispheres become very large, it is customary to give the name of brain to the whole medullary mass enclosed within the cranial cavity. Hence the whole of this medullary mass is generally regarded as consisting of only a single organ, whereas the fact is that it contains two, whose functions are essentially distinct.

Not only are the hemispheres special organs accessory to the brain, but they are in no way essential to the existence of the brain; this is placed beyond doubt by many known facts, showing that lesions may occur in them or that they may even be destroyed. With regard to the functions fulfilled by the hemispheres, there must be an emission of nervous fluid, which travels to these organs from its common reservoir or nucleus and enables them to carry out the functions for which they are adapted. Thus we may be sure that it is not the hemispheres themselves which dispatch to the nervous system the special fluid by which it works; for in that case the entire system would be dependent on them; which it is not.

From these principles it follows: that all animals which have a nervous system need not necessarily have a brain, since the latter is characterised by the faculty of giving immediate rise to some sense, at all events the sense of sight; that all animals which have a brain need not also have two wrinkled hemispheres, for the smallness of the brain in the last six classes of invertebrates shows that it can only serve for the production of muscular movement and feeling, and not for acts of intelligence; lastly, that all animals, whose brains are provided with two wrinkled hemispheres, possess the power of muscular movement and of feeling, the faculty of experiencing inner emotions, and, in addition, that of forming ideas, making comparisons and judgments and, in short, of carrying out various acts of intelligence, corresponding to the degrees of development of the hypocephalon.

On paying careful attention to the matter we shall feel that the operations which give rise to thoughts, meditations, etc., occur in the superior and anterior part of the brain, that is, in the two wrinkled hemispheres. We can, moreover, make out that these operations are not carried out either in the base of the brain, or in its posterior and inferior part. The two cerebral hemispheres composing what I call the hypocephalon, are therefore really special organs in which acts of intelligence are produced. Thus when we are thinking and fix our attention too long on one subject, we feel a pain in the head, especially in the part that I have mentioned.

It follows from these various principles that among animals which have a nervous system:

1. Those which have no brain, and consequently no special senses nor single centre of communication for the nerves, do not possess feeling but only the faculty of moving their parts by true muscles;

2. Those which have a brain and special senses, but not the wrinkled hemispheres which constitute the hypocephalon, only derive two or three faculties from their nervous system, viz. those of performing muscular movements, of experiencing sensations, that is to say, simple and fugitive perceptions when any object affects them, and perhaps also of experiencing inner emotions;

3. Lastly, those which have a brain together with its accessory hypocephalon enjoy the capacity for muscular movement, feeling, and emotion, and can moreover, by means of an essential condition (attention) form ideas, which are impressed on the organs, compare these ideas together, and produce judgments; and if their hemispheres are developed and perfected, they can think, reason, invent, and perform various intelligent acts.

No doubt it is very difficult to imagine how the impressions are formed that correspond to ideas; nothing whatever can be seen to indicate their existence in the brain. But surely the only conclusion to be drawn from this, is the extreme fineness of the marks, and the limitations of our own faculties. Will any one say that nothing exists but what man can perceive? It is enough for us that memory is a certain testimony of the existence of these impressions in the brain.

If it is true that nature does nothing suddenly or at a single swoop, she must have created successively all the organs which give rise to the faculties observed in the most perfect animals; and this is just what she has done, with the help of time and favourable conditions.

This assuredly has been her procedure, and we cannot substitute any other for it unless we abandon the positive ideas that we derive from the observation of nature.

Thus in the animal organisation, the nervous system was created in its turn like the other special systems, and this can only have occurred when the organisation was sufficiently developed for the three sorts of substances composing this system to have been formed and deposited in their proper situations.

It is therefore absurd to expect to find this system with its dependent faculties in animals so simply organised and so imperfect as the infusorians and polyps; for it is impossible that such complex organs could exist in these creatures.

Let me repeat that just as the special organs in animals were formed one after the other, so too each of them was gradually compounded, completed and perfected in correspondence with the increasing complexity of organisation; so that the nervous system presents in different animals the three following principal stages.

At its origin, when it is in its highest imperfection, the system appears to consist merely of various separate ganglia, which have communicating threads and from which issue other threads to certain parts of the body: it then shows no brain and cannot give rise either to sight, hearing, or possibly any true sensation; but it already possesses the faculty of exciting muscular movement. Such apparently is the nervous system of the radiarians, if there is any truth in the observations cited in Part I. of this work (Chapter VIII., p. 138).

At its next stage, the nervous system presents a ganglionic longitudinal cord and nervous threads which terminate in the ganglia of that cord: henceforward the ganglion at the anterior extremity of the cord may be regarded as a rudimentary brain, since it gives rise to the organ of sight and subsequently to that of hearing; but this small brain is still simple and has no hypocephalon or wrinkled hemispheres with special functions. Such is the nervous system of insects, arachnids, and crustaceans,—animals which have eyes and in the latter case even some traces of hearing: such again is the nervous system of annelids and cirrhipedes, some of which possess eyes, while others are destitute of them for the reasons named in Chapter VII. of Part I.

The molluscs, although having a higher organisation than the animals just mentioned, are in the midst of a change of plan on nature's part, and have no ganglionic longitudinal cord nor spinal cord; but they have a brain and some of them seem to possess the most perfect of the simple brains, that is, of brains without a hypocephalon: since the nerves of several special senses terminate in them. If this is the case, then the nervous system produces muscular movement and feeling in all animals from the insects to the molluscs inclusive, but it does not permit of the formation of ideas.

Lastly, at a far more perfect stage, the nervous system of vertebrates presents a spinal cord, nerves and a brain, of which the superior and anterior part is provided with two accessory wrinkled hemispheres whose development is proportional to the stage of progress of the new plan. This system then gives rise not only to muscular movement, feeling and inner emotions, but also to the formation of ideas, the clearness and number of which are proportional to the development of the hemispheres.

How can any one suppose that nature, who in all her productions invariably proceeds by gradual stages, could have endowed a nervous system at once with all the faculties which it possesses when it has attained its ultimate completion and perfection?

Moreover, since the faculty of feeling is not the property of any substance of the body, we shall see that the mechanism necessary for its production is so complex that the nervous system in its extreme simplicity could have had no other faculty than that of exciting muscular movement.

I shall endeavour to ascertain in Chapter IV. what is the power that causes and directs the emissions of nervous fluid to the hemispheres or other parts of the body. I shall merely say here that the dispatch of this fluid to the cerebral hemispheres arouses in them functions very different from those aroused in the muscles and vital organs.

I have now given a brief general account of the nervous system, the nature of its parts, the conditions that were required for its formation, and the four kinds of functions that it fulfils when it has attained its perfection.

Without undertaking any enquiry as to how nervous influence may set the muscles in action and cause the performance of their functions by various organs, I may observe that the explanation is probably to be found in a stimulus to the irritability of the parts.

But in the case of that function of the nervous system through which feeling is produced and which is rightly regarded as the most astonishing and difficult to conceive, I shall endeavour to describe the mechanism in Chapter III. I shall then do the same for the fourth function of the nervous system by which it produces ideas, thoughts etc., a function still more extraordinary than that which gives rise to feeling.

Since, however, I do not wish to set forth in this work anything that is not based on adequate facts or observations, I shall first consider the nervous fluid, and show that, so far from being a product of imagination, this fluid is made manifest by effects which cannot be otherwise produced, and which leave not the slightest doubt as to its existence.

CHAPTER II.

OF THE NERVOUS FLUID.

A SUBTLE substance, remarkable for the rapidity of its movements and receiving little attention because it cannot be directly observed, collected, nor experimentally examined; a substance of this character is the very strange and wonderful agent that nature employs for producing the muscular movement, feeling, inner emotions, ideas, and acts of intelligence, which many animals are able to carry out.

Now since we can only know this substance through the effects that it produces, we must begin by discussing it at the outset of the third part of this work; this fluid is the only substance capable of causing the phenomena which so much excite our wonder; and if we refuse to admit its existence and powers, we shall be forced to abandon all search for physical causes for these phenomena and to have recourse once more to vague and baseless theories for the satisfaction of our curiosity.

With regard to the necessity for investigating this fluid by means of its effects, is it not now an admitted fact that there exist in nature various kinds of substances, imperceptible to our senses, that we cannot take hold of, nor collect and examine as we should like; substances so attenuated and so subtle that they can only manifest their existence under certain circumstances, and through the medium of some of their results, which we succeed by careful attention in identifying; substances, in short, whose nature we can only ascertain up to a certain point by means of inductions and analogies, guided by a large number of observations? The existence of these substances is however proved by certain effects which can be produced in no other way; effects which we have to study carefully in various phenomena whose causes we seek.

It may be said by some that since we possess so few means for determining with precision the nature and qualities of these substances, every wise man who is concerned only with exact knowledge should leave them out of account.

I may be mistaken, but I confess that I am of quite a different opinion; I am firmly convinced that these same substances play an important part in most of the physical facts that we observe, and especially in the organic phenomena presented by living bodies; and hence that their investigation is of the greatest importance for the progress of knowledge on these subjects.

Thus, although we cannot know directly all the subtle substances existing in nature, yet if we were to abandon all enquiry with regard to some of them, we should in my opinion be rejecting the only clue that can lead to a knowledge of natural laws; we should be giving up the hope of real progress in our knowledge of living bodies, as also of the causes of the phenomena that we observe in their functions; we should at the same time be relinquishing the only path that can lead to the perfection of our physical and chemical theories.

It will soon be clear that these remarks are not irrelevant to my purpose, and indeed that they are entirely applicable to that nervous fluid about which we so greatly desire information.

Since our observations are now too advanced to permit of any real doubt as to the existence of a subtle fluid which circulates and moves about in the substance of the nerves, let us see how far we are led, on so delicate and difficult a subject, by the actual state of knowledge.

But before speaking of the nervous fluid, it is very important to establish the following proposition:

All the visible fluids contained in an animal's body, such as the blood or substitute for the blood, the lymph, secreted fluids, etc., move too slowly in the canals or parts which contain them, to be capable of conveying with sufficient rapidity the movement or cause of movement which produces actions in animals; for these actions are carried out in many animals with an amazing rapidity and vivacity, and the animals can interrupt them, start them again, and vary them with all possible degrees of irregularity. The slightest reflection should suffice to convince us that it is absolutely impossible that fluids so gross as those just mentioned, and whose movements are usually so regular, can be the cause of the various actions of animals. Yet everything that passes in them results from relations between their contained fluids, or such as penetrate into them, and their containing parts, or the organs affected by these contained fluids.

Assuredly it can only be a fluid, moving almost with the swiftness of light, that could work such effects as those I have named; now we have some knowledge of fluids which possess this faculty.

All action is the product of movement, and the nerves certainly act by a movement of some sort; the opinion of those who regarded the nerves as vibrating cords has been discussed and effectually refuted by M. Richerand in his *Physiologie* (vol. ii., p. 144 *et seq.*). "This hypothesis," he says, "is so absurd that there is reason for astonishment at the favour that it has long enjoyed."

The same thing might well be said of the hypothesis of vibration among molecules so soft and inelastic as those of the medullary substance of nerves, if anyone were to suggest it.

"It is much more reasonable," M. Richerand continues, "to believe that the nerves act by means of a subtle, invisible, impalpable fluid to which the ancients gave the name of *animal spirits*."

Farther on, when examining the special properties of the nervous fluid, this physiologist adds: "Have not these conjectures acquired some degree of probability, since the analogy between galvanism and electricity, originally suggested by their discoverer, has been confirmed by those remarkable experiments of Volta, which are at this moment being repeated, discussed and expounded by all the physicists in Europe?"

However manifest may be the existence of the subtle fluid by means of which the nerves work, there will be for a long time and perhaps for ever, men who dispute it because it cannot be proved except by effects which could not be produced in any other way.

Yet it seems to me that when all its effects unite to demonstrate its existence, it is wholly unreasonable to deny it on the mere grounds that we cannot see the fluids. It is particularly unreasonable to do so, seeing that we know that all organic phenomena result exclusively from relations between moving fluids and the organs concerned. It is still more unreasonable when we remember that the visible fluids (blood, lymph, etc.) which travel to the nerves and brain and penetrate their substance are too gross and move too slowly to be capable of giving rise to actions of such swiftness as those involved in muscular movement, feeling, ideas, thought, etc.

As a result of these reflections, I recognised that in every animal which possesses a nervous system there exists in the nerves and in the medullary nuclei where the nerves terminate, a very subtle, invisible, containable fluid, that is but little known since there are no means for examining it directly. This fluid, which I call nervous fluid, moves with extraordinary rapidity in the substance of the brain and nerves, and yet does not form any visible channels in them for the transmission of its movements.

It is by means of this subtle fluid that the nerves work, that muscular movement is set going, that feeling is produced, and that the cerebral hemispheres carry out those acts of intelligence to which they give rise in proportion to their development.

Although the actual nature of the nervous fluid is little known to us since we can only appreciate it by its effects, yet since the discovery of galvanism, it has become increasingly probable that it is closely analogous to the electric fluid. I am convinced even that it *is* electric fluid, which has been modified in the animal economy and to some extent animalised by its residence in the blood, and which has there undergone sufficient change to have become containable and to remain entirely within the medullary substance of the nerves and brain, to which it is incessantly provided by the blood.

I base this statement, viz. that the nervous fluid is only electricity modified by its residence in the animal economy, on the fact that this nervous fluid, although its effects closely resemble some of those produced by the electric fluid, is yet distinguished from it by some peculiar properties, among which that of being retained within an organ and moving about there, sometimes in one direction and sometimes in another, appears to be characteristic.

The nervous fluid is therefore quite distinct from the ordinary electric fluid, since the latter passes through every part of our body at its usual velocity and without any pause, when we complete a circuit in the discharge of a Leyden jar or electric conductor.

It is different even from the galvanic fluid obtained from Volta's pile: the latter indeed, which is still only electric fluid acting in a smaller quantity, density and activity than the electric fluid of a Leyden jar or charged conductor, derives from its special circumstances certain properties or faculties which distinguish it from the electric fluid collected and condensed by our ordinary methods. This galvanic fluid therefore has more action on our nerves and muscles than the ordinary electric fluid: yet since it is not animalised, that is to say, has not undergone the influence imparted by a residence in the blood (especially of warm-blooded animals), it does not possess all the qualities of the nervous fluid.

The nervous fluid of cold-blooded animals, being less animalised, is more allied to the ordinary electric fluid and especially the galvanic fluid. This is the reason why our galvanic experiments produce very energetic effects on the tissues of cold-blooded animals like frogs; and also why in certain fishes such as the torpedo, the electric eel, and the trembling catfish, a large electrical organ generates electricity which is completely adapted to the animal's needs. (See the interesting Memoir of M. Geoffroy on these fishes in the *Annales du Muséum d'Histoire naturelle*, vol. i., p. 392.)

In spite of the modifications by which the electric fluid is converted into nervous fluid, it still preserves to a great extent its extreme subtlety and rapidity of movement; qualities which render it suitable for performing its functions in the animal.

This electric fluid is incessantly penetrating into the blood, either by respiration or some other method, and is there gradually modified and animalised so as to acquire at length the properties of the nervous fluid. Now we may probably regard the ganglia, spinal cord, and especially the brain with its accessories, as the organs which secrete this animal fluid.

There is indeed reason to believe that the substance of the nerves, which, on account of its albumino-gelatinous nature, is a better conductor of the nervous fluid than any other substance of the body, and far better than the aponeurotic membranes which invest the nervous threads and cords, continually draws off the subtle fluid in question from the minute arteries circulating in it. It is no doubt these minute arteries and veins that give rise to the grey colour of the external or cortical layer of the medullary substance.

In this way, there is incessantly produced in animals with a nervous system the subtle and invisible fluid which moves in the substance of their nerves and in the medullary nuclei where these nerves terminate.

This fluid works in the nerves by two kinds of movements in opposite directions; and the activities of the cerebral hemispheres suggest that it carries out many and varied movements in them that are beyond our powers of ascertaining.

In the nerves which give rise to sensations, we know that the fluid moves from the periphery of the body towards the centre, or rather towards the nucleus that produces these sensations; and since the individuals which have a nervous system may also experience internal impressions, the fluid in these cases moves in the nerves of the internal regions, but still in the direction of the nucleus of sensations.

In the nerves which produce muscular movement, on the contrary, whether voluntary or involuntary, the nervous fluid moves from the centre or common nucleus towards the parts which have to be moved.

In both these cases where the nervous fluid moves in the nerves, and also in the case of its various movements in the brain, the activity of this fluid consumes a certain part of it, which is thereby dissipated and lost. This loss necessitates a restoration, which is continually being made by healthy blood.

The following is a very important observation for understanding the phenomena of organisation:

Individuals, which use their nervous fluid only for the production of muscular movement, make good their losses with interest, so that their strength is increased; since this muscular movement accelerates the circulation and other organic movements, and the secretions which make up for the used fluid are copious at the periods of rest.

Individuals, on the other hand, who use their nervous fluid only in the production of acts depending on the hypocephalon, such as continuous thought, deep meditation, mental agitation produced by passions, etc., make good their losses slowly and often incompletely; for muscular movement is then slight, the organic movements are enfeebled, the organic faculties lose their energy, and the secretions, which make up for the nervous fluid used, become less copious, and mental repose very difficult.

The nervous fluid in the brain does not merely convey sensations from their nucleus, and move about in various ways, but it also stamps impressions on the organs, and these impressions last longer or shorter according to their depth.

This statement is not one of those monstrous products begotten by the imagination: in a brief review of the chief acts of intelligence, I shall endeavour to show that it is well-founded, and that we are obliged to recognise it as one of those truths, that can still only be attained by irrefutable inductions.

I shall conclude my remarks on this singular fluid by some observations which may throw much light over the various organic functions fulfilled by means of this fluid.

All parts of the nervous fluid are in communication, in the system of organs which contains them; so that, according to the causes which excite it, either a part of the fluid may be set in motion, or nearly the whole fluid, or at least such part of it as is free.

The occasions when motion occurs only in parts of it, which may even be quite small, are as follows:

1. When it stimulates muscular activity, either dependent or independent of the will of the individual;

2. When it performs some act of intelligence.

The same fluid, on the contrary, moves as a whole (so far as it is free):

1. When it produces any sensation by some general movement of reaction;

2. Whenever it causes emotions of the inner feeling by a general agitation that does not constitute a reaction.

These details as to the movements of the nervous fluid cannot be proved by individual experiments; at least I do not see how; but the reader will probably think them justified, if he closely follows my observations on the functions of the nervous system in this third part of my *Zoological Philosophy*. In particular, the following facts should be considered:

1. That the nervous influence which moves the muscles only demands a simple emission of a part of the nervous fluid towards the muscles which have to act, and that the fluid here acts only as a stimulus;

2. That, in acts of intelligence, the organ of understanding is only passive and is prevented from reacting by its extreme softness; it acquires no activity from the nervous fluid but merely impressions, of which it preserves the tracings; the part of the fluid, which works in the various portions of this organ, is modified in its movements by the tracings already present, and at the same time traces more; so that the organ of understanding, which has only a narrow channel of communication with the rest of the nervous system, uses only a part of the whole fluid of the system. Moreover this narrow channel of communication preserves that part of the nervous fluid, contained in the organ of intelligence, from the general agitation which occurs in emotions of the inner feeling, and in the formation of sensations, except when that agitation is of great intensity, in which case nearly all the functions and faculties of the system are disturbed.

On these grounds, it appears probable that the entire mass of nervous fluid secreted and contained in the system is not at the disposal of the inner feeling of the individual, but that some of it is held in reserve to provide for the continuance of the vital functions. Hence, just as there are muscles independent of the will, and others which only enter into activity when excited by the inner feeling driven by the will or some other cause; in the same way, no doubt, one part of the nervous fluid is less at the service of the individual than the other, lest the whole fluid should be drained away from the vital functions.

Indeed, since the nervous fluid is never used without a proportional amount of loss, it follows necessarily that the individual is only free to use up a certain part of it: untoward effects ensue even when this part is run too low, for then some of that held in reserve becomes available and the vital functions suffer accordingly.

I shall have further opportunities later on for extending these various remarks on the nervous fluid; but let us first enquire what is the mechanism of sensations, and how the marvellous faculty of feeling is produced.

CHAPTER III.

OF PHYSICAL SENSIBILITY AND THE MECHANISM OF SENSATIONS.

How are we to conceive that certain parts of a living body can possess the faculty of feeling, when no kind of matter whatever does or can enjoy any such faculty!

It was indeed a great mistake to imagine that animals, even the most perfect, were endowed with feeling in certain of their parts. Assuredly the various humours and fluids of living bodies can no more possess the faculty of feeling than the solid parts.

It is only by a real hallucination that the separate parts of our body appear to be sensitive; for it is our entire being that feels or rather undergoes a general effect, on the stimulus of some affective cause. Since this effect is always referred to the part affected, we promptly derive from it a perception to which we give the name of sensation; and we are misled into the belief that it is the affected part of our body that feels an impression, whereas it is a commotion throughout the entire sensitive system which conveys to that part its general effect.

These considerations may appear strange and even paradoxical, for they are far removed from the common opinion on this subject. Yet if the reader will suspend his judgment until he has examined the grounds on which I base my opinion, he will doubtless abandon the idea of attributing the faculty of feeling to any individual part of a living body. But before stating my views, it must first be determined what animals possess the faculty of feeling and what animals have no such faculty.

Let me first enunciate the following principle: Every faculty possessed by animals is necessarily the product of an organic act and consequently of some movement; if the faculty is special, it results from the function of some special organ or system of organs; but no part of the animal body that remains motionless could possibly give rise to any organic phenomenon, not even to the smallest faculty. Hence feeling, which is a faculty, is not a property of any individual part, but the result of an organic function.

I infer from the above principle that every faculty, that arises exclusively from the functions of some especial organ, is confined to animals which possess that organ. Thus just as no animal can see unless it has eyes, so no animal can feel unless it has a nervous system.

It is useless to object that light does make impressions on certain living bodies that have no eyes, and affects them in spite of that deficiency : it still remains true that plants and a number of animals, such as polyps and many others, do not see although they move towards the light ; and that all animals are not endowed with feeling, although they may perform movements when irritated in special ways.

No sort of sensibility (conscious or latent) can, then, be ascribed to animals that have no nervous system, on the mere grounds that they have irritable parts ; I have already shown in Part II., Chapter IV., that feeling and irritability are very different in character, and are due to quite unlike causes. Indeed the conditions required for the production of feeling are of altogether another nature from those necessary for the presence of irritability. The former demand a special organ which is always distinct, complex, and extended throughout the animal's body, whereas the latter demand no special organ and give rise only to an isolated and local phenomenon.

But animals, which possess a sufficiently developed nervous system, possess at the same time their natural irritability and also the faculty of feeling ; they have, without being conscious of it, the intimate feeling of their existence, and though they are still liable to excitations from without, they act by an internal power that we shall shortly examine.

In some, the activities of this internal power are guided by instinct, that is, by inner emotions produced by their needs and habits ; while in others they are guided by a will that is more or less free.

Thus the faculty of feeling is exclusively the property of animals which have a *sensitive* nervous system ; and since it gives rise to the intimate feeling of existence, we shall see that this latter feeling endows animals with the faculty of acting by emotions which cause them internal excitations, and permit them to produce for themselves the movements and actions necessary for the satisfaction of their needs.

But what is physical sensibility or the faculty of feeling ? What again is the inner feeling of existence ? What are the causes of these wonderful phenomena? Lastly, how can the feeling of existence or general inner feeling give rise to a force that produces action?

I have conscientiously considered these matters and the wonders that flow from them, and now state my opinion on the first of these interesting questions.

The faculty of receiving sensations constitutes what I call *physical sensibility*, or feeling properly so called. This sensibility must be distinguished from *moral sensibility* which is quite a different thing, as I shall show, and which is only excited by emotions raised in us by our thoughts.

Sensations arise in us, on the one hand, from the impressions that external objects make on our senses; and, on the other hand, from those made on our organs by disordered internal movements which have an injurious effect and produce internal pains. Now these sensations affect our physical sensibility or faculty of feeling, bring us into communication with the outer world, and acquaint us vaguely with what is happening within us.

Let us now enquire into the mechanism of sensations, and let us begin by showing the harmony which exists in all the parts concerned of the nervous system, and afterwards the result on the entire system of any impression on a single part of it.

Mechanism of Sensations.

Sensations, which by an illusion we refer to the actual places where the impressions that cause them are made, are based upon a system of special organs belonging to the nervous system and called the *system of sensations* or of sensibility.

The system of sensations is composed of two essential and distinct parts, viz.:

1. A special nucleus, that I call the nucleus of sensations, and must be regarded as a centre of communication to which are conveyed all the impressions which act upon us;
2. A large number of simple nerves which start from all the sensitive parts of the body and proceed to their destination in the nucleus of sensations.

It is with this system of organs, whose harmony is such that all or nearly all parts of the body share equally in an impression made on any one part, that nature succeeds in giving, to all animals with a nervous system, the faculty of feeling both what affects them internally and also impressions made upon their senses by external objects.

It may be that the nucleus of sensations is broken up and multiple in animals which have a ganglionic longitudinal cord; yet we may suspect that the ganglion at the anterior extremity of the cord is a small rudimentary brain, since it gives rise immediately to the sense of sight. But in the case of animals which have a spinal cord, there can be no doubt that the nucleus of sensations is simple and indivisible; this nucleus is apparently situated at the anterior extremity of the spinal cord, at the base of what is called the brain and therefore below the hemispheres.

The sensitive nerves arriving from all parts and terminating in a centre of communication, or in several such nuclei connected together, constitute the harmony of the system of sensations, for they spread throughout the system all the impressions, whether isolated or generalised, that the individual may experience.

To form a clear conception of the wonderful mechanism of this sensitive system, we must recall what I said before, viz. that an extremely subtle fluid, whose movements of translation or oscillation are nearly as rapid as light, is contained in the nerves and their nucleus, where it moves freely without escaping.

If we now consider that harmony which unites all parts of the system of sensations, we see that every impression, internal or external, that any individual receives, immediately causes an agitation throughout the system or the subtle fluid contained in it, and consequently throughout the entire body, although it may pass unperceived. Now this sudden agitation promptly gives rise to a reaction, which is brought back from all parts to the common nucleus, and there sets up a singular effect, in short, an agitation, which is thereafter propagated through the one nerve, that does not react, to the point of the body that was originally affected.

Man, who possesses the faculty of forming ideas out of what he experiences, has formed one out of this singular effect produced at the nucleus of sensations and propagated to the point affected, and has given it the name of sensation, in the belief that every part that receives an impression possesses in itself the faculty of feeling. Feeling, however, does not exist anywhere but in the actual idea or perception which constitutes it, since it is not a faculty belonging to any part of our body nor any of our nerves, nor even to the nucleus of sensations, but is purely the result of an emotion of the whole sensitive system which becomes perceptible in some point of our body. Let us examine in further detail the mechanism of this singular effect of the system of sensibility.

In the case of animals which have a spinal cord, there start from every part of their body, the most deeply situated as well as the most superficial, nervous threads of extreme fineness, which without any division or anastomosis proceed to the nucleus of sensations. Now these threads, in spite of junctions which they form with others, travel without any discontinuity to their nucleus, always retaining their individual sheaths. This does not prevent the nervous cords, which arise from the junctions of several such threads, from having their own sheath as well; and the same applies to the still larger cords formed by the union of several of these.

Each nervous thread may thus be distinguished by the name of the part where it starts, for it only transmits impressions made on that part.

We are here dealing exclusively with nerves serving for sensations: those destined for muscular movement start apparently from some other nucleus, and constitute a special system within the nervous system, distinct from that of sensations; in the same way that the latter is distinct from the system providing for the formation of ideas and acts of the understanding.

It is true that in consequence of the close connection between the system of sensations and that of muscular movement, paralysis usually extinguishes both feeling and movement in the parts affected; nevertheless cases are seen where sensibility is quite extinct in certain parts of the body which still possess freedom of movement,[1] and this proves that the systems of sensation and movement are really distinct.

The special mechanism which constitutes the organic act giving rise to feeling, consists therefore in the following process:

When an impression is received at the extremity of a nerve, the movement thereupon set up in the subtle fluid of that nerve is transmitted to the nucleus of sensations, and from there to all the nerves of the sensitive system. But the nervous fluid immediately reacts from all the nerves together, and brings back this general movement to the common nucleus, where the only nerve which brought no reaction receives the entire product of all the rest and transmits it to the point of the body originally affected.

For greater clearness let us take a special example of the details of this mechanism.

[1] M. Hébréard records in the *Journal de Médecine, de Chirurgie et de Pharmacie* that a man fifty years of age had suffered, ever since he was fourteen, from an absolute insensibility in the right arm. Yet this limb retained its activity, size and usual strength. A phlegmon grew upon it, causing heat, swelling and redness, but no pain even when squeezed or pressed.

While working, this man fractured the bones of his fore-arm in their lower third. At first he only heard a crack, and thought that he had broken the spade which he held in his hand; but it was intact, and he only discovered his accident, because he could not continue his work. The next day the site of the fracture had swollen, and the temperature of the fore-arm and hand had risen: yet the patient experienced no pain even during the extensions necessary for reducing the fracture, etc.

The author concludes from this fact and from similar experiences by other doctors, that sensibility is absolutely distinct and independent from contractility, etc., etc. (*Journal de Médecine pratique*, 15th June, 1808, p. 540).

If I am pricked in the little finger of one of my hands, the nerve of the part affected runs in its special sheath and without communication with others to the common nucleus, where it delivers the agitation it has received; this agitation is immediately propagated thence to the fluid in all the other nerves of the sensitive system: then by a true reaction or repercussion this agitation flows back from all points to the common nucleus, where it produces a shock and a compression of the agitated fluid on all sides save one. This total effect produces a perception, the result of which is carried back by the single nerve which did not react.

In point of fact, the nerve which brought in the original impression, and thus set up the agitation of fluid in all the rest, is the only one which gives no reaction; for it is the only one that is active while all the rest were passive. The whole effect of the shock produced in the common nucleus and passive nerves, as also the resulting perception, must therefore be carried off by this active nerve.

An effect of this kind resulting from a general movement throughout the individual, necessarily advertises him of an event passing within; and this individual, though he cannot make out any details, derives from it a perception to which he gives the name of sensation.

The strength of this sensation is presumably proportional to the intensity of the impression, and its character would correspond with the actual nature of the impression received; it appears to be produced in the very part affected, simply because the nerve of that part is the only one affected by the general disturbance that is set up.

Thus every shock produced in the nucleus or centre of communication of the nerves, due to an impression received, is felt throughout the body though always appearing to us to take place in the very part which received the impression.

With regard to this impression, there is necessarily an interval between the moment at which it is made and that when sensation is produced, but this interval is so short, on account of the rapidity of the movements, that it is impossible for us to perceive it.

Such, in my opinion, is the wonderful mechanism underlying physical sensibility. Let me repeat that it is not matter that feels, for matter has no such faculty; it is not even a part of the individual's body, for the sensation experienced in any such part is only an illusion, as many facts combine to prove; but it is a general effect, produced throughout the body, which is entirely concentrated on the same nerve that first caused it, and which is necessarily felt by the individual at the extremity of that nerve where the impression was made.

All that we perceive is within ourselves: this is now a well-established truth. For a sensation to arise, it is absolutely necessary that the impression received by the part affected should be transmitted to the nucleus of the system of sensations; but if the whole action ended there, there would be no general effect, and no reaction would be conveyed to the point which received the impression. As regards the transmission of the original movement impressed, it doubtless only takes place through the nerve which was affected and by means of the nervous fluid moving in its substance. We know that, if by a ligature or tight compression of the nerve, we intercept communication between the portion arriving from the part affected and that passing on to the nucleus of sensations, no transmission of movement is effected.

The ligature or tight compression interrupts the continuity of the soft substance of the nerve by binding together the walls of its sheath, and thus suffices to intercept the passage of the nervous fluid; but, as soon as the ligature is removed, the elasticity of the nervous substance permits of restoration of continuity in the nerve, and sensation can then be produced again.

Although therefore it is true that we only feel what is within ourselves, yet the perception of the objects which affect us does not occur, as has been held, in the nucleus of sensations, but at the extremity of the nerve which received the impression; and all sensation is thus actually felt in the part affected, because it is there that the nerve of this part terminates.

But if this part no longer exists, the nerve, which ran there, continues to exist although it is shortened; if therefore this nerve receives an impression a sensation is experienced which appears by illusion to be in the part that is no longer possessed.

It has been observed that people who have lost a leg, feel, when the stump has healed, at changes of the weather, pains in the foot or leg which they no longer possess. It is obvious that in these individuals there is an error of judgment as to the actual site of their sensation; but this error is due to the fact that the nerves affected were just those which were originally distributed in the foot or leg of these individuals; in reality the sensation is produced at the extremity of the shortened nerve.

The nucleus of sensations only serves for the production of the general disturbance set up by the nerve which received the impression, and for bringing back into this nerve the reaction from all the rest; hence there results at the end of the affected nerve an effect, which all parts of the body combine to produce.

Cabanis seems to have had some notion of the mechanism of sensations, for although he did not work out the principles of it clearly, and although he wrongly suggests a mechanism analogous to that by which the nerves excite muscular action, yet he obviously had a general idea of what actually occurs in the production of sensations; he expresses himself as follows:

"The operations of sensibility may be regarded as being made up of two phases. In the first place the extremities of the nerves receive and transmit the original impression either to the entire sensitive organ, or, as we shall see later, to one of its isolated systems; the sensitive organ then reacts on the nerve endings, so that the sensibility, which in the first phase seems to have travelled from the periphery to the centre, returns in the second phase from the centre to the periphery; and the nerves, to put it briefly, exert a real reaction on themselves for producing feeling, just as they do on the muscular parts for producing movement." (*Rapports du Physique et du Moral*, vol. i., p. 143.)

The only deficiency in the above statement is the omission to state that the nerve at whose extremity the original impression was received is the only one which does not subsequently react; and that for this reason the general reaction from the other nerves of the system, on reaching the common nucleus, is necessarily transmitted into the only nerve that is at the moment in a passive state, and thus conveys to the point first affected the general effect of the system, that is to say, sensation.

As to the statement of Cabanis that the nerves exert a similar reaction on the muscular parts for setting them in motion, I believe that this comparison between two very different acts of the nervous system has no true foundation; but that a simple emission of the fluid of the nerves from its reservoir to the muscles is a sufficient explanation: there is no necessity to assume any nervous reaction.

I shall conclude my observations on the physical causes of feeling by the following reflections, for the purpose of showing that it is a mistake to confuse the perception of an object with the idea that may be called up by the sensation of that object, and also to imagine that every sensation necessarily yields an idea.

To experience a sensation and to distinguish it are two very different things: the former without the latter constitutes only a simple perception; whereas the latter, which is never found apart from the former, alone gives rise to the idea.

When we feel a sensation from some external object and distinguish that sensation, although we only feel what is within ourselves and although we have to make one or more comparisons in order to differentiate the object from our own existence and form an idea of it, we carry out almost simultaneously by means of our organs two acts of essentially different kinds, one of which makes us feel while the other makes us think. We shall never succeed in disentangling the causes of these organic phenomena, so long as we confuse the entirely distinct facts which constitute them, and fail to recognise that they cannot both have a common origin.

A special system of organs is certainly needed for producing the phenomenon of feeling, since this is a faculty peculiar to certain animals and not general for all. So too a special system of organs is necessary for carrying out acts of the understanding; for thought, comparison, judgment, reasoning are organic acts of a very different character from those producing feeling. Hence, when we think, we do not feel any sensation, although our thoughts impress the inner feeling or ego of which we are conscious. Now since all sensation arises from a special sense, it follows that the consciousness of one's thoughts is not a sensation, but differs radically from it, and must be kept distinct. In the same way, when we feel a simple sensation that constitutes perception, and thus passes unnoticed, no idea of it is formed and no thought produced, so that the sensitive system alone is in action. We may therefore think without feeling, and feel without thinking. Hence for each of these two faculties there is a separate system of organs; just as there is a separate system for movements which is independent of the other two, although one or other of them is the remote cause which sets the latter in action.

Thus it is wrong to confuse the system of sensations with the system that produces acts of the understanding, and to imagine that the two kinds of organic phenomena arising from them can be the result of a single system of organs. This is why men of the highest capacity and knowledge have been mistaken in their arguments on subjects of this nature.

"A creature," says M. Richerand, "absolutely destitute of sensitive organs would have a purely vegetative existence; if it acquired one sense it still would not possess any understanding, since, as Condillac has shown, the impressions produced on this single sense could not be compared; it would have nothing more than an inner feeling of its existence, and it would believe that all things which affect it are part of itself." (*Physiologie*, vol. ii., p. 154.)

We see from this quotation that the senses are considered, not merely as sensitive organs, but also as organs of the understanding; since, if instead of a single sense the creature had several, then according to the received opinion the mere existence of these senses would endow it with intellectual faculties.

There is even a contradiction in the passage above cited; for it is there stated that a creature which had only one sense would still not possess understanding, and farther on it is said, with reference to the impressions which it would experience, that the whole effect would be limited to an inner feeling of its existence, and it would believe that all things which affect it are part of itself. How can this being think or form judgments, if it possesses no understanding? For to believe anything is to form a judgment.

As long as we fail to distinguish the facts connected with feeling from those connected with intelligence, we shall often make mistakes of this character. It is an established fact that there are no innate ideas, but that every simple idea arises exclusively from a sensation. But I hope to show that not every sensation produces an idea; indeed that it need cause no more than a perception, and that for the production and impression of a permanent idea a special organ is needed, as well as compliance with a certain condition not involved in the organ of sensations.

It is a long way from a simple perception to an impressed and permanent idea. Indeed no sensation, which causes only a simple perception, makes any impression on the organ; it does not need the essential condition of *attention*, and can do no more than excite the inner feeling of the individual and give it momentary perceptions of objects, without the production of any thought. Moreover memory, whose seat can only be in the organ where ideas are traced, can never bring back a perception which did not penetrate to this organ, and therefore left no impression on it.

I regard perceptions as imperfect ideas, always simple, not graven on the organ and needing no condition for their occurrence; and this is a very different state of affairs from what prevails in the case of true and lasting ideas. Now these perceptions, by means of habitual repetitions which cut out certain channels for the nervous fluid, may give rise to actions which resemble those of memory. Examples are furnished us by the manners and habits of insects.

I shall hereafter revert to this subject; all that I had here to remark was the necessity of distinguishing perception, which results from every unnoticed sensation, from an idea, which, as I hope to show, requires a special organ for its formation.

From the foregoing principles I think we may conclude:

1. That the phenomenon of feeling is not more miraculous than any other phenomenon in nature, that is, any phenomenon produced by physical causes;

2. That it is not true that any part of a living body, or any substance composing it, has in itself the faculty of feeling;

3. That feeling is the result of an action and reaction, which become general throughout the nervous system, and are performed with rapidity by a very simple mechanism;

4. That the general effect of this action and reaction is necessarily felt by the individual's indivisible ego, and not by any separate part of his body; so that it is only by illusion that he thinks that the effect is entirely wrought at the point where the impression was received;

5. That whenever an individual notices a sensation, identifies it, and distinguishes the point of its body on which it takes effect, the individual then has an idea, thinks, carries out an act of intelligence, and must therefore possess the special organ for producing it;

6. Lastly, that, where the system of sensations exists without the system of understanding, the individual performs no act of intelligence, has no ideas, and only derives from its senses simple perceptions which it does not notice, although they may arouse its inner feeling and make it act.

Let us now endeavour if possible to form a clear idea of the emotions of the inner feeling of any individual possessing physical sensibility, and to ascertain what power this individual gains from it, for carrying out its actions.

CHAPTER IV.

OF THE INNER FEELING, THE EMOTIONS THAT IT MAY EXPERIENCE, AND THE POWER WHICH IT THENCE DERIVES FOR THE PRODUCTION OF ACTIONS.

My purpose in the present chapter is to treat of one of the most remarkable faculties conferred by the nervous system on all animals in which it is highly developed; I mean that singular faculty, with which certain animals and man himself are endowed, consisting in the capacity to experience inner emotions called forth by the needs and various causes external or internal; from this faculty arises the power for performing diverse actions.

Nobody, so far as I know, has yet considered the interesting subject with which I am about to deal; yet, unless we fix our attention on this subject, we shall never be able to account for the numerous phenomena presented by animal organisation, which have their origin in the aforementioned faculty.

We have seen that the nervous system consists of various organs which are all in communication with one another; consequently every part of the subtle fluid contained within this system is in communication with every other part, and is therefore liable to undergo a general agitation in the presence of certain causes adapted for exciting it. We have here an essential principle that we must keep in mind throughout our enquiries; its accuracy cannot be questioned since proofs of it are furnished by the observed facts.

Yet the whole of the nervous fluid is not always sufficiently free to take part in the agitation under consideration, for in ordinary cases it is only one portion of the fluid (no doubt a large portion) that is capable of undergoing this agitation when aroused by certain emotions.

It is certain that in various circumstances, the nervous fluid undergoes movements in more or less isolated parts of its mass: portions of this fluid are thus dispatched to the various parts for muscular action, and for the vivification of organs, without the whole fluid being set in motion ; in the same way portions of a fluid may be agitated in the cerebral hemispheres, while the rest of the fluid remains still : these are truths that cannot be called in doubt. But if it is manifest that the nervous fluid may receive movements in a certain part of its mass, it is no less so that special causes may agitate almost the entire mass of this fluid, since all its parts are in communication. I say *almost* the entire mass, because in the ordinary inner emotions, that portion of the nervous fluid which serves for the excitation of muscles independently of the individual, and often also that portion contained in the hemispheres of the brain, are sheltered from the agitations which constitute emotions.

The nervous fluid may therefore undergo movement in its separate parts, or as a whole ; it is these latter movements which constitute the general agitations of the fluid, that we are about to discuss.

General agitations of the nervous fluids are of two kinds, viz. :

1. Partial agitations which become general and end by reactions ; it is this kind of agitation which produces feeling. We have dealt with it in Chapter III. ;

2. Agitations which are general from the first and form no reaction : it is these which constitute the inner emotions, and it is exclusively with them that we shall now deal.

But a word must first be said about the feeling of existence, since it is in this feeling that the inner emotions take their rise.

Of the Feeling of Existence.

The feeling of existence, which I shall call *inner feeling*, so as not to imply a universality which it does not possess, is a very obscure feeling possessed by animals whose nervous system is sufficiently developed.

This feeling, obscure as it is, is none the less very powerful ; for it is the origin of the inner emotions, and consequently of that singular force which enables individuals to produce for themselves the movements and actions which their needs demand. Now this feeling, regarded as a very active motive power, works simply by dispatching to the required muscles the nervous fluid which is to excite them.

The feeling in question is now well recognised, and results from the confused assembly of inner sensations which are constantly arising throughout the animal's life, owing to the continual impressions which the movements of life cause on its sensitive internal parts.

Indeed, as a result of the organic or vital movements taking place in every animal, those which possess an adequate nervous system derive from it physical sensibility and are incessantly being affected by impressions throughout their sensitive internal parts; the animal feels these all confused together, without being able to distinguish any one.

As a matter of fact, all these impressions are very weak; and although they vary in intensity according to the health of the individual, they are usually very difficult to distinguish, because they are not liable to any sudden interruption or renewal. Yet the sum-total of these impressions, and the confused sensations resulting from them, constitute in all animals subject to them a very obscure but real inner feeling that has been called the feeling of existence.

This continuous and intimate feeling, which is not appreciated because it is not noticed, is general, in that all the sensitive parts of the body share in it. It constitutes that ego with which all animals that are merely sensitive are imbued without knowing it, while those which also possess an organ of intelligence may notice it, since they have the facuties of thought and attention. Finally, it is in both cases the source of a power, which the needs can evoke, which only acts effectively by emotion, and from which movements and actions derive their motive force.

The inner feeling may be considered under two quite different aspects viz.:

1. In so far as it is the result of obscure sensations which continue without cessation in all sensitive parts of the body: from this aspect I call it simply inner feeling;

2. In its faculties: for the general agitations of the subtle fluid confer upon animals the power of producing movements and actions for themselves.

This feeling, which forms a very simple whole, is susceptible of being stirred by various causes. When it is aroused, it can excite movements in the free portions of the nervous fluid, direct those movements, and convey the stimulating fluid to one or another muscle or to any part of the cerebral hemisphere; it then becomes a power for exciting action or thought. From this second aspect therefore we may regard the inner feeling as the fountain from which the force that produces actions derives its energy.

It is necessary for a comprehension of the phenomena produced, to examine this feeling from the two aspects named above; for by its very nature as a feeling of existence it is always in action throughout the waking period; and by its faculties it gives rise to a force which causes action.

Finally, the inner feeling only manifests its power and produces actions when there exists a system for muscular movement: this system is always dependent on the nervous system and cannot exist without it.

It would therefore be inconsistent to look for muscles in animals which obviously had no nervous system.

Let us now turn to a discussion of the chief facts about the emotions of the inner feeling.

Of the Emotions of the Inner Feeling.

We now have to investigate one of the most important phenomena of animal organisation, viz. those emotions of the inner feeling which lead to action in animals and even in man, sometimes without any effort of their will; emotions long ago recognised, but as to the origin or causes of which no one seems to have paid any attention.

Observation leaves no room to doubt that the general inner feeling experienced by animals which possess the requisite nervous system, is susceptible of being stirred by causes which affect it; now these causes always consist in the need either of assuaging hunger or of flying from danger, or avoiding pain, or seeking pleasure, or what is agreeable to the individual, etc.

The emotions of the inner feeling can only be known by man, since he alone can notice and mark them; but he only perceives those which are strong, and which agitate, so to speak, his whole being; close attention and thought is necessary before he can recognise that he experiences them in all degrees of intensity, and that it is exclusively the inner feeling that under various circumstances stirs up in him those inward emotions which lead him to the execution of some action.

I have already said, at the beginning of this chapter, that the inner emotions of a sensitive animal consist in certain general agitations of all the free parts of the nervous fluid, and that since the agitations are followed by no reactions they produce no distinct sensation. Now we may easily imagine that, when these emotions are weak or moderate, the individual may dominate them and control their movements; but that when they are sudden and powerful he is then mastered by them: this is a very important consideration.

The positive fact of the existence of these emotions is no mere hypothesis. Who has not noticed that a loud and unexpected noise makes us start or give a sort of jump, and execute corresponding movements that our will had not determined?

Some time ago I was walking in the street with my left eye covered by my pocket-handkerchief, because it was in pain and irritated by the sunlight. A horse and rider that I did not see suddenly fell, quite close to me on my left: I instantly found myself transported two steps towards the right by a movement or bound, in which my will had not the smallest share, and before I had any idea of what was happening close to me.

Everybody is acquainted with these kinds of involuntary movements, for everybody has experienced the like; and they are only noticed because they are extreme and sudden. But we pay no attention to the fact that everything, which affects us, stirs our emotions proportionally, that is to say, stirs to some extent our inner feeling.

We are moved by the sight of a precipice, of a tragic scene, either real or on the stage or even on a picture, etc., etc.: and where is the power of a fine piece of music well executed, if not in producing emotions of our inner feeling? Consider again the joy or sorrow that we suddenly feel on hearing good or bad news about something that interests us: what is it but the emotion of that inner feeling, which is very difficult to master on the spot?

I have several times seen pieces of music played on the piano by a young lady who was deaf and dumb: her playing was far from brilliant and yet it was passable; she kept good time, and I perceived that her entire personality was stirred by regular movements of her inner feeling.

I gathered from this that the inner feeling in this young person took the place of the organ of hearing, which was of no use. Her music-master too told me that he had practised her in keeping time by measured signs, and I soon became convinced that these signs had stirred within her the feeling in question; hence I inferred that what we attribute entirely to the highly trained and delicate ears of good musicians belongs rather to their inner feeling, which from the first bar is stirred by the kind of movement necessary for the performance of a piece.

Our habits, temperament, and even education, modify this faculty of undergoing emotion; so that in some individuals it is very weak while in others it is very strong.

A distinction should be drawn between the emotions aroused by the sensation of external objects and those which come from ideas, thoughts, or acts of our intellect; the former constitute *physical* sensibility, whereas the latter characterise the *moral* sensibility, that we shall now turn to consider.

Moral Sensibility.

Moral sensibility is very different from the physical sensibility that I have already mentioned; the former is only excited by ideas and thoughts which move our inner feeling, while the latter only arises from impressions produced on our senses; for these can likewise stir our inner feeling.

The seat of moral sensibility has been wrongly supposed to be in the heart, on the ground that the functions of that viscus are more or less affected by the different manifestations of sensibility. But in reality moral sensibility is nothing more than a very delicate susceptibility to emotion, which is aroused in the inner feeling of certain individuals on the sudden presentation of suitable ideas and thoughts. These individuals are then said to be highly sensitive.

This sensibility, when developed to the extent attainable to a perfected intellect, appears to me a product and even a benefaction of nature. It then forms one of the finest qualities of man, for it is the source of humanity, kindness, friendship, honour, etc. Sometimes, however, circumstances make this quality almost as baneful to ourselves as its presence in others is beneficial: in order to derive from it the benefits without the disadvantages, we simply have to moderate its transports by methods which nothing but a good education can indicate.

A good education shows us the necessity on innumerable occasions for repressing our sensibility up to a certain point, in order not to fall short in the respect which man in society owes to his fellows, as also to the age, sex, and rank of those with whom he consorts: hence result that decorum and amenity in the expressions used in conversation, in short, that careful restraint in the expression of ideas which gives pleasure without ever wounding, and confers a quality of high distinction to those that possess it.

Up to this point our conquests can only turn to the general advantage. But the limits are sometimes passed; and we abuse the power, given us by nature, of stifling to some extent the finest of the faculties that we draw from her.

Indeed, many men possess certain propensities which lead them constantly to resort to dissimulation; it has become necessary for them habitually to restrain the emotions of their inner feeling, and carefully to hide their thoughts and such of their actions as may lead to the end they have in view.

Now since every faculty, that is not used, gradually degenerates until it almost becomes extinct, the moral sensibility which we are here discussing is almost absent in them; and they do not even esteem it in persons who still possess it in a moderate degree.

Just as physical sensibility is only brought about by sensations which, when they give birth to some need, immediately produce an emotion in the inner feeling and thus drive the nervous fluid to the muscles which have to act; so too moral sensibility is only brought about by the emotions which thought produces in this inner feeling; and when the will, which is an intellectual act, determines some action the feeling aroused by that act guides the nervous fluid towards the suitable muscles.

The inner feeling acquires its various emotions through one or other of two very different routes, viz.: that of thought and that of physical feeling or sensations. The emotions of the inner feeling may therefore be divided:

1. Into moral emotions, such as those raised by certain thoughts;
2. Into physical emotions, such as those derived from certain sensations.

Since, however, the results of the first kind of emotion belong to the sphere of moral sensibility, while those of the second kind depend on physical sensibility, it is enough to abide by the first distinction that we established.

Nevertheless, I shall take this opportunity of making a few remarks which seem to me not without interest.

A moral emotion when very powerful may temporarily extinguish physical feeling, disturb the ideas and thoughts, and cause some enfeeblement in the functions of several of the most essential organs.

It is known that distressing news, when unexpected, as also news which causes extreme joy, produces emotions whose consequences may be of this kind.

It is also known that among the minor effects of these emotions are digestive troubles or pains; and that, in the case of elderly people, when the emotions are at all strong, they may be dangerous and even fatal.

Lastly, the power of the moral emotions is so great as often to dominate physical feeling. Fanatics, for instance, are people whose moral feeling is so exalted as to overcome the impressions of the tortures, which they are forced to undergo.

Although as a general rule the moral emotions are more powerful than the physical emotions, yet the latter when very strong may also disturb the intellectual faculties; they may even cause delirium and throw the organic functions out of order.

I shall conclude these remarks with a reflection that I believe to be well-founded, viz.: that the moral feeling exercises in course of time a greater influence on the organisation than the physical feeling is capable of working.

Indeed, almost any disturbance may be produced in the organic functions, especially in the abdominal viscera, by deep and prolonged sorrow.

Cabanis held that individuals, who are always sad and melancholy, often without any real cause, are victims of a sort of degeneration in the abdominal viscera; whence he concluded that the melancholy of these individuals must be attributed to this kind of degeneration, and therefore that these viscera contribute to the formation of thought.

It seems to me that this savant has pushed too far the conclusions to be drawn from his observations.

No doubt any degeneration of organs, and especially of the abdominal viscera, frequently corresponds to degeneration of the moral faculties and does actually contribute to it. But this condition, in my opinion, cannot be regarded as assisting in the formation of thought; it merely gives the individual a propensity towards some one kind of thoughts rather than some other.

Now it is known that the moral feeling may act strongly on the state of the organs and may affect them for a long period in some particular way; it thus appears to me that constant and real griefs in any individual may set up degeneration of the abdominal viscera; and that these degenerations, once started, may in their turn perpetuate the inclination to melancholy, even when there is no longer any real cause for it.

Reproduction, indeed, may transmit a tendency of the organs or a state of the viscera adapted for giving rise to any special temperament, inclination, or characteristic; but it is essential that circumstances should favour the development of this tendency in the new individual; for otherwise the individual would acquire another temperament, inclinations, and characteristics. It is only in animals of very low intelligence that reproduction transmits almost without variation the organisation, inclinations, habits and special peculiarities of each race.

I should travel too far afield if I were to discuss these matters further; I now therefore return to my subject.

I may sum up my observations on the inner feeling, by saying that this feeling is the source of movements and actions for beings that possess it; these occur, either when sensations, which awake needs, arouse emotions in it, or when thought similarly awakes a need or points out a danger, etc., and thus stirs it more or less strongly. These emotions, however aroused, promptly agitate the available nervous fluid: and since every need that is felt, causes the resultant emotion to affect the parts which have to act, movements are invariably carried out by this method, and are always in relation to the individual's needs.

Lastly, since these inner feelings are very obscure, the individual in whom they are working does not perceive them; yet they are real; and if man, who has a highly developed intellect, were to pay them some attention, he would soon recognise that all his acts are due to emotions of his inner feeling, some of which are inspired by ideas, thoughts, and judgments, thus making him feel certain needs, and exciting his will to action; while the others result immediately from sudden and pressing needs, and cause him to carry out actions quite independently of his will.

I may add that, since the inner feeling may produce the agitations in question, it follows that if the individual governs the emotions of that inner feeling, he may repress or moderate them and even prevent their effects. This is how the inner feeling of any individual is a power, which leads him to act according to his needs and customary inclinations.

But when the emotions in question are so powerful as to cause an agitation in the nervous fluid sufficient to affect the movements of the portion contained in the cerebral hemispheres and also that which controls the involuntary muscles, the individual then loses consciousness, and suffers from syncope ; and the functions of his vital organs are more or less deranged.

Here then we seem to have reached great truths, which the philosophers could never discover because they have not sufficiently observed nature, and which the zoologists did not perceive because they were too much occupied with matters of detail. We may at all events affirm that the physical causes, indicated above, would be capable of accounting for the phenomena of organisation into which we are enquiring.

In compliance with the ordinary rules for the exposition of ideas, I now have to establish a fundamental distinction of the first importance : I have already said that the inner feeling derives its emotions from two quite different causes :

1. From some intellectual operation culminating in an act of will ;

2. From some sensation or impression, which causes a need to be felt or a propensity to be followed, independently of the will.

These two kinds of causes, which stir the inner feeling of the individual, show that there is really a distinction to be drawn among the factors which control the movements of the nervous fluid in the production of actions.

In the first case, the emotion arises from an act of intelligence, that is to say, from a judgment, which determines the will to act ; and the emotion then directs the movements of the available nervous fluid in the direction impressed on it by the will.

In the second case, on the contrary, the intelligence has no share in the emotion of the inner feeling ; and this emotion directs the movements of the nervous fluid in the direction demanded by the needs, to which the sensations have given rise, and in the direction of the acquired propensities.

There is another fact to be named no less important than the foregoing: it is that the inner feeling may be suspended either entirely or in part.

During sleep, for instance, the feeling is suspended or nearly so; the free part of the nervous fluid is in a kind of repose, and no longer undergoes a general agitation; nor does the individual any longer enjoy his feeling of existence. The system of sensations is not exercised, and none of the actions dependent on the individual are carried out; the muscles necessary for their production are no longer excited, but fall into a kind of relaxation.

If sleep is imperfect, and if there exists some cause of irritation which agitates the free part of the nervous fluid, particularly that in the cerebral hemispheres, the inner feeling (being suspended) no longer directs the movements of the fluid in the nerves, and the individual is then abandoned to dreams, that is, to involuntary recurrences of his ideas following one another in characteristic disorder and confusion.

In the waking state, the functions of the inner feeling may be greatly disturbed either by too strong an emotion which cuts off the nervous fluid as it flows towards the involuntary muscles, or by some deep irritation affecting chiefly the fluid of the brain. It then ceases to direct the movements of the nervous fluid; syncope supervenes if the trouble comes from a strong emotion, or delirium if it is caused by severe irritation or some act of lunacy, etc., etc.

In accordance with the above principles, it seems to me obvious that the inner feeling is the only factor that produces actions in man and such animals as possess it; that this feeling only works when prompted by its emotions; that it is moved sometimes by acts of intelligence, sometimes by a need or sensation acting suddenly upon it; that in men, its weak emotions may be controlled since intelligence is highly developed, but that in animals this can only be done with great difficulty and never in those that lack intelligence; that its functions are suspended during sleep, and that then it no longer directs the movements of the free part of the nervous fluid; that its functions may also be interrupted and disturbed during the waking state; finally, that it is the product, on the one hand, of the individual's feeling of existence, and, on the other hand, of the harmony in the parts of the nervous system, as a result of which the free portions of the subtle nervous fluid are all in communication and capable of undergoing a general agitation.

On the same principles, it also appears to me obvious that moral sensibility only differs from physical sensibility, in that the former results exclusively from emotions prompted by acts of the intelligence; while the latter is produced only by emotions aroused by sensations and the needs which they evoke.

If these principles are well founded, they appear to me to be of the highest importance; for not only do they correct our mistakes as to the phenomena of life and organisation and the faculties to which these phenomena give rise, but they set a limit to the marvels created by our imagination; and they give a truer and higher idea of the Supreme Author of all existing things, by disclosing to us the simple method that he has adopted for working the wonders of which we are witnesses.

Hence the intimate feeling of existence possessed by animals which have the faculty of feeling but not that of intelligence, confers upon them at the same time an inner power, which only works through the emotions called up by the harmony of the nervous system, and which causes them to carry out actions without any co-operation of the will. But such animals as possess both the faculty of feeling and that of intelligence, have this advantage over the first: that the inner power, which inspires their actions, is susceptible of receiving its driving emotions either through sensations produced by internal impressions and wants that are felt, or through a will which, though more or less dependent, is always the result of some act of intelligence.

We shall now consider more attentively this singular internal power, which confers the faculty of acting on such animals as possess it: the next chapter, which is devoted to this purpose, may be regarded as a completion of the present.

CHAPTER V.

OF THE FORCE WHICH PRODUCES THE ACTIONS OF ANIMALS, AND OF CERTAIN PECULIAR FACTS RESULTING FROM THE USE OF THIS FORCE.

ANIMALS, independently of their organic movements and of the functions essential to life, carry out yet other movements and actions of which it is very important to ascertain the cause.

We know that plants can satisfy their needs without changing their position or making any sudden movement: the reason of this is that every plant in a favourable situation finds in its neighbourhood the substances which it requires for food; so that it only has to absorb them and profit by some among them.

The case is different with animals: for, with the exception of the most imperfect animals at the beginning of the animal chain, the food on which they live is not always at hand, and they are obliged to carry out movements and actions in order to procure it. Moreover, most of them have other wants to be satisfied as well, which in their turn demand other movements and other actions.

Now the question at issue was how to identify the source from which animals derive the faculty of making comparatively sudden movements of their parts;—in short, of carrying out the various activities by means of which they satisfy their needs.

I noted, in the first place, that every action is a movement, and that every movement necessarily has some cause which produces it: the question was therefore reduced to ascertaining the nature and origin of this cause.

Thereupon, when I reflected that the movements of animals are never communicated nor transmitted, but merely exited, their cause appeared to be revealed to me in the clearest and most decisive manner; and I became convinced that these movements are really in every case the product of some power which excites them.

Indeed in certain animals, muscular activity is a force fully adequate for the production of such movements; and nervous influence is no less adequate for exciting that activity. Now I recognised that in animals, which possess physical sensibility, the emotions of the inner feeling constitute the power which drives the exciting fluid to the muscles, and the problem thus seemed to be solved as far as these animals are concerned; in the case of animals so imperfect as to possess no physical sensibility, stimuli arriving from without are obviously sufficient for their movements, for they are just as irritable or even more irritable in their parts than the others.

In my opinion, this clears up a mystery which seemed indeed difficult to penetrate; nor does this solution seem to me to rest on mere hypothesis: for in the case of sensitive animals, muscular power and the need for nervous influence to excite that power are not hypothetical; and the emotions of the inner feeling, which I regarded as being capable of driving to the muscles the fluid that excites their activity, appeared to me too manifest to be regarded as conjectural.

If now we closely consider all existing animals, including the state of their organisation, the consistency of their parts, and the various circumstances under which they live, it will be difficult not to admit that the most imperfect of them, which have no nervous system and hence no muscular activity to help their movements and actions, move by a force that is outside them, that is to say, one which they do not possess themselves and is not available at will.

It is true that it is in the interior of these delicate bodies that the subtle fluids, entering from without, set up the agitations witnessed in their parts; nevertheless it is impossible for these fragile creatures, on account of their weak coherence and extreme softness, to contain within themselves any power capable of producing their movements. It is only as a result of their organisation that these imperfect animals act in a methodical manner, which they could in no wise originate for themselves.

Now, nature has wrought her various productions by slow and gradual stages; she has created the various organs of animals in turn, varying the shape and situation of these organs according to circumstances and progressively improving their faculties. Hence we feel that she must have begun by borrowing from without, that is, from the environment, the force which produces the organic movements and those of the external parts; that she afterwards transferred that force within the animal itself, and that finally, in the most perfect animals, she made a great part of that internal force available to their will, as I shall shortly show.

If we do not carefully note the gradual order followed by nature in the creation of the various animal faculties, I believe that we shall find it hard to explain how she could have brought feeling into existence, and still harder to conceive how simple relations between material substances can give rise to thought.

We have just seen that animals, which do not yet possess a nervous system, cannot contain within themselves the force which produces their movements, but that this force must be outside them. Now seeing that the intimate feeling of existence is entirely absent in these animals, and seeing that this feeling is the source of that internal power which produces movements and actions in those that possess it, its privation and the privation of the power resulting from it, necessitate the existence of a force to excite movements, from sources that are altogether external.

Hence in imperfect animals the force which produces both vital and bodily movements is wholly outside them: they do not even control it; but they do control to some extent, as I said above, the movements which it impresses on them, and this they do by means of their organisation.

This force is the result of subtle fluids (such as caloric, electricity, and perhaps others) which incessantly penetrate these animals from the environment, set in motion the visible and contained fluids of their bodies, and by exciting the irritability of their containing parts, give rise to the various movements of contraction which they produce.

Now these subtle fluids moving incessantly in the interior of their bodies soon carve out special routes which they always follow till new ones are open to them. Hence the similarity of the movements observed in these animals; and hence again the appearance of an irresistible propensity to carry out movements which, when continued or repeated, give rise to habits.

As a mere statement of principles is not enough, let us endeavour to elucidate the arguments on which they rest.

The most imperfect animals, such as the infusorians and especially *Monas*, only feed by means of absorption which they carry out through the pores of their skin, and by an internal imbibition of the absorbed substances. They have no power of seeking their food nor even of seizing it, but they absorb it because it is in contact with every point of their surface, and the water in which they live provides it in sufficient abundance.

These fragile animals, in which the subtle fluids of the environment are the stimulating cause of orgasm, irritability, and organic movements, carry out, as I have said, movements of contraction which, being incessantly prompted and varied by this stimulating cause, facilitate and hasten the absorptions. Now in these animals the movements of the visible contained fluids are still very slow, and the absorbed substances make good the losses which they suffer as a consequence of life, and serve in addition for the individual's growth.

I have said that the subtle fluids which move in the interior of these living bodies cut out special routes which they continually follow, and thus begin to establish definite movements which become habits. Now if we reflect that organisation develops as life goes on, we shall understand how new routes must have been cut out, increased in number, and progressively varied for the furtherance of the movements of contraction; and how the habits to which these movements give rise then become fixed and irresistible to a corresponding extent.

Such, in my opinion, is the cause of movements in the most imperfect animals; movements that we are led to attribute to their own initiative, and to regard as a result of their faculties, because we know that in other animals the source of them is within their bodies; movements, in short, which are carried out without any will on the part of the individual, and which yet, from being very irregular in the most imperfect of these living bodies, gradually become more regular and are always the same in animals of the same species.

Finally, just as reproduction transmits acquired forms both internal and external, so too it transmits at the same time an aptitude for certain specialised types of movement and corresponding habits.

Of the Transference of the Force, which produces Movements, into the Interior of Animals.

If nature had confined herself to her original method, that is, to a force entirely external and foreign to the animal, her work would have remained very imperfect; animals would have been simply passive machines, and nature never would have produced in such organisms the wonderful phenomena of sensibility, the intimate feeling of existence, the power of acting and, lastly, ideas, by means of which she has created the most astonishing of all, viz. thought or intelligence.

With the intention of reaching these great results, she gradually prepared the way by increasing the coherence of the internal parts of animals, by diversifying their organs, and by multiplying and compounding their contained fluids, etc.; thereafter she was able to transfer into the interior of these animals that force productive of movements and actions, which in truth they did not originally control, but which was in great part rendered available for them as their organisation became more perfect.

Indeed, when the complexity of animal organisation had gone so far as to establish a nervous system of a certain development, as among the insects, then animals provided with this organisation were endowed with the intimate feeling of their existence ; and henceforward the force productive of movements was transferred into the interior of the animal itself.

I have already shown that this internal force, which produces movements and actions, finds its origin in the intimate feeling of existence possessed by animals which have a nervous system, and that this feeling when aroused by the needs sets in motion the subtle fluid contained in the nerves, and drives it to the muscles which have to act ; thus are produced the actions which the needs demand.

Now every need that is felt, produces an emotion in the individual's inner feeling ; and from this emotion springs the force which gives rise to the movements required. I furnished evidence of this in setting forth the communication and harmony existing throughout the nervous system, and the way in which the inner feeling on being stirred may excite muscular activity.

Hence, in animals which contain within them the force productive of movements and actions, the inner feeling which gives rise to that force puts it in action, whenever some need is felt, and thus excites movements in the subtle fluid of the nerves (called by the ancients animal spirits) ; the force directs that fluid towards such of the organs as have to act, and finally causes a reflux of the fluid into its usual reservoirs, when the needs no longer require the organ to act.

The inner feeling then takes the place of will ; for it now becomes important to remember that every animal which does not possess the special organ in which or by means of which thoughts, judgments, etc., are produced, has really no will, does not choose, and therefore cannot control the movements, prompted by its intimate feeling. Instinct directs these movements, and we shall see that this direction always results from emotions of the inner feeling, in which the intellect has no share, and from the organisation as modified by habit ; so that the needs of such animals being necessarily limited and always the same in the same species, the intimate feeling always produces the same actions.

The case is different in animals which nature has endowed with an additional special organ (two wrinkled hemispheres surmounting the brain), for the performance of acts of intelligence, and which consequently carry out comparisons, judgments, thoughts, etc. These animals control their power of acting in proportion to the perfection of their organ of intelligence ; and although they are still strongly subordinated to the habits which have modified their organisation, they yet enjoy a will that is more or less free ; they can choose, and introduce variation into their actions or at least into some of them.

We shall now say a word about the consumption of the nervous fluid which takes place in proportion as it works towards the production of animal actions.

Of the Consumption and Drainage of the Nervous Fluid in the Production of Animal Actions.

The nervous fluid, set in motion by the animal's inner feeling, is thus the motive power of the actions of these living bodies; hence it is used up in working its effects, and it would ultimately be exhausted and incapable of producing any further activity, if the individual's will kept it continually at work.

The nervous fluid, which is constantly being formed throughout life, is as constantly being consumed by the use which the individual makes of it.

One part of this fluid is kept occupied independently of the animal's will in the maintenance of vital movements and of the functions necessary to life.

The other part of the fluid is at the service of the individual and can be used either for the production of its acts and movements, or for the performance of its various acts of intelligence.

The individual thus uses up its invisible fluid in a degree proportional to the duration of the action which it produces, or to the effort which that action requires; and it would exhaust the whole of the available part, if it continued too long to carry out actions which required much of it.

Hence arises the need of rest after a certain period of action: the individual falls asleep while the exhausted fluid is recuperated.

The loss of strength, and consequently of the nervous fluid which gives strength, becomes manifest therefore in all actions that are too prolonged or are painful, and for that reason called fatiguing.

If you take too long a walk, you become fatigued after a period relative to your strength; if you run, you become fatigued much sooner, for you dissipate the principle of your strength much more rapidly and copiously; lastly, if you take a weight of fifteen or twenty pounds, and hold it out with your arm extended horizontally, you will at first find the action quite easy because you have the wherewithal to do it, but as your motive principle is rapidly used up, the weight soon seems heavier and harder to hold, and in a short time the action becomes altogether impossible.

Your organisation, however, remains the same throughout, for, if it were to be examined, no change would be found in it from the first moment, when the action was started, to the moment when you could no longer support the weight in question.

Is it not obvious that the real difference is only in the dissipation of an invisible fluid, which we cannot perceive owing to the limitation of our available resources ?

The consumption and exhaustion of the subtle fluid of the nerves in prolonged or painful actions, is certainly a fact that can never be disputed, since reason and organic phenomena afford the strongest proof of it.

Although it is true that one part of an animal's nervous fluid is constantly employed, independently of its will, in the maintenance of its vital movements and of such functions as are necessary to its existence, yet when the individual uses up too much of the free fluid available to it, the functions of its vital organs suffer. For in these circumstances the unavailable part of the nervous fluid fills the place of the available fluid that has been dissipated. It is therefore run too low, and provides inadequately for the operations of the vital organs ; thereupon the functions of these organs flag to some extent, and are imperfectly fulfilled.

It is chiefly man—who is allied to animals by his organisation—that uses up his physical strength in this way ; for, of all his actions, those that use the most nervous fluid are the prolonged acts of his understanding, thoughts, meditations, and, in short, the continued efforts of his intellect. The digestion is then impaired, and the physical strength is proportionally diminished.

The consumption of nervous fluid in the movements and actions of animals is so well known that I need not dwell upon it longer ; but I may say that this by itself would be a sufficient proof of the existence of the fluid in the most perfect animals, if there were not many others of equal cogency.

Of the Origin of the Propensity towards Repeating the Same Actions, and also of Instinct in Animals.

It is assuredly worth while to investigate the cause of the well-known phenomenon which compels nearly all animals always to repeat the same actions, and which gives rise even in man to a propensity towards any action that has become habitual.

If the principles expounded in this work are really well-founded, the causes which we seek may be very easily deduced from them ; so that we shall cease to be astonished by phenomena seemingly so mysterious, when once we have recognised the simplicity of their origin.

Let us see, then, how we can account for the phenomena in question, in accordance with the principles enunciated above.

Every action is caused by some movement in the fluid of the nerves. Now when this action has been several times repeated there is no doubt that the fluid cuts out a route which becomes specially easy for it to traverse, and that it acquires a readiness to follow this route in preference to others.

How great is the light shed by this simple and fertile principle on the power of habit—a power from which even man can only escape with great difficulty by the help of his highly developed intellect!

Who can now deny that the power of habits over actions is inversely proportional to the intelligence of the individual, and to the development of his faculty of thinking, reflecting, combining his ideas, and varying his actions?

Animals, which are only sensitive, and have not yet acquired the organ in which are produced comparisons between ideas, as also thoughts, reasonings and the various acts constituting intelligence, have nothing but perceptions (often very confused), do not reason, and can make little variation in their actions. They are therefore permanently subject to the power of habit.

Thus insects, which have the least developed nervous system of any animals that possess feeling, experience perceptions of the objects which affect them, and seem to possess a memory gained through a repetition of these perceptions. They can, however, neither vary their actions nor alter their habits, since they possess no organ to give them this power.

Of Instinct in Animals.

By instinct is meant the fixed tendencies displayed by animals in their actions; and many people have held that these tendencies are the produce of a reasoned choice, and therefore the fruit of experience. Others, as Cabanis said, prefer to think, in common with observers of all ages, that many of these tendencies cannot be referred to any sort of reasoning, and that, although taking their origin in physical sensibility, they are usually formed without any share of the individual's will, beyond that of improving their execution.

If attention had been paid to the fact that all animals which can feel have an inner feeling capable of being stirred by their needs, and that the resulting movements of their nervous fluid are always controlled by this inner feeling and by habit, it would then have been apparent that in all such animals that are devoid of intelligence, the tendencies of action can never be the product of a reasoned choice or judgment or experience turned to account, or will, but that they are subject to needs, excited by certain sensations, which then awake irresistible propensities. Even in animals which possess some degree of intelligence, the actions are still commonly controlled by the inner feeling and by habit.

Lastly, although the motive power of movements and actions, as also their directive cause, are exclusively internal, we must not, as has been done,[1] limit the original or prompting cause of these acts to internal impressions, with the view of attributing all acts of intelligence to external impressions; a slight knowledge of the facts is enough to show that in both cases the causes, which stimulate actions, are sometimes internal and sometimes external, and yet that they do actually give rise to impressions which only act internally.

According to the common notion attached to the word instinct, it is regarded as a torch which lights and guides animals in their actions, and which is for them what reason is for us. Nobody has shown that instinct may be a force that induces action, nor that it does induce action without any intervention of the will but only under the control of acquired propensities.

The view of Cabanis that instinct arises from internal impressions, while reasoning is the produce of external sensations, cannot be justified. It is in ourselves that we feel; our impressions can only be internal; and the sensations of external objects, derived from our special senses, cannot produce in us any but internal impressions.

When my dog is out for a walk and sees in the distance another animal of his own species, he undoubtedly experiences a sensation from that external object through the intermediary of sight. Thereupon his inner feeling, aroused by the impression received, guides his nervous fluid in the direction of a tendency acquired in all the individuals of his race; by a kind of involuntary impulse, his first movement is towards the dog that he sees. This is an act of instinct, excited by an external object, and many others of the same kind may similarly be carried out.

With regard to these phenomena, of which we find so many examples in animal organisation, it seems to me that a true and clear idea of their cause can only be found when we have recognised: (1) That the inner feeling is a very powerful generalised feeling, which has the faculty of exciting and controlling the movements of the free part of the nervous fluid and of making the animal carry out various actions; (2) that this inner feeling is capable of being moved either by acts of intelligence culminating in a will to act, or by sensations which evoke needs which in their turn immediately excite it, so that it guides the productive force of actions in the direction of some acquired propensity without the co-operation of any act of will.

There are thus two kinds of causes which may move the inner feeling, viz. those depending on intellectual operations, and those

[1] Richerand, *Physiologie*, vol. ii., p. 151.

which excite it immediately to guide its power of acting in the direction of the acquired propensities.

It is purely these last kind of causes that constitute instinct; and seeing that they are not the produce of any deliberation, thought, or judgment, instinctive actions always satisfy surely and unerringly the wants felt and the habits acquired.

Hence instinct in animals is a compelling tendency, prompted by sensations in giving rise to needs; and it causes the execution of actions without the intervention of any thought or act of will.

This propensity is a part of the organisation acquired by habit; it is excited by impressions and needs, which move the individual's inner feeling, and cause it to dispatch nervous fluid to the appropriate muscles as required by the propensities.

I have already said that the habit of using any organ or any part of the body for the satisfaction of constantly recurring needs, gives the subtle fluid so great a readiness for moving towards that organ where it is so often required, that the habit becomes inherent in the nature of the individual.

Now the needs of animals with a nervous system vary in proportion to their organisation, and are as follows:

1. The need of taking some sort of food;
2. The need for sexual fertilisation, which is prompted in them by certain sensations;
3. The need for avoiding pain;
4. The need for seeking pleasure or well-being.

For the satisfaction of these needs they acquire various kinds of habits, which become transformed in them into so many propensities; these propensities they cannot resist nor change of their own accord. Hence the origin of their habitual actions and special inclinations, which have received the name of instinct.[1]

This propensity of animals to the preservation of habits, and to the repetition of the resulting actions when once it has been acquired, is propagated to succeeding individuals by reproduction so as to preserve the new type of organisation and arrangement of the parts; thus the same propensity exists in new individuals, before they have even begun to exert it.

Hence it is that the same habits and instinct are handed on from

[1] Just as all animals do not possess the faculty of will, so too instinct is not a property of all existing animals; for those which have no nervous system have no inner feeling, and cannot therefore have any instinct for their actions.

These imperfect animals are entirely passive, do nothing of their own accord, feel no needs, and are provided for by nature in everything just as in the case of plants. Now since their parts are irritable, nature causes them to carry out movements, which we call actions.

generation to generation in the various species or races of animals, without any notable variation so long as no alteration occurs in their environment.

Of Skill in certain Animals.

In animals which have no special organ for intelligence, the name skill, applied to certain of their actions, is scarcely deserved; for it is only by illusion that we ascribe to them a faculty they do not really possess.

Propensities acquired and transmitted by reproduction, together with the resulting habits of carrying out complicated actions, in the face of various difficulties which are regularly and habitually overcome by emotions of the inner feeling: these constitute the sum-total of actions, always the same in individuals of the same race, to which we carelessly give the name of skill.

The instinct of animals consists in the habit of satisfying the four kinds of needs named above, and results from propensities acquired long previously, which constrain to act in a way predetermined for each species. It subsequently happened, in the case of some animals, that a complication in the actions required for the satisfaction of these four kinds of needs or of some of them, combined with various difficulties which had to be overcome, gradually forced the animal to increase and develop its methods, and led it to the performance of various new activities, without any choice or act of intelligence, but solely by emotions of the inner feeling.

Hence the origin in certain animals of various complicated actions, which people call skill, and which they are never tired of enthusiastically admiring because they always imagine, tacitly at least, that these actions are the result of thought and calculation; this, however, is an obvious mistake. They are merely the result of a necessity which has developed and guided the habits of animals, until they reach the point at which we find them.

What I have just said is especially applicable to the invertebrates, which can carry out no act of intelligence. None of these animals indeed has any power to vary its actions, or to abandon its so-called skill in favour of some other method.

There is then nothing more wonderful in the alleged skill of the ant-lion (*Myrmeleon formica leo*) which digs out a hole in loose sand and then waits until some victim falls into the bottom of a hole by the slipping of the sand, than there is in the operation of the oyster, which for the satisfaction of all its needs has only to open slightly and close its shell. So long as their organisation remains unchanged they will both continue to do just what they do now, without any intervention of will or reasoning.

It is only among the vertebrates, and particularly among the birds and mammals, that we find the characteristics of a true skill; for in difficult situations their intelligence may assist them to vary their actions notwithstanding their subjection to habit. These characteristics however are not common and it is only in a few races that we often witness them.

Let us now enquire what constitutes the determining motive of action, called will: and let us consider if it is actually, as has been supposed, the motive of all the actions of animals.

CHAPTER VI.

OF THE WILL.

In the present chapter I propose to show that the will, which has been regarded as the source of all actions in animals, can only be present in those which possess a special organ for intelligence; and that even in these, which include man himself, it is not always the motive of the actions performed.

On considering the matter, we find indeed that the will is the immediate result of an active intelligence: for it is always the effect of a judgment and hence of an idea, thought, comparison or choice; lastly, we perceive that the faculty of willing is nothing else than that of determining by thought, that is by an operation of the organ of understanding, to perform some action, combined with the faculty of exciting an emotion of the inner feeling which can produce that action.

Thus the will is a determination towards some action, and is achieved by the intelligence of the individual: it always results from a judgment, and this judgment itself necessarily springs from an idea or thought or from some impression giving rise to such idea or thought, so that it is purely by an act of intelligence that the will, which determines an individual to some action, can be developed.

But if the will is nothing else than a determination following upon a judgment, and hence only the result of an intellectual act, it becomes clear that animals which have no organ for intelligence cannot carry out any acts of will. Yet all these animals carry out movements, which constitute their actions. There are then several different sources from which the actions of animals may be derived.

Now since the movements of all animals are excited and not communicated, there must be different kinds of exciting causes for these movements. Indeed we have seen that in certain animals these causes come entirely from without, that is to say, from the environment; while in others, which possess an inner feeling, there is a motive power sufficient for the production of the requisite movements.

But the inner feeling, which only becomes a power when moved by a physical cause, acquires its emotions by two very different routes: in animals which have no organ for the will, the inner feeling can only be moved by means of sensations; whereas in those which have an organ for intelligence, the emotions of the inner feeling may either be the exclusive result of sensations, or that of a will born from an operation of the understanding.

Hence we have three distinct sources for the actions of animals, viz.: (1) The external causes which excite the irritability of animals; (2) the inner feeling moved by sensations; (3) the inner feeling moved by the will.

The actions or movements arising from the first of these three sources take place without the intermediary of muscles; for the muscular system does not exist in animals of this character; and when it begins to be developed, excitations from without still take the place of the non-existent inner feeling; but the actions or movements which originate from the emotions of the individual's inner feeling are only carried out through the intermediary of muscles excited by nervous fluid.

Thus when the will determines an individual towards some action, its inner feeling promptly receives an emotion, and the resulting movements are directed in such a way that the nervous fluid is immediately dispatched to the required muscles.

As to the animals which are endowed with physical sensibility, while yet possessing no organ for intelligence, and which therefore cannot carry out any act of will, their needs always result from some sensation, that is, from some perception that makes itself felt and not from any idea or judgment; and this need or perception immediately affects the individual's inner feeling. From this it follows that these animals, before acting, do not deliberate or form a judgment, and come to no preliminary determination. Their inner feeling is directly stirred by the need, and its movements are thereafter controlled by the nature of this need, so that it promptly puts in action those parts which have to move. The actions therefore which arise from this source are not preceded by a true will.

Whereas this is a necessity in the case of the animals under consideration, it is still the most frequent factor in those that are endowed with intellectual faculties; for nearly all the needs of these latter are due to sensations, which revive certain habits and thus immediately stir the inner feeling, causing these animals to act without any previous thought. Even man carries out actions of this character when the needs which prompt them are pressing. If, for instance, you absent-mindedly take up a piece of iron which unexpectedly turns out to be very hot, the pain caused by the heat of the iron promptly moves your inner feeling; and before you have had time to think what you ought to do, your muscles have already carried out the action which consists in letting go the hot iron that you were holding.

From these observations it follows that actions carried out in consequence of needs prompted by sensations which immediately affect the individual's inner feeling, are in nowise the result of any thought, judgment, or act of will; while those which are wrought in consequence of needs aroused by ideas or propensities are exclusively the result of those acts of intelligence which also affect immediately the inner feeling and enable the individual to act by a manifest will.

This distinction between actions of which the immediate cause is in some sensation and those which result from a judgment or act of intelligence, is of great importance for the avoidance of confusion and error in our investigations of these wonderful phenomena. It is because the distinction had not been drawn that animals have been usually credited with a will in the performance of their actions; and people have supposed, on the analogy of man and the most perfect animals, that all animals have the faculty of voluntary movement. Yet this is not so, even in those that possess a nervous system, and still less in those that do not.

It is certain that animals which have no nervous system cannot possess the faculty of will; so far from it are they, that they cannot even have the feeling of their existence: the infusorians and polyps are in this position.

Those which have a nervous system capable of yielding the faculty of feeling, but which have no hypocephalon or special organ for intelligence, possess indeed an inner feeling—the mainspring of their actions—and form confused perceptions of the objects which affect them; but they have no ideas, do not think, compare or judge, and hence carry out no act of will. There is reason to believe that this is the case in the insects, arachnids, crustaceans, annelids, cirrhipedes, and even molluscs.

The inner feeling, aroused by some need, is the source of all the actions of these animals. They act without any deliberation or preliminary determination, and always in the one direction to which the need drives them; and when they encounter some obstacle while acting, if they avoid it or turn aside and seem to choose, it is only because a new need has aroused their inner feeling. Thus the new action does not result from any combination of ideas, or comparison between objects, or judgments, since these animals cannot carry out any operations of the intelligence, having no organs for the purpose;

the new action is purely a sequence of some emotion of their inner feeling.

It is therefore only the animals which, in addition to a nervous system, possess the special organ for complex ideas, thoughts, comparisons, judgments, etc., that possess the faculty of will and can carry out acts of will. This apparently is the case with the vertebrates: and since the fishes and reptiles have so imperfect a brain as not completely to fill up the cranial cavity—an indication that their acts of intelligence are extremely limited—it is mainly in the birds and mammals that the faculty of will is to be found; for they obviously carry out various acts of intelligence, and certainly have the special organ which renders them capable of such acts.

But as I have already observed: in the case of animals which have a special organ for intelligence, all the actions do not result exclusively from a will, that is, from a preliminary intellectual determination which excites the motive power of actions. Some of them are no doubt the product of the faculty of will, but many others simply arise from a direct emotion of the inner feeling, which is excited by sudden needs and causes the animal to carry out actions that are not preceded by any determination or thought.

Even in man, how many actions there are that are exclusively prompted and carried out by an emotion of the inner feeling, without any intervention of the will! Do not many of our actions owe their origin to instantaneous and uncontrolled movements; and what are these movements but results of the inner feeling?

If, as I said above, there is no true will in animals which have a nervous system without any organ for intelligence, and if this is the reason why such animals only act by emotions produced by their sensations, this truth applies still more to animals that have no nerves. It appears therefore that these latter only move by an excited irritability, and as an immediate result of external excitations.

It is easy to imagine from the above exposition that, after nature had transferred the power of acting into the interior of animals, that is, after she had created the inner feeling based on the nervous system as the origin of the force for producing actions, she then perfected her work by creating a second internal power, that, viz. of the will, which arises from acts of intelligence and is the only faculty that can cause any variation in the habitual actions.

For this purpose, nature was obliged to add a new organ to the nervous system, in which acts of intelligence might be carried on; and she had to separate the nucleus of sensations or perceptions from the organ of ideas, comparisons, judgments, reasonings, thoughts.

Thus in the most perfect animals the spinal cord provides for the muscular movement of the parts of the body, and for the maintenance of the vital functions; while the nucleus of sensations, instead of being situated throughout the length, or in any isolated part of the spinal cord, becomes concentrated at its superior or anterior extremity in the inferior part of the brain. This nucleus of sensations is therefore quite close to the organ in which are carried out the various acts of the intelligence, and is yet separate from it.

When animal organisation has reached the limit of perfection implied by an organ for acts of intelligence, the individuals which possess such an organisation have simple ideas, and may form complex ideas. They possess a will, free in appearance, which determines some of their actions; they have passions, that is, heightened inclinations which draw them towards certain kinds of ideas and actions beyond their control; lastly, they are endowed with memory and have the faculty of calling up ideas previously traced in their organ, a process which is due to the nervous fluid passing over, and being acted on by, the impressions or traces implanted by these ideas.

It seems probable that the disturbed agitations of the nervous fluid in contact with these traces are the cause of dreams, which often call up ideas in animals during sleep.

Animals which have intelligence yet carry out most of their actions by instinct and habit, and in this respect they never make a mistake; when they act by will, that is, as the effect of a judgment, they still make no mistakes, or, at least, very rarely so; because the elements which enter into their judgments are few in number and usually furnished by sensations, and still more because in any one race there is no inequality in the intelligence and ideas of individuals. Hence it follows that their acts of will are determinations which lead them always unerringly to the satisfaction of their needs. For this reason it has been said that instinct is for animals a torch which lights them better than our reason.

The truth is that animals are less free than ourselves to vary their actions and more subject to habit, and hence that they find in their instinct a compelling necessity, and in their acts of will only a cause the elements of which are invariable, unmodified, quite simple and always the same in individuals of the same race; so that it has in all an equal power and extent. Finally, since there is no inequality in the intellectual faculties of individuals of the same species, their judgments and their will, resulting from these judgments, cause them always to carry out the same actions in the same circumstances.

I shall conclude these theories on the origin and results of the will by some remarks on this faculty in man. We shall then see that things are quite different in his case from what they are in the case of animals;

for although his will seems to be much freer than theirs, it is not really so; and yet, for a reason that I shall try to explain, individuals of his species act very differently under similar circumstances.

Since the will is always dependent on some judgment it is never really free; for the judgment which gives rise to it is a necessary result of the union of its component elements, like the quotient in an arithmetical operation. But the produce of a judgment must vary in different individuals for the reason that the elements which enter into the formation of this judgment are apt to be quite different in different individuals.

In fact, so many and various elements enter into the formation of our judgments, so many are present which ought not to be present, and of the proper elements so many are unnoticed or rejected from prejudice, or are affected by our disposition, health, age, sex, habits, propensities, state of our knowledge, etc., that the union of these elements gives rise to very different judgments on the same subject in different individuals. The fact that our judgments depend on so many inappreciable elements has given rise to the belief that our determinations are free, although in reality they are not so, seeing that the judgments which produce them are not free themselves.

The diversity of our judgments is so remarkable that it often happens that a subject gives rise to as many individual judgments as there are persons to discuss it. This variation has been taken for a freedom of determination, and a mistake has thus been made; for it is only the result of the difference between the elements entering into the judgment of each person.

Yet there are subjects so simple and straightforward that the judgments passed upon them are in almost universal agreement. But these subjects are almost exclusively confined to what is outside of us, and only known by the sensations which they excite or have excited on our senses. Our judgments with regard to them involve scarcely any elements other than those furnished by sensation together with the comparisons which we draw between these and other known bodies. Lastly, for this class of judgment there is very little call upon our understanding.

One result of the immense number and variety of the causes which affect the elements entering into the formation of our judgments, especially those which require various intellectual operations, is that these judgments are usually erroneous and inaccurate; and that on account of the inequality between the intellectual faculties of individuals these same judgments are commonly as varied as the people who form them, each one importing different elements into them. Hence it follows, moreover, that disorders of these acts of intelligence necessarily involve others in our wills and consequently in our actions.

If the scope of the present work permitted, I might adduce many instances in support of these observations; I might even make remarks that would not be without interest.

I might show, for instance, that while man derives great advantages from his highly developed intellectual faculties, the human species in general suffers from them at the same time considerable disadvantages; since these faculties confer the means for doing harm as easily as good, and their general effect is always to the disadvantage of those individuals who make least use of their intelligence, and this is necessarily the case of the greater number. It would appear therefore that the main evil in this respect resides in the extreme inequality of intelligence between individuals, an inequality that cannot be entirely destroyed. Nevertheless it may be inferred with still greater certainty that the thing most important for the improvement and happiness of man is to diminish as far as possible this enormous inequality, since it is the origin of most of the evils to which he is exposed.

Let us now endeavour to ascertain the physical causes at work in acts of the understanding: we shall try at least to determine what are the necessary conditions of the organisation in order that these wonderful phenomena may be produced.

CHAPTER VII.

OF THE UNDERSTANDING, ITS ORIGIN, AND THE ORIGIN OF IDEAS.

THIS is the most curious and interesting, and at the same time the most difficult subject presented to man in his studies of nature; it is the subject in which it is most important for him to possess positive knowledge, while yet offering the fewest facilities for its acquisition.

The question is, how purely physical causes, that is to say, simple relations between different kinds of matter, can produce what we call ideas; how, out of simple or direct ideas, complex ideas may be formed; how, in short, out of ideas of any kind, faculties can arise, so astonishing as those of thought, judgment, analysis and reasoning.

He indeed is more than bold who undertakes such a research, and flatters himself that he has found in nature the origin of these wonderful phenomena.

Assuredly, I have not the presumption to suppose that I have discovered their causes; but I started with the conviction that all acts of intelligence are natural phenomena and hence derive their source exclusively from physical causes. Seeing that the most perfect animals enjoy the faculty of producing them, I thought that by careful observation, attention, and patience, I might reach, especially through induction, ideas of great moment on this important subject; I now have to present my conclusions.

Under the title of understanding or intelligence I include all known intellectual faculties, such as those of forming ideas of various kinds, of comparing, judging, thinking, analysing, reasoning, and, lastly, of recalling ideas previously acquired as well as past thoughts and reasonings; this faculty constituting memory.

All these faculties unquestionably result from acts peculiar to the organ of intelligence, and each such act is necessarily the product of relations subsisting between that organ and the nervous fluid moving in it.

The special organ in question, to which I have given the name of hypocephalon, consists of two wrinkled pulpy hemispheres, investing or covering that medullary region which I call more strictly the brain, and which contains the nucleus or centre of communication of the nervous system and gives rise to the nerves of the special senses ; the cerebellum is a mere appendage of it.

This region (the brain properly so-called, including the cerebellum) and the hypocephalon are quite distinct from each other, especially as regards their functions, although they are commonly confused together under the general name of brain or encephalon. Now it is in the functions of the hypocephalon alone that I shall seek the physical causes of the various intellectual faculties, because this organ is the only one that has the power of giving rise to them.

Though not easily recognised, there is a real differentiation in the parts of this organ. This differentiation and the varied movements of the contained subtle fluid are the only origin of the different intellectual acts which I have named. This is the general idea, which I propose briefly to develop.

At the outset, in order to introduce some order into our discussions we must posit or recall the two following principles which are at the base of all legitimate hypotheses on this subject.

First principle : All intellectual acts whatever originate from ideas, acquired either at the time or previously, for these acts may always be reduced to ideas, or relations between ideas, or operations on ideas.

Second principle : Every idea of any kind originates either directly or indirectly from a sensation.

Of these two principles, the first is fully confirmed by an investigation as to what the various acts of the understanding really consist of ; in all such acts, ideas are invariably the subject or material of the operations carried out.

The second of these principles had been recognised by the ancients, and is perfectly expressed in that axiom, which Locke has so successfully sustained, viz. : *That there is nothing in the understanding which was not previously in sensation.*

Hence it follows that every idea must in the last analysis be resolved into an image perceived, and that since whatever is in our understanding came there *viâ* sensation, all that issues from it and can find no perceptible object to fasten upon is absolutely chimerical. This is the obvious conclusion drawn by M. Naigeon from Aristotle's axiom.

The axiom however is not yet universally admitted, for many people hold, on account of certain facts whose causes they do not perceive, that there really are innate ideas. They persuade themselves that they have found a proof of it in the child which a few minutes after birth tries to suck, and seems to look for its mother's breast, although it can still have no knowledge of it through freshly acquired ideas. I shall refrain from citing the alleged case of a kid which when dragged from its mother's breast selected laburnum from several plants offered it. It is well known that this was a mere imagination, which cannot have been well-founded.

When once we have recognised that habits are the source of propensities, that continued exercise of these propensities modifies the organisation in a corresponding direction, and that they are then transmitted to new individuals by reproduction, we shall see how the newly born child may try to suck by pure instinct, and may take the breast offered it without having the slightest idea of it, or carrying out any thought, judgment, or act of will; and we shall further see that the child performs this action exclusively through the slight emotion which the need raises in its inner feeling, so as to make it act in the direction of an acquired propensity without any previous experience; we shall see in the same way that the duckling just out of the egg, if it is in the neighbourhood of water, immediately runs to it and swims about on the surface, without having any idea of it or knowing what it does. The animal does not perform this action by any intellectual deliberation, but by a transmitted propensity evoked by its inner feeling without any call upon the intelligence.

I lay down, then, as a fundamental principle and unquestionable truth, the proposition that there are no innate ideas, but that all ideas whatever spring either directly or indirectly from sensations which are felt and noticed.

From this principle it follows that the organ of intelligence, being the ultimate perfection which nature has bestowed on animals, can only exist in those which already possess the faculty of feeling. Hence the special organ in which are carried out ideas, judgments, thoughts, etc., only begins to be formed in animals with a highly developed system of sensations.

All intellectual acts occurring in an individual are due therefore to a combination of the following causes, viz.:

1. The faculty of feeling;
2. The possession of a special organ for intelligence;
3. The relations subsisting between this organ and the nervous fluid moving freely about in it;
4. Lastly, the fact that the results of these relations are always conveyed to the nucleus of sensations, and therefore to the individual's inner feeling.

This is the chain which is in harmony throughout, and constitutes the developed physical cause of the most wonderful of natural phenomena.

To set up a reasonable opposition to the views stated above, it would be necessary to show that the harmony existing throughout the nervous system is not capable of producing sensations and the individual's inner feeling; that intellectual acts, such as thoughts, judgments, etc., are not physical acts, and do not result immediately from relations between a subtle agitated fluid and the special organ containing it; lastly, that the results of these relations are not transmitted to the individual's inner feeling. Now the physical causes named above are the only ones that can possibly give rise to the phenomena of intelligence. If therefore the existence of these causes is denied, and if consequently it is denied that the resulting phenomena are natural, it will then be necessary to seek another source of these phenomena outside nature. It will be necessary to substitute for the physical causes rejected, fantastic ideas of our own imagination, ideas that are always baseless since it is quite obvious that we can have no other positive knowledge than that derived from the actual objects which nature presents to our senses.

Since the marvels which we are investigating, and whose causes we are seeking, are based upon ideas, and since in acts of intelligence we are dealing only with ideas and operations on ideas; before enquiring what ideas themselves are, let us illustrate the gradual formation of the organs which give rise firstly to sensations and the inner feeling, then to ideas and, lastly, to the operations performed upon them.

The very imperfect animals of the earlier classes, having no nervous system, are simply irritable and merely have habits without feeling any sensations or ever forming ideas. But the less imperfect animals, which have a nervous system without, however, the organ of intelligence, have instinct, habits and propensities and feel sensations, while yet forming no ideas. I venture to affirm that where there is no organ for a faculty, that faculty cannot exist.

Now if we admit that every idea originates from a sensation, which indeed cannot seriously be disputed, I hope to show that it does not follow that every sensation necessarily yields an idea. The organisation must have reached a state favourable to the formation of the idea, and, moreover, the sensation must be accompanied by a special effort of the individual, in short, by a preparatory act which renders the special organ of intelligence capable of receiving the idea, that is, of retaining impressions.

Indeed, if it is true that in creating organisation, nature necessarily began by forming it in extreme simplicity without the intention of giving living bodies any other faculties than those of feeling and reproducing, those bodies which were endowed by her with organisation and life could have no other organs than those necessary for the maintenance of life. This fact is confirmed by observation of the most imperfect animals, such as the infusorians and polyps.

But when she subsequently complicated the organisation of these earliest animals and created, with the help of long periods of time and an infinite diversity of circumstances, the multitude of different forms which characterise the later classes, nature formed successively the various organs which animals possess and the different faculties to which these organs give rise. She produced them in an order that I have determined in Part I., Chapter VIII., and from this order it may be seen that the hypocephalon, consisting of the two wrinkled hemispheres covering the brain, is the last organ which she brought into existence.

Long before creating the hypocephalon, or special organ for the formation of ideas and of all the operations carried out upon them, nature had established in a great number of animals a nervous system, which gave them the faculty of exciting muscular activity, and afterwards of feeling and acting by the emotions of their inner feeling. Now although for this purpose she had multiplied and scattered the nuclei for muscular movement either by establishing separate ganglia, or by distributing these nuclei throughout the length of a ganglionic longitudinal cord or spinal cord, yet she concentrated the nucleus of sensation in a special locality and fixed it in a small medullary mass, which gives direct origin to the nerves of some special senses and which has received the name of brain.

It was therefore only after having wrought these various perfections of the nervous system that nature put the finishing touch on her work by creating, in close proximity to the nucleus of sensations, the hypocephalon, that remarkable and interesting organ in which ideas are graven and where all the operations constituting intelligence are carried out.

It is exclusively these operations that we shall study, in our endeavours to determine their most probable physical causes, by careful inductions and a knowledge of the conditions required.

Let us now investigate how an idea comes to be formed, and under what conditions a sensation can produce it; let us even enquire, at least in outline, in what way acts of the intelligence are carried out in the hypocephalon.

It is a very singular circumstance that the special organ now under consideration never exerts any action itself in any of the acts or phenomena to which it gives rise, and that it does nothing more than receive and preserve for a longer or shorter period the images transmitted to it and the impressions graven upon it. This organ, together with the brain and nerves, differs from all the other organs of the animal body in that it is not active, and does no more than provide the means for the nervous fluid to carry out its various phenomena.

Indeed, when I remember the extreme softness of the medullary pulp constituting the nerves, brain, and hypocephalon, I cannot think that in the relations of the nervous fluid with the medullary parts, the latter are capable of exerting the slightest action. Beyond doubt these parts are entirely passive and unable to react upon anything that affects them. Hence it results that the medullary parts, of which the hypocephalon consists, receive and preserve the traces of all the impressions made upon them by the movements of the nervous fluid; so that the only active element in the functions of the hypocephalon is the nervous fluid itself; or to express the matter more precisely, the organ carries out no function, the nervous fluid by itself carries out all; but this fluid could in nowise give rise to them without the existence of the organ in question.

I may be asked how it is possible to conceive that any fluid, however subtle and varied its movements, can by itself give rise to that astonishing variety of acts and phenomena characteristic of the intellectual faculties. To this I reply, that the entire marvel is in the composition of the hypocephalon itself.

The medullary mass constituting the hypocephalon, that is, the two wrinkled hemispheres which cover over the brain,—this mass, I say, which seems to be only a pulp, whose parts are continuous and coherent throughout, consists, on the contrary, of an inconceivable number of separate and distinct parts, from which result a vast quantity of cavities of infinitely varied size and shape and appearing to occupy distinct regions, equal in number to the intellectual faculties of the individual; lastly, however it may come about, the composition of the organ is different in each region, for each is devoted to some individual faculty of the intelligence.

The examination of the white medullary part of the hypocephalon has disclosed numerous fibres in it: now it is probable that these fibres are not, as elsewhere, organs of movement; their consistency would not permit of it: there is more reason to believe that they are so many individual canals, each terminating in a cavity which would be in the form of a cul-de-sac unless they communicate together by lateral paths. These cavities, which are imperceptible to us, are as innumerable as the tubular threads leading to them; and it may be presumed that it is on their internal walls that the impressions brought by the nervous fluid are engraved; there may also be little medullary plates or leaves for the same purpose.

Since we cannot know what is actually the case on this subject, I believe I have attained my purpose by showing what may be the case or even probably is the case: for I can do no more.

The wonderful composition of the hypocephalon, both in its main features and in its individual regions, which are all duplicated, one being in each hemisphere, is more than a mere hypothesis, in spite of the fact that we lack the means for seeing it and assuring ourselves of it. The organic phenomena constituting the intelligence, and the fact that each of these phenomena requires a special locality in the organ and indeed a special organ, so to speak, for its production, should give us a moral conviction that the composition of the hypocephalon is as I have represented it.

Certainly, individuals are not born in possession of all their possible intellectual faculties, for the organ of intelligence, like the rest, develops according to its use. The same applies with each special kind of intellectual faculty: it is brought into existence in the appropriate region of the hypocephalon by the needs that are felt; and in correspondence with the frequency of the repetition of these acts, the special organ adapted to them develops and enlarges its faculty.

It is therefore not true that any of our intellectual faculties are innate; and this applies also to those of our propensities which depend on the faculty of thought. These faculties and propensities grow and strengthen according to the exercise which we give to their underlying organs. We may indeed derive certain tendencies from our inherited organisation: but unless we exercise these faculties and inclinations ourselves we gradually lose the aptitude for them.

Dr. Gall having noticed that some individuals had particular faculties, more developed and conspicuous than others, conceived the idea of enquiring whether any part of their body might not present some external signs by which this faculty could be recognised.

He does not seem to have studied any faculties but those connected with the intelligence; for if he had he would have met with abundant evidence that any part which is much exerted and has acquired a highly developed faculty, always exhibits manifest signs of it in its size, shape, and vigour. We cannot look at the posterior extremities and tail of a kangaroo, without being convinced that these much used parts possess great power of action, a fact which also applies to the posterior legs of grasshoppers, etc. In the same way, when we reflect upon the immense growth of the elephant's nose, which has been transformed into an enormous trunk, we cannot help recognising that the habitual exercise of this organ, which is used by the animal as a hand, is accountable for its size, strength, and wonderful suppleness, etc., etc.

But M. Gall seems to have particularly devoted himself to a search for the external signs which might indicate a high eminence of intellectual faculties in certain individuals. He recognised that all these faculties are based upon functions of the cerebral organ, and therefore directed his attention to the encephalon. After some years of research, he became convinced that those of our intellectual faculties, which are greatly developed and have acquired a high degree of complexity, may be recognised by external signs consisting of certain protuberances of the cranium.

M. Gall certainly founded his theory on a sound principle: for if it is true, as I have sufficiently proved in Part I., Chapter VII., that all parts of the body which are vigorously and constantly used, acquire a characteristic development and strength of function, the same must hold good equally for the organ of understanding in general, and even for each of the special organs which compose it: so much is certain and easy to demonstrate by many established facts.

M. Gall's underlying principle therefore is beyond criticism; but the results of enquiries into the doctrines taught by this savant give rise to the belief that he misapplied his theory, as regards most of the inferences which he drew from it.

As a matter of fact, in the case of the special organs which enter into the composition of the two cerebral hemispheres and give rise to the different kinds of intellectual faculties, the principle seems to me to be much less applicable than M. Gall supposes; and indeed it can only be in a very small number of extreme cases that certain faculties which have acquired extraordinary power may present indubitable external signs of their existence. In such cases I should not be in the least surprised at the discovery of some such signs, since they have a true natural cause. But with regard to our intellectual faculties, M. Gall has lost sight of the larger distinctions, to embark upon a crowd of details, including even the gradations between these faculties; and this in my opinion is an instance of a very ordinary abuse of the imagination, which destroys the value of our discoveries in the study of nature. And now because M. Gall wanted to prove too much, the public have fallen into the opposite extreme and rejected the whole. Such is the usual progress of the human mind in its various acts; the good that has been done is most often spoilt by excess and abuse. The exceptions to this rule are only found in a small minority of persons who by the aid of a powerful reason set limits to the imagination which tends to carry them away.

To regard certain propensities which have become altogether dominant as innate in the human species, is not only a dangerous belief but also a genuine mistake. We may no doubt possess at birth special tendencies transmitted from the organisation of our parents, but it is certain that if we did not strongly and habitually exercise the faculties favoured by these tendencies, the special organs concerned would never have developed.

In truth, every individual, from the moment of his birth, is in the midst of a set of circumstances altogether peculiar to himself, which to a large extent contribute to make him what he is at the different stages of his life, and which put him in the way of exerting or not exerting one or other of his faculties or inherited tendencies ; so that it may be said in general that we only have quite a moderate share in bringing about the condition in which we find ourselves throughout our existence, and that we owe our tastes, propensities, habits, passions, faculties, and knowledge to the infinitely varied but special circumstances in which each of us has been placed.

From our earliest infancy, those who bring us up sometimes leave us entirely at the mercy of surrounding circumstances, or themselves create circumstances highly disadvantageous to us by their mode of life, thought, and feeling; and sometimes by ill-advised weakness they spoil us and let us acquire many pernicious faults and habits, whose consequences they do not foresee. They laugh at what they call our tricks, and make jokes over all our follies, in the belief that they will be able later on easily to change our vicious inclinations and correct our faults.

It is difficult to conceive how great is the influence of early habit and inclinations on the propensities which will some day dominate us, and on the character which we form. The very pliable organisation of early youth yields readily to the habitual movements of our nervous fluid in one or other direction, in correspondence with our inclinations and habits. The organisation thence acquires a modification which may increase under favourable circumstances, but which can never be entirely obliterated.

It is in vain that after our infancy, we make efforts to guide our inclinations and actions by means of education towards whatever may be useful to us, in short, to give ourselves principles to form our reason and manner of judging, etc. We meet with so many circumstances that are difficult to master, that we are each of us, so to speak, constrained by them and gradually acquire a mode of life in which we ourselves have only had a very small share.

I need not here enter upon the numerous details which form a special environment for each individual, but I must observe, since I am convinced of it, that everything tending to make any of our actions habitual modifies our internal organisation in favour of that action ; so that as a result this action becomes a sort of necessity for us.

Of all parts of our organisation, that which most readily undergoes modifications, due to some acquired habit of thought, idea, or action, is our organ of intelligence. Now the region which is modified is necessarily that concerned with the ideas or thoughts which habitually occupy us. I repeat, then, this region of our intellectual organ when vigorously exercised acquires a development which may finally become noticeable by external signs.

We have dealt with the general principles concerning the organ which gives rise to intelligence; we shall now pass on to a enquiry into the formation of ideas.

Formation of Ideas.

My purpose here is not to undertake any analysis of ideas, nor to show how they become compounded and increased, nor, in short, how the understanding is perfected. Many celebrated men since Bacon, Locke, and Condillac, have dealt with these matters and greatly illuminated them. I need not therefore stay to consider them.

My purpose is simply to indicate by what physical causes ideas may be formed, and to show that comparisons, judgments, thoughts, and all operations of the understanding are at the same time physical acts, which result from the relations between certain kinds of matter in action, and which are carried out in a special organ which has gradually acquired the faculty of producing them.

All that I have to say on this important subject is entirely a matter of probabilities. All is a product of imagination; limited however by the necessity for admitting nothing but physical causes compatible with the known properties of matter, nothing, in short, but causes which may be and probably are correct. With regard to the physical acts which I shall endeavour to analyse, none of them can be witnessed and none therefore can be proved.

I should mention that ideas are of two distinct kinds, viz.:

Simple or direct ideas; complex or indirect ideas.

By simple ideas I mean those which spring immediately and exclusively from the noticed sensations that are impressed upon us by objects either within or without us.

By complex ideas I mean those which are found within us as a result of some operation of our understanding, performed on previously acquired ideas; they therefore require no immediate sensation.

Ideas of all kinds are the result of images or special outlines of objects which have affected us; these images or outlines only become ideas for us, when they have been traced on some part of our organ; and the nervous fluid then passes over them and carries back the result to our inner feeling, by which we become conscious of it.

Not only do ideas originate in two different ways, but a distinction has also to be drawn between those which only become perceptible to us in company with the sensation which produced them, and those which are presented to our consciousness without any accompanying sensation.

The former I call *physico-moral* ideas, and the latter *moral ideas* simply.

Physico-moral ideas are clear, vivid, and sharply defined, and are felt with all the force that they derive from their accompanying sensation. Thus the sight of a building or any other object under my eyes to which I pay attention, gives rise in me to one or several ideas by which I am vividly affected.

Moral ideas, on the other hand, both simple and complex, of which we only become conscious as the result of an operation of our understanding excited by our inner feeling, are very vague and ill-defined and do not affect us with any vividness, although we are sometimes stirred by them. Thus, when I recall an object that I have seen and noticed, a judgment that I have formed, a reasoning which I have carried out, etc., the idea only affects me in a weak and vague manner.

We must then beware of confusing what we experience when we have the consciousness of some idea, from what we experience when a sensation affects us and we pay attention to it.

All that we are conscious of only comes to us through the organ of intelligence, and whatever causes sensation in us works firstly through the sensitive organ and afterwards through the idea that we form of it, if we happen to pay attention to it.

It is essential, therefore, to distinguish moral from physical feeling, since experience of the past teaches us that a failure to recognise that distinction has led men of the highest ability to draw up theories which now have to be destroyed.

No doubt both feelings are physical, but the difference in the terms that I use to distinguish them is sufficient for the purpose that I have in view, and moreover they are the terms in common use.

By moral feeling I mean the feeling that we experience when an idea or thought or any act of our understanding is transmitted to our inner feeling, so that we then have consciousness of it.

By physical feeling I mean the feeling which we experience from some sensation due to an impression on any of our senses and compelling our attention.

These clear and simple definitions must show that the two things are quite different from one another, both as regards their origin and their effects.

It is however through confusing them, as Condillac had done, that M. de Tracy has said :

"Thinking is only feeling, and feeling is for us the same thing as existing; since it is sensations that tell us of our existence. Ideas or perceptions are either true sensations or recollections, or affinities that we perceive, or indeed the desire that may be raised in us by these affinities: the faculty of thought is thus sub-divided into true sensibility, memory, judgment, and will."

In all this there is clearly a confusion between sensations and the consciousness of our ideas, thoughts, judgments, etc. It is a similar confusion between moral feeling and physical feeling that has led to the belief that every creature, which possesses the faculty of feeling, also has that of intelligence; for this is certainly ill-founded.

Sensations no doubt tell us of our existence, but only when we pay attention to them. That is to say, we have to think of them, and this is an act of intelligence.

Thus, in the case of man and the most perfect animals, sensations that are noticed acquaint them with existence and give ideas; but in the case of the more imperfect animals, such for instance, as the insects, in which I recognise no organ for intelligence, sensations cannot be remarked, nor yield ideas; and they can only form simple perceptions of the objects which affect the individual.

Yet the insect possesses an inner feeling capable of emotions which make it act; but since there is no idea connected with it, it cannot be conscious of its existence; in short, it never experiences any moral feeling.

In the case of all creatures endowed with intelligence, we must therefore say: to think is to feel morally, or to have consciousness of one's ideas and thoughts, and also of one's existence; but this is not the same as physical feeling, for the latter is a product of the system of sensations while the former comes from the organic system of intelligence.

Simple Ideas.

A simple idea, arising from a sensation of some object affecting one of our senses, can only be formed when the sensation is remarked, and when the result of the sensation is transmitted to the organ of intelligence and traced or graven on some part of it; this result is perceived by the individual because at the same moment it affects his inner feeling.

Indeed every individual which possesses the faculty of feeling and also an organ for intelligence, promptly receives in this organ the image or outlines brought by the sensation of any object, if the organ is prepared for it by attention. Now these outlines or image of the object reach the hypocephalon by means of a second reaction of the nervous fluid, which, after having produced sensation, carries into the intellectual organ the particular agitation which it derived from that sensation, impresses on some part of it the characteristic outlines of its movement, and finally renders them perceptible to the individual by transmitting their effects to his inner feeling.

The ideas formed on seeing a sky-rocket for the first time, on hearing a lion's roar, and on touching the point of a needle are simple ideas.

Now the impressions which these objects make upon our senses immediately excite in the fluid of the corresponding nerves an agitation which is different in each case; the movement is propagated to the nucleus of sensations; the whole system immediately shares in it; and sensation is produced by the mechanism that I have already explained.

Thus, if our attention has prepared the way, the nervous fluid instantly conveys the image of the object or some of its outlines to our organ of intelligence, and impresses that image or those outlines on some part of the organ; the idea traced is then immediately carried back by it to our inner feeling.

In the same way that the nervous fluid by its movements is the agent for carrying to the nucleus of sensations the impressions of external objects that affect our senses, so too this subtle fluid is the agent for conveying from the nucleus of sensations to the organ of intelligence the product of each sensation that is raised, for tracing its outlines there or impressing them on it by its agitations if the organ has been prepared by attention, and for subsequently carrying back the resultant to the individual's inner feeling.

Thus, in order that the outlines or image of the object which has caused the sensation may reach the organ of understanding and be impressed on some part of it, it is necessary that the organ should first be prepared by an act of attention for receiving the impression, or that this act should open the way by which the product of that sensation may travel to the organ where the outlines of the object may be impressed: in order that any idea may arise or be recalled in consciousness, it is necessary again by means of attention that the nervous fluid should convey its outlines to the individual's inner feeling; this idea then becomes perceptible to him and may be recalled at the will of the individual for a longer or shorter period.

The impression which forms the idea is thus actually traced and graven on the organ, since memory can recall it at the will of the individual and make it perceptible once more.

This in my opinion is the probable mechanism of the formation of ideas; the mechanism by which we recall them at will until time has obliterated or blurred their outlines so that we can no longer remember them.

To try to determine how the agitations of the nervous fluid trace or engrave an idea on the organ of understanding, would be to court the risk of committing one of those numerous errors to which the imagination is liable; all that we can be sure of is that the fluid in question is the actual agent which traces and impresses the idea; that each kind of sensation gives a special agitation to this fluid and consequently causes it to impress equally special outlines on the organ; and, lastly, that the fluid acts upon an organ so soft and delicate and finds its way into such narrow interstices and tiny cavities, that it can impress on their delicate walls traces more or less deep of every kind of movement by which it may be agitated.

Do we not know that in old age, when the organ of intelligence has lost some of its delicacy and softness, ideas are graven less deeply and with greater difficulty; that memory, which is gradually being lost, only recalls ideas graven long ago upon the organ, since they were then more easily and deeply impressed?

Moreover, with regard to the organic phenomenon of ideas, we are dealing exclusively with relations between moving fluids and the special organ which contains them. Now for operations so swift as ideas and all intellectual acts, what other fluid could produce them but the subtle and invisible fluid of the nerves, a fluid so analogous to electricity; and what organ could be more appropriate for these delicate operations than the brain?

Thus a simple or direct idea is formed, whenever the fluid of the nerves is agitated by some external impression or even by some internal pain, and conveys the agitation to the nucleus of sensations, and when attention has prepared the way for the further conveyance of that agitation to the organ of intelligence.

As soon as these conditions are fulfilled, the impression is immediately traced upon the organ, the idea comes into existence and is at once perceived because the individual's inner feeling is affected by it; lastly, the idea may be called up afresh by memory, although obscurely, whenever the individual directs the nervous fluid over the subsisting traces of that idea.

Every idea recalled by memory is therefore much vaguer than when it was formed; because the fact which renders it perceptible does not then result from a present sensation.

Complex Ideas.

By a complex or indirect idea, I mean one that does not arise immediately from the sensation of some object, but is the result of an act of intelligence working on ideas already acquired.

The act of understanding which gives rise to the formation of a complex idea is always a judgment; and this judgment itself is either a consequence or a determination of a relationship. Now this act appears to me to be due to the resultant movement of the nervous fluid which has been broken up by the inner feeling into separate streams, each of which traverses previously made impressions of certain ideas, and thence undergoes special kinds of modifications in its movements. When the streams reunite their individual movements are then combined into this resultant movement.

It is then by means of this movement of the nervous fluid, which is really due to compared ideas or to relations between them, that the subtle fluid makes its impressions on the organ and at the same moment transfers the effect to the individual's inner feeling.

Such is in my opinion the physical cause and precise mechanism which give rise to the formation of all kinds of complex ideas. These complex ideas are quite distinct from simple ideas, since they do not result from any immediate sensation or impression on any of our senses, but originate from several ideas already impressed, and, further, are exclusively due to an act of the understanding in which the sensitive system has no share.

There is this difference between the act of understanding which forms a judgment whence arises a complex idea, and that called a recollection or act of memory, which merely consists in recalling ideas to the individual's inner feeling: that in the former case the ideas employed take part in an operation resulting in a new idea, whereas in the second case the ideas employed take part in no operation, give rise to no new idea, but merely become present to the individual's consciousness.

If it is true that the emotions of our inner feeling give us the faculty and power of acting, and that they enable us to put our nervous fluid in motion and direct it over the impressions of various ideas made upon different parts of the recipient organ, it is obvious that this subtle fluid, while passing over the tracings of any idea, will undergo a special modification in its mode of agitation. We may suppose that, if the nervous fluid simply brings back this special modification of its agitation to the individual's inner feeling, it only makes the idea perceptible or present to the individual's consciousness; but if the fluid, instead of merely passing over the tracings or image of a single idea, divides into several streams, each of which travels over some individual idea, and if the streams then all reunite, the resultant movement of the combined fluid will impress on the organ a new and complex idea, and will then transfer its effect to the individual's consciousness.

If we form complex ideas out of pre-existing simple ideas, we shall have, as soon as they are impressed on our organ, complex ideas of the first order: now it is obvious that if we compare together several complex ideas of the first order by the same organic means with which we compared simple ideas, we shall obtain a resultant judgment of which we form a new idea ; this will be a complex idea of the second order, since it arises from several previously acquired complex ideas of the first order. In this way complex ideas of various orders may be multiplied almost to infinity, as indeed we see in most of our reasonings.

Thus there are carried out in the organ of intelligence various physical acts which give rise to the phenomena of comparison, judgment, analysis of ideas, and reasoning, and these different acts are only operations on ideas already traced, due to the resultant movement acquired by the nervous fluid when it impinges on their tracings or images : and since these operations on pre-existing ideas and even on series of ideas, taken in turn or all together, are only relations sought out by thought through the inner feeling between the various kinds of ideas, these same operations culminate in results which we call judgments, inferences, conclusions, etc.

In the same way, intellectual phenomena are physically produced in the most perfect animals. They are no doubt of a very inferior order, but they are altogether analogous to those described above ; for these animals do receive ideas and have the faculty of comparing them and drawing judgments from them. Their ideas are therefore actually traced or impressed on the organ on which they are formed, since they evidently have a memory, and when asleep may often be seen dreaming, that is, experiencing involuntary recurrences of their ideas.

As regards the signs, so necessary for the communication of ideas, and so useful for increasing their number, I am compelled to confine myself to a simple explanation, with reference to the double service that they render us.

"Condillac," said M. Richerand, "has acquired an immortal glory, in being the first to discover and to prove irrefutably that signs are as necessary to the formation as to the expression of ideas."

I am sorry that the limits of the present work do not permit me to enter into the detail necessary for showing that there is an obvious error in a use of terms which suggests that signs are necessary to the direct formation of ideas; for this cannot have the slightest foundation.

I do not yield to M. Richerand in admiration of the genius and the profound thoughts and discoveries of Condillac, but I am quite convinced that the signs which have to be used for the communication of ideas are generally only necessary to their formation, because they furnish an indispensable means for increasing their number and not because they actually contribute to their formation.

No doubt a language is as useful for thinking as for talking; we have to attach conventional signs to the conceptions we acquire, in order that these conceptions may not remain isolated, but may be associated and compared, and their relationships determined. But these signs are artificial aids, infinitely useful in helping us to think, and not immediate causes of formation of ideas.

Signs of whatever kind do no more than assist our recollection of conceptions acquired recently or long ago, furnish us means of bringing them back into consciousness, and thereby facilitate the formation of new ideas.

Condillac has successfully proved that without signs man could never have extended his ideas as he has done and still does; but it does not follow that signs are themselves elements in ideas.

I greatly regret that I cannot continue the important discussion which this subject demands; but probably some one will see the mistake to which I have drawn attention, and furnish a complete proof of it. Then while recognising all that we owe to the art of signs, we shall recognise at the same time that it is only an art and consequently outside nature.

From the observations and considerations set forth in this chapter I conclude:

1. That the various acts of the understanding require a special organ or system of organs, just as one is required for feeling, another for movement, and a third for respiration, etc.;

2. That in the performance of acts of intelligence the movement of the nervous fluid is the only active factor, while the organ remains passive though contributing to the diversity of the operations by a corresponding diversity of its parts and of the impressions preserved upon them;—a diversity that cannot be calculated since it increases up to infinity according as the organ is used;

3. That acquired ideas are the material of all operations of the understanding; that with this material the individual who habitually exerts his intellect may be continually forming new ideas, and that the only means open to him for this extension of his ideas consists in the art of signs to assist his memory, an art which man alone can use, which he makes more perfect every day, and without which his ideas would inevitably remain very limited.

To throw further light on the subject under discussion, I shall now pass to an examination of the principal acts of the understanding, that is to say, those of the first order from which all the rest are derived.

CHAPTER VIII.

OF THE PRINCIPAL ACTS OF THE UNDERSTANDING, OR THOSE OF THE FIRST ORDER FROM WHICH ALL THE REST ARE DERIVED.

THE subjects which I propose to treat in the present chapter are so vast as to make it impossible for me, within the limits which I have set myself, to exhaust all the problems and topics of interest which they present. I shall therefore confine myself to an attempt to show how all the acts of the understanding, and all the phenomena that result from them, originate in the physical causes which I have expounded in the previous chapter.

The special organ which gives rise to the wonderful phenomena of intelligence is not limited to a single function; it clearly performs four essential functions; and in proportion to its development, these functions acquire more capacity and energy, or are subdivided into many others; so that when the organ is highly developed the intellectual faculties are numerous and some of them attain an almost infinite capacity.

Hence man, who alone furnishes instances of this latter event, is the only kind of being who is enabled by his lofty intellectual faculties to give himself up to the study of nature, to perceive and wonder at the invariable order, even to discover some of her laws, and finally to ascend in thought to the Supreme Author of all things.

The principal functions performed in the organ of intelligence are four in number, and consequently give rise to four very different kinds of acts, viz.:

1. The act which constitutes attention;

2. That which gives rise to thought, from which spring complex ideas of all orders;

3. That which recalls acquired ideas and is named recollection or memory;

4. Lastly, that which constitutes judgments.

We shall enquire therefore what really are the acts of the understanding which constitute attention, thought, memory, and judgments. We shall find that these four acts are the principal ones, the type or source of all the rest, and that it is wrong to place in the first rank *will*, which is only a result of certain judgments, *desire* which is only a moral need, and *sensations* which have nothing to do with intellect.

In saying that desire is only a need or the consequence of a need that is felt, I rely on the fact that needs may be divided into physical needs and moral needs.

Physical needs are those which arise in consequence of some sensation, such as those for escaping from pain or discomfort and for satisfying hunger, thirst, etc.

Moral needs are those which arise from thoughts, and in which sensations have no share, such as those for seeking pleasure or comfort, of fleeing from danger, of indulging one's interests or vanity, or any passion or inclination, etc., etc. ; desire is of this order.

Both these kinds of needs arouse the individual's inner feeling in proportion as he feels them, and this feeling promptly sets in movement the nervous fluid so as to produce actions, either physical or moral, suitable for satisfying them.

Let us now examine each of the faculties of the first order which when combined constitute the understanding or intellect.

OF ATTENTION.

The First of the Principal Faculties of the Intellect.

I now come to one of the most important subjects of study for understanding how ideas and all intellectual acts come to be formed, and how they result exclusively from physical causes ; I refer to *attention*.

Let us then enquire what attention is, and whether the definition of it that I am about to give is confirmed by the known facts.

Attention is a special act of the inner feeling ; it takes place in the organ of intelligence and enables that organ to carry out its functions, for which indeed it is indispensable. Attention is therefore not in itself an operation of the intelligence but of the inner feeling, and prepares the organ of thought or some part of it for carrying out its acts.

It may be described as an effort of an individual's inner feeling, sometimes prompted by a need arising in consequence of some sensation, and sometimes by a desire called up by the memory of an idea or thought. This effort, by transporting and directing the available part of the nervous fluid towards the organ of intelligence, prepares some part of that organ, so that it is ready either to re-awake ideas already impressed upon it or to receive the impression of new ideas.

It seems to me manifest that attention is not a sensation, as Senator Garat has said,[1] and that it is not an idea, or any operation upon ideas ; consequently that it is not an act of will, since this is always the result of a judgment ; but that it is an act of the individual's inner feeling, by which some part of the organ of understanding is prepared for an intellectual operation, and by which that part becomes fitted to receive impressions of new ideas, or recall to the individual ideas which had previously been traced.

Indeed I can prove that when the organ of understanding is not prepared by that effort of the inner feeling called attention, no sensation can arise ; or if one does arise, it leaves no impression but merely skims over the organ without producing any idea, or recalling to consciousness any that had been previously traced.

I was justified in the statement that although every idea is derived in the first instance from a sensation, every sensation does not necessarily yield an idea. The citation of certain well-established facts will suffice to justify this proposition.

When you are reflecting, or your thoughts are occupied with something, although your eyes are open and external objects are constantly affecting your vision by the light which they emit, you do not see any of these objects or at least you do not distinguish them, because the effort of your attention guides the available portion of your nervous fluid over the outlines of the ideas which are occupying you, and because the part of your intellectual organ that is adapted to the reception of sensations of external objects is not at the time prepared to receive these sensations. Thus the external objects which affect your senses from all sides produce no idea within you.

The fact that your attention is directed to the other points of your organ, where the ideas that occupy you are traced, and where perhaps you are still tracing new and complex ones by your reflections, puts these other points in the state of tension or preparation necessary for the working of your thoughts. Hence, under these circumstances, although your eyes are open and receive the impression of external objects, yet you form no idea of them, because the sensations ensuing from them cannot penetrate to your organ of intelligence, which is not prepared to receive them. In the same way you do not hear, or at least do not distinguish, the sounds which strike your ear.

Finally, if somebody speaks to you, although distinctly and in a loud voice, at a moment when you are engrossed in some particular

[1] Course of lectures on the analysis of the understanding for the *École normale*, p 145.

subject, you hear everything and yet grasp nothing, and you are entirely ignorant what has been said because your organ was not prepared by attention to receive the ideas communicated to you.

How often it happens that we read an entire page of a work, when thinking of something different from what we are reading, and without taking in anything of what we have completely read. To this state of intellectual preoccupation is given the name of *distraction*.

But if your inner feeling, on being aroused by some need or interest, suddenly causes your nervous fluid to flow towards the point of your intellectual organ that corresponds to the sensation of some object before your eyes, or some sound that affects your ear, or some body that you are teaching, your attention then prepares this point of your organ to receive the sensation of the object in question and you immediately acquire some idea of that object; if you pay it enough attention, you even acquire all the ideas that its shape, size, and other qualities can impress on you through the medium of the various sensations.

It is, then, only *noticed sensations*, that is, those which arrest attention, that give rise to ideas: thus every idea of whatever kind is the real produce of a noticed sensation, or of an act which prepares the intellectual organ to receive the characteristic outlines of that idea; and every sensation, that is not noticed, but finds the organ of intelligence unprepared by attention, fails to form any idea.

Mammals have the same senses as man and, like him, receive sensations of whatever affects them. But since they do not dwell on most of these sensations, nor fix their attention upon them, but only notice those that are immediately related to their usual needs, these animals have but a small number of ideas, which are always more or less the same with little or no variation.

Hence except for objects which may satisfy their needs and give rise to ideas in them through being noticed, everything else is non-existent for these animals.

Nature offers to the eyes of the dog or cat, horse or bear, etc., nothing that is wonderful, curious, or interesting, but only what ministers directly to their needs or comfort; these animals see everything else without noticing it, that is to say without fixing their attention on it; consequently they can acquire no idea of it. Nor could this be otherwise so long as circumstances do not compel the animal to vary its intellectual acts, to develop the organ which produces them, and to acquire by necessity ideas different from those that their usual needs produce. The results of the education forced upon certain animals are well known.

I am therefore justified in the statement that these animals distinguish scarcely anything of what they perceive, and that everything that they do not notice is for them non-existent, although most of the surrounding objects act upon their senses.

How great is the light thus thrown over the question why animals which possess the same senses as man have yet so few ideas, think so little, and are always subjected to the same habits!

Shall I go on to say how many men there are also who remain unconscious of nearly all that nature presents to their senses? Now in consequence of this method of employing their faculties, and limiting their attention to a small number of objects that interests them, these men only exert their intellects very little, make small variation in the subjects of their thoughts, have very few ideas, like the animals that we have spoken of, and are strongly subjected to the force of habit.

In the case of men who have not been compelled by early education to exert their intellects, the needs are confined to what they think necessary for their preservation and physical well-being, but are extremely limited as regards their moral well-being. The ideas which they form are almost entirely reduced to ideas of self-interest, property, and a few physical enjoyments; their whole attention is absorbed by the few subjects which promote the satisfaction of these needs. Whatever is irrelevant to the physical needs of such a man, to his ideas of self-interest, and his very limited physical and moral enjoyments, is as good as non-existent for him, since he never does nor can notice it, having acquired no habit of varying his thoughts.

Finally education, which so wonderfully develops the human intellect, only achieves this result by imbuing a habit of thinking and of fixing attention on the numerous and varied objects which may affect the senses, on all that can increase his physical and moral well-being, and, consequently, on the true aspects of men's relations with one another.

By fixing attention on the various objects which may affect our senses, we establish distinctions between these objects and determine their differences, affinities, and individual qualities: hence the origin of the physical and natural sciences.

In the same way, by fixing attention on the interests of men in relation to one another, we form moral ideas both as to right conduct in the situations that arise in the course of social life, and as to the progress of useful knowledge: hence the origin of the political and moral sciences.

Education thus inculcates a habit of exciting the intellect and varying the thoughts, and this greatly increases the power of giving attention to numerous different objects, of forming comparisons, of carrying out judgments with a high degree of accuracy, and of multiplying ideas of every kind, but especially complex ideas. Lastly, this habit of exerting the intellect, when the circumstances of life are favourable, enables a man to extend his knowledge, to enlarge and guide his genius, in short, to see things on a large scale, to embrace in thought almost an infinity of objects, and to obtain from the intellect the most solid and permanent enjoyments.

I shall conclude this subject with the remark that although the attention works only by means of the individual's inner feeling when aroused by a need, usually a moral need, it is nevertheless one of the essential faculties of the intellect, and is only carried on in the organ which produces these faculties. Hence there is justification for the belief that no being which is destitute of this organ could give attention to any object.

This section on attention might well be extended, for the subject seems to me very important to investigate, and I am firmly convinced that without a knowledge of the necessary condition under which a sensation can produce an idea, we should never have been able to grasp the matters connected with the formation of ideas, thoughts, judgments, etc., nor the cause why most animals which have the same senses as man only form very few ideas, vary them with difficulty, and are dominated by the influence of habit.

There is therefore good ground for the belief that no operation of the organ of the understanding can take place unless the organ is prepared by attention; and that our ideas, thoughts, judgments, and reasonings, only continue so long as the organ, in which they are carried on, is maintained in a proper condition for producing them.

Since the nervous fluid is the chief instrument in an act of attention, a certain quantity of it is consumed when that act is in progress. Now if it lasts too long, the individual becomes so fatigued and exhausted, that the other functions of his organs suffer proportionally. Hence men who think much, and are constantly meditating, and have acquired the habit of straining their attention almost incessantly on things that interest them are much enfeebled in their digestive faculties and muscular power.

Let us now pass on to an examination of thought, the second of the principal intellectual faculties, though the earliest and most universal of its operations.

OF THOUGHT.

The Second of the Principal Faculties of the Intellect.

Thought is the most universal of intellectual acts, for if we exclude attention, which is the condition of thought, and the other acts of the understanding, thought really embraces all the rest and yet deserves to be especially distinguished.

Thought must be regarded as an action carried out in the organ of intelligence by movements of the nervous fluid. It works on ideas already acquired, either by restoring them unchanged to the consciousness of the individual, as in *memory*; or by comparing some of these ideas together so as to draw judgments from them, or to ascertain their relations, which are also judgments, as in *reasoning*; or by methodically dividing and decomposing them as in *analysis*; or lastly, in creating new ideas on the model of the old, or in contrast to them, and thence new ideas again, as in the operations of the *imagination*.

Is every thought either an act of memory or a judgment? I at first thought so; in that case thought would not be a special intellectual faculty, distinct from recollections and judgments. I believe, however, that we should classify this act of the understanding as one of its special primary faculties, for the thought which constitutes *reflection* and consists in the inspection or examination of an object, is more than an act of memory and yet is not a judgment. Indeed comparisons and investigations of relations between ideas are not mere recollections, nor are they judgments, although these thoughts nearly always terminate in one or more judgments.

Although all acts of the understanding are thoughts, we may yet regard thought itself as the result of a special intellectual faculty, since some of its acts cannot be ranked either as memory or judgments.

If it is true that all intellectual operations are thoughts, it is also true that ideas are the raw material of these operations and that the nervous fluid is the sole agent which gives immediate rise to them, as I have already explained in the previous chapter.

Thought, being an operation of the understanding wrought on previously acquired ideas, can alone give rise to judgments, reasonings, and acts of the imagination. In all this, ideas are the raw material of the operation and the inner feeling is always the stimulating and controlling cause, for it sets the nervous fluid in motion in the hypocephalon.

This act of understanding is sometimes produced as a result of some sensation, which gives rise to an idea and thence to a desire; but it is usually carried out without any immediately preceding sensation, for the recollection of an idea, giving rise to a moral need, is enough to stir the inner feeling and incite it to stimulate the act.

The organ of intelligence thus sometimes carries out its functions as a result of an external cause, which evokes some idea and stirs the individual's inner feeling; while sometimes the organ enters into activity of its own accord, as when some idea recalled by memory gives rise to a desire, that is, to a moral need, and subsequently to an emotion of the inner feeling which leads it to produce some act of intelligence or several such acts in succession.

As in every other bodily activity, thought is achieved only by an excitation of the inner feeling, so that, except for organic movements essential to the preservation of life, acts of intelligence and those of the muscular system are always excited by the individual's inner feeling and should really be regarded as the product of that feeling.

Seeing that thought is an action, it follows that it can only be carried out when the inner feeling excites the nervous fluid of the hypocephalon to produce it, and that, considering the passive condition of the cerebral pulp, the fluid in question must be the only active body that takes a share in this action.

Since a being, endowed with an organ of intelligence, has the faculty of setting its nervous fluid in motion, and of guiding that fluid over the impressed outlines of some previously acquired idea, such a being immediately becomes conscious of this idea when the action is excited. Now this act is a thought, although a very simple one, and it is at the same time an act of memory. But if, instead of recalling a single idea, the individual recalls several, and carries out operations on these ideas, he then forms thoughts less simple and more prolonged, and he can thus carry out various intellectual acts and indeed a long succession of such acts.

Thought is therefore an action, which may be complicated by a great many others of the same kind carried out successively or sometimes almost simultaneously; it may also embrace a large number of ideas of all kinds.

Not only do the operations of thought include ideas already in existence or traced in the organ, but they may also produce ideas which did not previously exist. The results of comparisons, the relations ascertained between different ideas, and, lastly, the products of the imagination, are so many new ideas for the individual; they are generated by his thought, impressed on his organ, and subsequently transferred to his inner feeling.

Judgments, for example, which are also called inferences, because they result from comparisons or calculations, consist both of thoughts and of acts subsequent to thoughts.

The same thing holds good with regard to arguments, for we know that several judgments drawn in turn from compared ideas constitute what we call an argument; now arguments, being only series of inferences, likewise consist of thoughts and of acts subsequent to thoughts.

It follows from all this that no creature which is destitute of ideas could carry out any thought or judgment, and still less any argument.

To meditate is to carry out a succession of thoughts, to sift by successive thoughts either the affinities between several objects that are under consideration, or the different ideas that may be obtained from a single object.

A single object may indeed provide an intelligent being with a number of different ideas, such as those of its mass, size, shape, colour, consistency, etc.

If the individual becomes conscious of some of these ideas when the object is not present, he is said to be thinking of that object; and indeed he actually does carry out one or more successive thoughts with regard to it; but if the object is present he is then said to observe it and examine it, in order to derive from it all the ideas that his method of observation and capacity for attention permit of.

Just as thought works on direct ideas, that is, on ideas obtained by sensations that are noticed, so too it works on the complex ideas that the individual possesses, and may restore them to consciousness.

Hence the object of a thought or succession of thoughts may be material or include various material objects; but it may also be constituted out of a complex idea or several ideas of this character. Now, by means of thought, the individual may obtain still further ideas from these two different kinds, and so on indefinitely. Hence we get *imagination*, which originates in the habit of thinking and forming complex ideas, and which creates by similarity or analogy special ideas on the model of those yielded by sensations.

I shall now bring my remarks on this subject to a close: for I do not propose to undertake any analysis of ideas, as has already been done by abler men and more profound thinkers; I shall have attained my purpose if I have made clear the true mechanism by which ideas and thoughts are formed in the organ of intelligence, in response to excitations of the individual's inner feeling.

I shall merely add that thought is always accompanied by attention, so that when the latter ceases the former promptly comes to an end.

I shall further add that since thought is an action, it uses up the nervous fluid, and consequently, that when it is maintained too long, it causes fatigue, exhaustion and injury to all the other organic functions and especially that of digestion.

I shall conclude with the following remark, which I believe to be well-founded, viz.: that the available portion of our nervous fluid becomes larger or smaller in accordance with certain conditions, so that it is sometimes abundant and more than sufficient for the production of prolonged thought and attention, while sometimes it is sufficient and cannot provide for a succession of intellectual acts, except at the expense of the functions of the other organs of the body.

Hence those rises and falls in the activity of thought noted by Cabanis; hence that facility at certain times and difficulty at other times of fixing attention and following out a line of thought.

In one who is weakened as a result of disease or age, the functions of the stomach are carried out with difficulty, and use up a large part of the available nervous fluid. Now if during this labour of the stomach you divert nervous fluid from the digestion towards the hypocephalon, if, that is to say, you give yourself up to close study and a succession of thoughts which require profound and continued attention, you damage your digestion and endanger your health.

In the evening, when one is more or less exhausted by the various fatigues of the day, especially when one is no longer in the vigour of youth, the available portion of the nervous fluid is generally less abundant and less fitted to provide for continued thought: in the morning, on the contrary, after the recuperation of a good sleep, the available portion of the nervous fluid is very abundant, and can adequately meet the demands that are made upon it by intellectual operations or bodily exercises. Finally, the more you consume of your nervous fluid that is available for intellectual operations, the smaller is your capacity for bodily labour and exercise, and *vice versâ*.

In consequence of these causes and many others, there are remarkable fluctuations in our faculty for following a line of thought, meditating, reasoning, and especially exerting our imagination. Among these causes, variations in our physical condition and the influence upon it of atmospheric changes are not the least powerful.

Since acts of imagination are at the same time thoughts, this is the right place to speak of them.

IMAGINATION.

The imagination is that faculty for creating new ideas that the organ of intelligence acquires by means of its thoughts. It is dependent on the presence of many ideas, out of which new complex ideas are constantly being formed.

The intellectual operations, which give rise to acts of the imagination, are excited by the individual's inner feeling, carried out like other acts of thought by the movements of his nervous fluid, and controlled by judgments.

Acts of imagination consist in creating new ideas by comparisons and judgments of previous ideas, these being taken either as models or as contrasts; so that with this material the individual can form for himself a number of new ideas which are impressed on his organ, and out of these many more again, with no limit to this infinite creation beyond that suggested by his endowment of reason.

I have said that the previously acquired ideas which furnish the material for acts of the imagination are employed in these acts either as models or as contrasts.

In fact, if we consider all the ideas produced by human imagination, we shall see that some, including the larger number, are modelled on simple ideas which have arisen immediately from sensations, or on complex ideas based upon the simple ideas, and that the rest originate in contrast or opposition to the simple or complex ideas that had been acquired.

Since man cannot form any true idea, except of objects or things in the likeness of objects, his intellect would have been limited to the elaboration of this one kind of idea, if it had not possessed the faculty of taking these ideas either as models or contrasts, and forming from them ideas of another kind.

It is by contrast to simple or complex ideas, that man imagined the infinite, basing it on his idea of the finite; when he had conceived the idea of a limited duration, he imagined eternity or an unlimited duration; when he had formed the idea of a body or of matter, he imagined mind or an immaterial being, etc., etc.

It is not necessary to show that every product of the imagination, which does not present a contrast to some simple or complex ideas originally acquired through sensations, is necessarily modelled upon some such idea. How many citations I might make, if I wished to show that wherever man has tried to create ideas, his materials have always been in the likeness of previously acquired ideas or in contrast to them!

It is a truth borne out by observation and experience, that the intellectual organ is in the same case as all the other bodily organs; the more it is exercised the more it develops and the more its faculties extend.

Those animals, which are endowed with an intellectual organ, nevertheless lack imagination, because they have few needs, vary their actions but little, and hence acquire but few ideas, and especially because they rarely form complex ideas and then only of the first order.

But man, who lives in society, has so largely increased his needs that he has been obliged to increase his ideas to a corresponding extent; so that of all thinking creatures he is the one who can most easily exercise his intellect, vary his thoughts the most and, lastly, form the greatest number of complex ideas: hence we have reason to believe that he is the only creature that can have imagination.

If, on the one hand, imagination can only exist in an organ which already contains many ideas, and only originates from the habit of forming complex ideas; and if, on the other hand, it is true that the more the organ of intelligence is exercised, the more it develops, and the more its faculties extend and increase, we shall perceive that although all men might possess this fine faculty called imagination, yet there are only very few who could have it to any high degree.

How many men there are, even excluding those that have had no education, who are forced by their condition of life to occupy themselves daily during the chief part of their lives with the same kinds of ideas and to carry out the same work, and who as a result are scarcely at all able to vary their thoughts! Their habitual ideas revolve in a little circle which is nearly always the same and they make but few efforts to enlarge it, because they have no great interest in doing so.

Imagination is one of the finest faculties of man: it ennobles and elevates his thoughts and relieves him from the domination of minute details; and when it reaches a very high development, it makes him superior to the great majority of other people.

Now genius in an individual is nothing else but a high imagination, guided by exquisite taste and a well-balanced judgment, and nurtured and enlightened by a vast knowledge, and controlled in short by a high degree of reason.

What would literature be without imagination! It is useless for the man of letters to be a perfect master of his language; it is useless for him to cultivate a purified diction and faultless style; if he has no imagination, he is cold, lacks thought and images, rouses no emotion or interest, and all his efforts are futile.

How could poetry, that beautiful branch of literature, and how even could rhetoric dispense with imagination?

For myself, I hold that literature, that beautiful produce of the human intellect, is the noble and sublime art of arousing our passions, elevating and widening our thoughts, and transporting them out of their usual routine. This art has its rules and precepts, but imagination and taste are the exclusive source of its finest products.

Since literature arouses, animates, and charms every man who is able to appreciate it, science is to that extent inferior; for she teaches coldly and stiffly: but science is superior in this, that not only does she serve all the arts and furnish us with the best means of providing for all our physical needs, but that she also greatly broadens our thoughts by showing us everywhere what is really there and not what we want to find there.

The purpose of the former is to give pleasure; that of the latter is to collect all practicable positive knowledge.

This being so, imagination is as much to be feared in the sciences as it is indispensable in literature; for its aberrations in the latter are merely a lack of taste and reason, whereas its aberrations in the former are errors; for the imagination nearly always gives rise to errors, when it is not controlled and limited by learning and reason; and if these errors are captivating, they inflict upon science an injury which is often very difficult to repair.

Yet without imagination there is no genius, and without genius there is no possibility of discovering anything but facts, without drawing any satisfying inferences. Now since every science is a body of principles and inferences carefully deduced from observed facts, genius is absolutely necessary for stating these principles and drawing their inferences; but it has to be guided by a sound judgment, and kept within the limits imposed by a high degree of enlightenment.

Thus, although it is true that imagination is to be feared in the sciences, this only holds good when it is not controlled by a lofty and enlightened reason; when it is so controlled, it is one of the essential factors in the progress of science.

Now the only means of limiting our imagination, so that its aberrations may not affect the advancement of knowledge, is to allow it to work only on real natural objects; since such objects are all that we can possibly know positively; its various acts will then possess a reliability that is proportional to the number of facts considered in the object concerned, and to the excellence of our judgment.

I shall conclude this section with the remark that if it is true that we derive all our ideas from nature, and have none that do not come originally from her, it is also true that we can modify these ideas in various ways by means of our imagination, so as to create new ones entirely foreign to nature; but these latter are always either contrasts to acquired ideas, or else more or less distorted images of objects the knowledge of which we derive from nature alone.

Even in the most exaggerated and extraordinary ideas of man, it is impossible not to recognise their origin, by means of a close examination.

OF MEMORY.

The Third of the Principal Faculties of the Intellect.

Memory is a faculty of the intellectual organs; the recollection of an object or thought is an act of this faculty; and the organ of the understanding is the seat where this wonderful act occurs, while the nervous fluid by its movements in that organ is the sole agent of its occurrence.

This I propose to prove; but first let us consider the importance of the faculty in question.

Memory may be described as the most important and necessary of the intellectual faculties, for without memory what could we do? how could we provide for our various needs, if we could not recall the different objects, that we have come to know or to use for their satisfaction?

Without memory, man would have no kind of knowledge; he would be absolutely destitute of science; he could cultivate no art; he could not even have a language for the expression of his ideas; and seeing that, in order to think or even to imagine, he must, in the first place, have ideas and, in the second place, institute comparisons between these ideas, he would be altogether deprived of the faculty of thought and imagination if he had no memory. When the ancients said that the muses were daughters of memory, they proved that they were conscious of the importance of this intellectual faculty.

We saw in the preceding chapter that ideas spring from sensations which we have experienced and noticed, and that with the ideas thus impressed upon our organ we can form others which are indirect and complex. Since the time of Locke, it has been recognised that all ideas whatever originate from sensations and that none have any other origin.

We shall now see that memory can only come into existence after ideas have been acquired, and consequently that no individual could display any act of memory unless he had ideas impressed on the organ which is the seat of it.

If this is so, nature can have given to the most perfect animals and even to man nothing but memory; she cannot give prescience, that is to say, a knowledge of future events.[1]

Man would no doubt be very unhappy, if he knew definitely what was going to happen to him, the precise date of the end of his life, etc., but the real reason why he has not this knowledge is that nature could not give it to him; it was impossible for her. Seeing that memory is only the recollection of past events, of which we were able to form ideas, and seeing that the future will give rise to events which do not yet exist, we cannot form any idea of it, except in the case of such facts as belong to certain ascertained parts of the order followed by nature.

[1] With regard to future events, those which flow from comparatively simple causes and from the laws which man has discovered in his studies of nature, are capable of being foreseen by him, and up to a certain point of being referred in advance to more or less definite dates. Thus astronomers can prophesy the future date of an eclipse, or when some star will be in some particular position; but this foreknowledge of certain facts is confined to a very small number of objects. Yet many other future events of a different kind are also known to him: for he knows that they will occur without being able to specify precisely when.

Let us now enquire what the mechanism of this wonderful faculty may be; I shall endeavour to show that the operation of the nervous fluid, which gives rise to an act of memory, consists in the acquisition of a special movement by that fluid as it passes over the impressions of some acquired idea, and in the transference of that movement to the individual's inner feeling.

Since ideas are the material of all intellectual acts, memory presupposes ideas already acquired; and it is obvious that an individual who had never had an idea could not possess any memory. Hence the faculty called memory only begins to exist in an individual who possesses ideas.

Memory throws light on the nature of ideas, and even suggests the answer to the question as to what they really are.

Now I have pointed out that the ideas which we have formed through the medium of sensations, and those that we have acquired later by means of thought, consist of specific images or outlines graven or impressed more or less deeply on some part of our organ of intelligence. These ideas are recalled by memory, whenever our nervous fluid, aroused by our inner feeling, comes in contact with their images or outlines. The nervous fluid then transmits the effects to our inner feeling and we immediately become conscious of these ideas: that is how acts of memory take place.

The inner feeling, which controls the movement of the nervous fluid, may direct it over one only of the previously traced ideas or over several of them; hence memory may recall one idea alone, or several ideas in succession, according to the individual's desire.

It follows from the above that if our ideas, both simple and complex, were not impressed more or less deeply on our organ of intelligence, we should be unable to recall them and memory could not therefore exist.

Suppose that some object strikes our attention, a fine building, for instance, which has caught fire and is being burnt up before our eyes. Now for some time after, we can recall that object perfectly without seeing it; for this purpose an act of thought is quite sufficient.

This process must be due to the fact that our inner feeling sets our nervous fluid in motion, and drives it into our organ of intelligence over the outlines impressed by the sensation of the conflagration; and that the modification, acquired by our nervous fluid in its movements as it passes over these particular outlines, is promptly transmitted to our inner feeling and thereupon restores to clear consciousness the idea that we are seeking to recall; although the idea is less vivid than when the conflagration was actually taking place before our eyes.

We likewise recall any person or object that we have previously seen and noticed; and in the same way we recall complex ideas that we have acquired.

Our ideas, then, are specific images or outlines impressed on some part of our organ of intelligence, and we only become conscious of these ideas when our nervous fluid is set in motion and transmits to our inner feeling that modification of its movement which it acquired while passing over these outlines. So true is this, that if during sleep our stomach is disordered or we suffer from some internal irritation, our nervous fluid acquires an agitation which is propagated into our brain. It is easy to conceive how this fluid, when its movements are no longer controlled by our inner feeling, follows no order in passing over the outlines of the various ideas impressed upon it, but brings them into consciousness in the greatest confusion, usually distorting them by strange associations and unbalanced judgments.

During perfect sleep, the inner feeling undergoes no emotion, and for practical purposes ceases to exist; consequently it no longer controls the movements of the available portion of the nervous fluid. Thus a sleeping individual is as though he did not exist. He no longer possesses feeling, although the faculty of it is intact; he no longer thinks, although he still has the power to do so; the available portion of his nervous fluid is in a state of rest, and since the factor which produces actions (the inner feeling) is no longer active, the individual also can do nothing.

But if sleep is imperfect, owing to some internal irritation which stimulates an agitation of the free part of the nervous fluid, the movements of the latter are not controlled by the inner feeling; they therefore occasion disordered ideas and strange and motley thoughts, owing to the haphazard association of ideas that have no relation to one another. Thus are formed the various dreams which we have, when our sleep is not perfect.

These dreams, or the disordered ideas and thoughts which constitute them, are nothing but acts of memory occurring at random and in confusion; they are irregular movements of the nervous fluid in the brain, whereby consciousness is filled with disconnected ideas, since the inner feeling no longer exerts its functions during sleep nor guides the movements of the nervous fluid.

This is why we have dreams when digestion is very difficult, or when we have been much agitated by some great interest or by objects which have stirred us. These produce during sleep a great agitation of our spirits, that is, of our nervous fluid.

Now these disordered acts are always wrought upon ideas that have been acquired and impressed on the organ of intelligence: an individual could never have an idea in a dream, that he had not had when awake, nor recall an object of which he was previously ignorant.

If someone were to be confined from his childhood in a room where daylight was only admitted from above, and if all necessaries were supplied to him without communication, he would assuredly never see in his dreams any of those objects which affect men so strongly in society.

Thus dreams disclose to us the mechanism of memory, just as memory teaches us the mechanism of ideas; when I see my dog dreaming, barking in his sleep, and giving unequivocal signs of the thoughts which agitate him, I become convinced that he too has ideas, of however limited a kind.

It is not only during sleep that the functions of the inner feeling may be suspended or disturbed. While we are awake, a sudden strong emotion sometimes suspends altogether the functions of this feeling, and even all movement in the free part of the nervous fluid; we then suffer from syncope, that is to say, we lose all consciousness and power of action; sometimes also an extensive irritation, such as occurs in certain fevers, similarly suspends the functions of the inner feeling and yet agitates the free portion of the nervous fluid in such a way as to call up disordered ideas and thoughts, and lead to actions no less disordered: in such a case we suffer from what is called delirium.

Delirium therefore resembles dreams as regards the disorder of ideas, thoughts and judgments, and it is clear that this disorder in both cases arises from the fact that the functions of the inner feeling are suspended, so that it no longer controls the movements of the nervous fluid.[1]

But the violence of the nervous agitation causing delirium is the reason for believing that this phenomenon is not only the product of a strong irritation, but sometimes also of a powerful moral affection; so that individuals experiencing it then obtain very little advantage from their knowledge, for their inner feeling, whose functions are disturbed and suspended, no longer guides the nervous fluid in a way suitable for correct ideas.

Indeed, when moral sensibility is very great, the emotions produced in the inner feeling by certain ideas or thoughts are sometimes so considerable, that the functions of this feeling are disturbed, and it is unable to guide the nervous fluid towards the performance of the new

[1] With regard to the faint delirium or kind of dizziness, commonly experienced when we are falling off to sleep, it is probably due to the fact that the inner feeling, which is losing control of the movements still taking place in the nervous fluid, resumes and again gives up that control several times alternately, until complete sleep has supervened.

thoughts which ought to be produced; the intellectual faculties are then suspended or in disorder.

We shall see that lunacy is also due to a very similar cause, which prevents the inner feeling from directing the movements of the nervous fluid into the hypocephalon.

In fact, when any accidental injury has caused some disturbance in the organ of intellect, or when a powerful emotion of the inner feeling has left upon this organ traces deep enough to produce some degeneration in it, the inner feeling no longer controls the movements of the nervous fluid in that organ, and the ideas raised in the individual by the agitations of the fluid present themselves without any order or connection. He gives expression to whatever occurs to him, and his actions are of a corresponding kind. But we see from the acts of this individual that he is always affected by ideas previously acquired and then brought into consciousness. In point of fact memory, dreams, delirium, acts of lunacy, never bring out any ideas beyond those already possessed by the individual.

There are some acts of lunacy which follow from a disorder of certain special organs of the hypocephalon, while the others maintain their integrity; it is then only in these special organs that the inner feeling cannot control or direct the movements of the nervous fluid. People who are affected in this way only perform acts of lunacy with regard to certain subjects that never vary: they appear to be in possession of their reason on all other subjects.

I should travel beyond my province if I tried to follow all the distinctions observed in the disorder of ideas, and to ascertain their causes. It is enough to have shown that dreams, delirium, and lunacy in general are only disordered acts of memory which always work upon ideas previously acquired and impressed on the organs, but which are beyond the control of the individual's inner feeling, because the functions of this power are suspended or disturbed, or because the state of the hypocephalon does not permit of their performance.

Cabanis had no idea of the strength of the inner feeling, nor did he perceive that this feeling constitutes in us a power that can be stirred by any need, or by the smallest desire, or by a thought, and that it is then able to set in motion the free portion of the nervous fluid and to direct its movements either into our organ of intelligence, or towards muscles which require to be actuated. Yet he was forced to recognise that the nervous system often enters into activity of its own accord, without the stimulus of external impressions, and that it can even disregard these impressions and escape from their influence, since concentrated attention or deep thought suspend the activity of the external organs of feeling.

"It is thus," said Cabanis, "that operations of imagination and memory are performed. The motions of the objects recalled and represented are, it is true, usually provided by impressions received in the various organs: but the act which recalls their image, which presents them to the brain in their correct form, which puts that organ in a condition to form numberless new combinations, often depends entirely on causes situated within the sensitive organ." (*Histoire des Sensations*, p. 168.)

This appears to me quite true; it is all the result of the power of the inner feeling of the individual, for this feeling can be aroused by a simple idea which gives rise to that moral need called desire; and we know that desire includes and leads to the performance both of those actions which set up muscular movement, and of those which give rise to our thoughts, judgments, reasonings, philosophical analyses, and to the operations of our imagination.

Desire creates the will to act in one or other of these two ways: now this desire, together with the will which it evokes, arouses our inner feeling, so that it dispatches nervous fluid either into some part of the muscular system, or of the organ which produces acts of intelligence.

If Cabanis, whose work on the *Rapports du Physique et du Moral* is an inexhaustible mine of observations and interesting discussion, had recognised the power of the inner feeling; if he had guessed the mechanism of sensations, and not mistaken physical sensibility for the cause of intellectual operations; if he had recognised that sensations do not necessarily yield ideas, but mere perceptions, which is a very different thing; if, lastly, he had distinguished what is due to irritability from what is the product of sensations; how great would have been the light which his interesting work might have shed! As it is, this work presents the best means for advancing that sphere of human knowledge now in question, on account of the multitude of facts and observations which it comprises. But I am convinced that these facts can only be made useful by fixing our ideas on the essential distinctions drawn in the course of the present work.

By paying attention to what I have said in the present section, we shall probably reach the conviction:

1. That the seat of memory is the organ of intelligence itself, and that the operations of memory are simply acts which recall to consciousness ideas already acquired;

2. That the outlines or images of these ideas are necessarily already graven in some part of the organ of understanding;

3. That the inner feeling, when stirred by any cause, drives the available nervous fluid over such of the impressed outlines as may be selected

by the emotion which it has derived either from a need or inclination, or from an idea which awakes a need or inclination; and that it promptly brings them into consciousness, by carrying back to the sensitive nucleus the modifications of movement which the nervous fluid has acquired from these outlines;

4. That when the functions of our inner feeling are suspended or disturbed, it ceases to direct the movements which may then set our nervous fluid in motion; so that if some cause then agitates this fluid in our intellectual organ, its movements bring back to the sensitive nucleus disordered ideas, strangely mixed and without any connection or sequence; hence dreams, delirium, etc.

We thus see that the phenomena in question everywhere result from physical acts which depend on the organisation and its condition, and on the circumstances in which the individual is placed, as well as on the variety of causes, likewise physical, which produce these organic acts.

Let us pass to an examination of the fourth and last kind of the principal operations of the intellect, viz. those operations which constitute judgments.

OF JUDGMENT.

The Fourth of the Principal Faculties of the Intellect.

The operations of the intellect which constitute judgments are the most important to the individual of any that his understanding can perform; they are those which can least easily be dispensed with, and which he is most often called upon to use.

It is in this faculty of judgment that the will takes its origin; it is also this faculty which gives rise to moral needs such as desires, wishes, hopes, anxiety, fear, etc.; lastly, it is always as a result of judgments that those of our actions, in which our understanding has had some share, are performed.

We cannot carry out any series of thoughts without forming judgments; our reasonings and analyses are pure results of judgments; the imagination itself has no power, except through its judgments, with regard to the models or contrasts used in the creation of ideas; finally, any thought, which is neither a judgment nor accompanied by a judgment, is a mere act of memory or else constitutes a barren inspection or comparison.

How important it is then for every being endowed with an intellectual organ to accustom himself to use his judgment, and to endeavour gradually to improve it by means of observation and experience; for he is then exercising his understanding at the same time, and he increases its faculties to a proportional extent!

Yet if we consider the great majority of men, we find that whenever there is no pressing need or danger they rarely judge for themselves but rely on the judgments of others.

This obstacle to the progress of individual intelligence is not merely due to idleness, carelessness, or lack of means; it is due further to the habit impressed on individuals, during their childhood and youth, of believing what they are told and of always submitting their judgment to some authority.

Now that we have briefly indicated the importance of judgment, and especially of developing it by use and gradually improving it by experience, let us enquire what it is itself and by what mechanism it works.

A judgment is a very peculiar act, carried out by the nervous fluid in the intellectual organ; its result is then traced upon that organ and is immediately brought back to the individual's inner feeling; in other words, is brought into consciousness. Now this act is always the result of some comparison made, or some relation sought, between ideas previously acquired.

The following is probably the mechanism of this physical act: for it is the only mechanism which seems to me capable of giving rise to it, and is in harmony with the known effects of the law of united or combined movements.

Each idea that is graven doubtless occupies a special site in the organ: now when the nervous fluid is agitated and traverses the outlines of two different ideas at the same moment, as occurs in a comparison of two ideas, it is then necessarily divided into two separate streams, one of which passes over the first of the two ideas while the other passes over the second. In each case these two streams of nervous fluid undergo a modification of movement, which is caused by the outlines they pass over and which is peculiar to the idea in question. Hence we may imagine that, if these two streams are subsequently united into one, their movements will also be combined, so that the common stream will have a compound movement which is the resultant of the two kinds of movements brought into combination.

The physical act giving rise to a judgment is therefore probably constituted by the movements of the nervous fluid, when spread over the impressed outlines of ideas that are being compared; and it appears to consist of as many special movements of the fluid, as there are ideas compared and separate streams of fluid passing over the outlines of these ideas. Now these separate streams of the same fluid, each with its special movement, all unite to form a single stream whose movement is compounded of all the special movements named; and this compound movement then impresses new outlines on the organ, that is to say, a new idea which is the judgment in question.

This new idea is instantly brought back to the individual's inner feeling; he has a moral feeling of it; and if it awakes in him any need, which is likewise moral, it evokes his will to act in order to satisfy it.

In addition to inexperience, and to the consequences of the almost universal habit of judging in imitation of others, numerous factors combine to affect our judgments, that is to say, to make them less well-balanced.

Some of these factors are due to an imperfection in the comparisons that are made, and to the preference given by the individual to one idea over another, according to his knowledge, special tastes, and condition; so that the true elements which should enter into the formation of these judgments are incomplete. In all ages there are but few men who are capable of profound concentration, and who, being accustomed to think and to learn from experience, can escape from these factors which tend to affect their judgments.

The others, which it is difficult to avoid, take their origin: (1) In the actual state of our organisation, which affects the sensations originating ideas; (2) in the error in which some of our sensations frequently involve us; (3) in the influence that our inclinations or passions exert on our inner feeling, leading it to give to the movements, which it impresses on our nervous fluid, a different direction from what it would have given them without that influence, etc., etc.

Since I have already treated of the judgment in Chapter VI. of this part, I should travel beyond the limits of my plan if I were to enter into the details of the numerous factors which combine to affect our judgment. It is sufficient, for the purpose that I have in view, to observe that many factors are apt to affect the value of the judgments that we form; and that in this respect there is as great a diversity in the judgments of men, as there is in their physical condition, environment, inclinations, knowledge, sex, age, etc.

Let us not be astonished then at the permanent but not universal disagreement noticed in the judgments that are passed on some thought, argument, work, or any other subject, in which no one can see anything but what he has decided for himself and what he can imagine to himself as a result of the character and extent of his knowledge; nothing, in short, but what he can understand, in accordance with the amount of attention that he pays to the subjects presented to him. How many persons there are, too, who have formed a habit of judging scarcely anything for themselves, and consequently of falling back in almost everything on the judgment of others!

These considerations, which seem to me to prove that judgments are subject to different degrees of correctness, and that this correctness is relative to the individual's circumstances, naturally lead me to say a word about reason, to enquire what it is, and to compare it with instinct.

OF REASON.

AND ITS COMPARISON WITH INSTINCT.

Reason is not a faculty ; still less is it a torch or entity of any kind ; but it is a special condition of the individual's intellectual faculties ; a condition that is altered by experience, gradually improves and controls the judgments, according as the individual exercises his intellect.

Reason therefore is a quality that may be possessed in different degrees, and this quality can only be recognised in a being that possesses certain intellectual faculties.

In the last analysis, it may be said that for all individuals endowed with intelligence, reason is nothing more than a stage acquired in the rectitude of judgments.

No sooner are we born than we experience sensations, mainly from external objects affecting our senses ; we quickly acquire ideas, which are formed in us as a result of noticed sensations ; and we soon compare almost mechanically the objects we have noticed and thus form judgments.

But we are then new to the whole of our environment, destitute of experience, and deceived by some of our senses, so that we judge badly; we are mistaken as to the distances, shapes, colours, and consistency of the objects that we notice, and we do not grasp the relations existing between them. It is necessary that several of our senses should combine gradually to destroy our errors and rectify our judgments ; lastly, it is only with the help of time, experience, and attention paid to the objects which affect us, that rectitude is slowly attained in our judgments.

The same thing is true with regard to our complex ideas, and the useful truths, rules, or precepts communicated to us. It is only by means of much experience, and memory in collecting all the elements for an inference ; only by means of the greatest use of our understanding, that our judgment on these matters is gradually improved.

Hence the wide difference existing between the judgments of childhood and those of youth ; hence again the difference found between the judgments of a young man of twenty and those of a man of forty or more, when the intellect in both cases has always been regularly exercised.

Since the extent of our reason is proportional to the rectitude of our judgment on all things, and especially on the ordinary affairs of life and our relations with our fellows, it follows that this quality is only a certain stage acquired in the rectitude of judgments; and seeing that environment, habit, temperament, etc., involve a great diversity in the exercise of our understanding, that is to say, in our way of thinking, investigating, and judging, it follows that there are real differences between the judgments that are formed.

Thus reason is not an individual object or entity that we may or may not possess, but it is a condition of the organ of understanding, from which results a greater or smaller degree of rectitude in the individual's judgments; so that every being who possesses an organ of understanding, who has ideas and performs judgments, must possess a certain degree of reason corresponding to his species, age, habit, and the various circumstances which combine to retard or advance or to keep stationary the progress made in rectitude of judgments.

Seeing that it is only by paying attention to the objects producing sensations in us, that these sensations can give rise to ideas, it is clear that the more we use this faculty of attention, and the deeper and more sustained it is, the clearer become our ideas, the more accurately are they defined, and the more correct are the judgments that we form from them.

Hence it follows that the highest stage of reason is that which comes from extreme clearness in the ideas, and from an almost invariable soundness of judgments.

Man, who is much more capable than any other intelligent being of this profound and continuous attention, and can fix it on a great many different objects, is the only one who can have an almost infinite variety of clear ideas and whose judgments consequently possess the highest rectitude; but for this purpose he has to exercise his intellect vigorously and perpetually, and his circumstances have to be favourable for it.

It follows from the above, that since reason is only a stage in the rectitude of judgments, and since every intelligent being can carry out judgments, they must all possess a certain degree of reason.

Indeed, if we compare the ideas and judgments of an intelligent animal, which is still young and inexperienced, with the ideas and judgments of the same animal when it has reached the age of acquired experience, we shall find that the difference between these ideas and judgments is quite as discernible in the animal as it is in man. A gradual improvement in the judgments, and an increasing clearness of the ideas, fill up in both cases the interval between the time of their childhood and that of their maturity. The age of completed experience and development is clearly distinguished from that of inexperience and low development of the faculties in such an animal, just as it is in man. In both cases the same features are to be recognised, and an analogous progress in the acquirements; it is only a matter of more or less in the different species.

In animals possessing a special organ for intelligence, there are therefore various degrees in the rectitude of judgments and hence various degrees of reason.

Doubtless the highest degree of reason gives man a perception of the propriety or impropriety both of his own ideas and beliefs, and of the ideas and beliefs of others; but this perception, which is a judgment, is not a property of all men. Those who do not possess it substitute a false perception in place of that just perception which arises from a highly trained intellect; and, since the former is the best they can attain, they believe it to be just. Hence arises that diversity of opinions and judgments in individuals of the human species; a diversity which will always stand in the way of a real agreement between the ideas and judgments of individuals, owing to the fact that men are situated in very different circumstances and therefore cannot attain the same degree of reason.

If we now compare reason with instinct, we shall see that the former to some extent gives rise to determinations to act, originating in the intellect, that is to say, in ideas, thoughts, and judgments; whereas instinct, on the contrary, is a force which impels towards an action, without any previous determination or intellectual act.

Now, since reason is only a stage acquired in rectitude of judgments, the determinations of action which spring from it may be wrong or unsuitable when the judgments producing them are erroneous; they may be false in whole or in part.

But instinct, which is simply an impelling force and arises from the inner feeling when stirred by some need, never makes a mistake in the action to be performed; for it does not choose, nor does it result from any judgment, and there are not really different degrees of it. All action caused by instinct is therefore invariably due to the kind of excitation produced by the individual's inner feeling, just as all movement communicated to a body is made up, both in direction and strength, by the power which communicated it.

There is nothing either clear or really exact in Cabanis' idea of attributing reasoning to external sensations, and instinct to internal impressions. All our impressions are invariably internal, although the objects causing them may be either external or internal. Observation of the facts should convince us that it is more just to say:

That the reasonings and determinations, following from judgments, take their origin from intellectual operations; whereas instinct, in causing some action, takes its origin from the needs and propensities which arouse immediately the individual's inner feeling and make him act, without any choice or deliberation, or, in short, any participation of the intellect.

The actions of certain animals are therefore sometimes based on rational determinations, but more often on an instinctive force.

If we pay attention to the facts and arguments presented in the course of the present work, we shall perceive that there must be animals which have neither reason nor instinct, such as those which are destitute of the faculty of feeling; that there must be others which have instinct but possess no degree of reason, such as those which have a sensitive system, but lack the organ for intelligence; lastly, that there must be others again which have instinct together with some degree of reason, such as those which possess a system for sensations and another for acts of the understanding. The instinct of these last is the source of nearly all their actions, and they rarely make use of such degrees of reason as they possess. Man, who comes next, also has instinct which in certain circumstances makes him act; but he is capable of acquiring much reason, and of using it to control the greater part of his activities.

Besides the *individual reason* of which I have been speaking, there is established, in every country and region of the earth, in proportion to the knowledge of the men who live there and to certain other factors, a *public reason*, which is almost universal, and which is upheld until new and sufficient causes operate to change it. In both cases, the individual and the public reason are always constituted by a certain degree of rectitude in the judgments.

It is true that in a society or nation, errors and false beliefs may be as much matters of general assent as ascertained truths; so that various errors, prejudices, and truths go to make up the degree of rectitude of judgment, both in individuals and in the received opinions of societies, groups, and nations, according to the age or period considered.

We have therefore to recognise different stages of advance in the reason of a people or society, as in that of an individual.

Men who strive in their works to push back the limits of human knowledge know well that it is not enough to discover and prove a useful truth previously unknown, but that it is necessary also to be able to propagate it and get it recognised; now both individual and public reason, when they find themselves exposed to any alteration, usually set up so great an obstacle to it, that it is often harder to secure the recognition of a truth than it is to discover it. I shall not dwell on this subject, because I know that my readers will see its implications sufficiently, if they have any experience in observation of the causes which determine the actions of mankind.

In concluding this chapter on the principal acts of the understanding, I terminate at the same time all that I propose to present to my readers in the present work.

In spite of the errors into which I may have been led, the work may possibly contain ideas and arguments that will have a certain value for the advancement of knowledge, until such time as the great subjects, with which I have ventured to deal, are treated anew by men capable of shedding further light upon them.

THE END.

INTRODUCTORY LECTURE FOR 1800

J. B. LAMARCK

PREFACE

On the Aims and the Plan of this Work.

In writing this work I have aimed to offer to the students who attend my lectures at the Museum, and to those of the central schools of the Republic, a brief account of the invertebrate animals together with a systematic classification of these animals based mainly upon consideration of their organization.

Although very restricted, this work will, I believe, be of use not only to those studying this great part of the animal kingdom, but also to those studying that kingdom as a whole. At the least it may help those intending to arrive at an exact and general view of these interesting creatures.

It has above all been made necessary since consideration of animal organization has shown that the class of insects and that of worms described by Linnaeus in the *Systema naturae* are extremely badly determined, classified, and delimited. It must certainly become more and more so in times like the present when the study of Natural History is so widely cultivated and even made part of our national education.

The determination of the classes, orders, and principal genera, of the invertebrate animals being once achieved, and in a way that conforms to the natural order of these animals, it will henceforth be easy to develop all that concerns their history, characters, and everything of interest shown by their species, either in books devoted to these things or in particular studies. And it is likely that less time will be spent on those arbitrary

Footnotes to this Introductory Lecture that are preceded by numbers are the author's; those following asterisks are the translator's.

systematic classifications in which one sees everywhere the most incongruous juxtapositions.

The Opening Lecture printed at the beginning of this work may serve to characterize in a general way the animals with which it deals, to give some idea of the astounding gradation which exists in their organization, and lastly to make felt all the kinds of interest which knowledge of these singular creatures can inspire. In it I have just touched upon several important philosophical matters, which the nature and limits of this work have not permitted me to develop, but to which I intend to return elsewhere in the detail necessary to make their basis known, and with such explanations as will prevent any misinterpretation.

Despite the brevity forced on me by my plan, I have permitted myself some expositions at the head of the treatment of each class and order, limited in length, but necessary for the adequate recognition of the objects mentioned in these groups. I have further prefaced the accounts of each of these classes with a general chart of the divisions and genera that it comprises. These charts may be consulted with advantage because they give an instantaneous picture of the size of the class, the nature and sequence of its divisions, and a list of its genera with their serial numbers.

I have not followed slavishly the characters given in other works because, having at my disposal the magnificent collection of the Museum and another fairly rich, which I have myself made in the course of nearly thirty years' work, I have been able to verify those which I have used from this material, and this practice I have followed as much as possible.

The usage generally adopted by Petrologists and Palaeontologists of ending the names of all remains of living bodies found in the fossil state uniformly, and in this way changing the name of *Pecten* to *pectinite*, of *Turbo* to *turbinite* etc., has forced me to change the names of some genera of the testate molluscs and the coral-forming polyps since their names have been inappropriately ended as if they applied to objects known only in the fossil state, which, however, was not so.

In order to identify with certainty the genera whose characters I give, I have named under each a single, or very rarely several, known species. In addition I have supplied some synonyms for which I can vouch. This should be sufficient to make me understood.

Finally, I hope later to offer to the public a *Table of Species* for each of the genera established in this work. Naturalists will realize that this is a considerable enterprise and a very difficult one. But Citizen Latreille*, who is most learned in entomology, and to whom we will owe the determination of all the species of insects in the Museum, has promised to undertake the part of this concerning the classes of crustaceans, arachnids, and insects. The public, who already know the distinction of this Naturalist, will certainly look forward to his contribution to the new work on which I hope soon to congratulate him.

DISCOURS D'OUVERTURE

Prononcé le 21 Floreal An 8†

Citizens,

If it is true that to study Natural History profitably, even in its smallest details, it is first necessary to have a mental picture of the whole of nature's products, and in this way to view them from such a distance that the great groups into which they appear to fall are made clear, may be compared the one with another, and lastly that their principal characters may be determined; if indeed this be true, then I must start by reminding you briefly of the great divisions that nature itself has apparently established in the immense range of natural objects, of the sequence or order in which it appears to have created them, and of the peculiar relationship which it has brought about between their actual natures and the ease or difficulty with which they multiply their numbers.

Thus, in order that I may be able to give you clear and useful ideas about the things with which I shall deal in this course, I must first run quickly over the principal sections which arise from the distinctions that nature itself has established between its numerous products, pointing out that which is especially inportant in them and serves to distinguish them, and finally I must indicate the place occupied by the natural beings with which I attempt to acquaint you in the table of affinities and in the systematic classification that I have made.

* Pierre André Latreille, (1762–1833), French naturalist who joined the Museum in 1796 where he was charged with arranging the entomological collection. Collaborated with Lamarck in the production of the last two volumes of the *Histoire naturelle des Animaux sans Vertèbres.*

† 11 May 1800.

You are aware that all natural objects that we can observe have for many years been divided by Naturalists into three kingdoms, known as the *animal kingdom,* the *vegetable kingdom,* and the *mineral kingdom.* By this division the objects within each one of these kingdoms have suffered comparison with each other on equal terms, although some have quite a different origin from others.

I have found it more convenient to use a different primary division, whose object it is to enable the things in question to be better known. Thus I divide all the natural objects comprising the three kingdoms into two principle branches* :

1st. Into organized bodies, living.

2nd. Into inert bodies, without life.

The beings, or living things, such as the animals and plants thus form the first of these two branches of natural objects. They have, as is well known, the ability to feed, to develop, to reproduce, and they are necessarily subject to death.

But what is not equally realized is that they fashion their substance themselves as a result of the actions and the faculties of their organs : and what is still less known is that they give rise, through their remains, to all the formed inert substances seen in nature. The diversity of these substances increases as a result of changes, more or less rapid according to circumstances, by which they are completely dissociated into their constituents. In a great stretch of land where the soil has been denuded of plant and animal life for many centuries, such as the deserts of Africa, one may seek in vain for any but almost purely vitreous substances ; the mineral kingdom is here seen very greatly reduced. The contrary holds in all countries long covered by an abundant vegetation and with a varied fauna. Here a vegetable or organic earth, thick, succulent, and fertile, covers in one place or another mineral matter of nearly every kind, sometimes saline, bituminous, sulphurous, pyritic, sometimes stony, etc. etc. etc. I have developed the proofs of these important facts in a work published under the title *Mémoires de Physique et d'Histoire Naturelle* (see the 7th memoir) etc.

These varied inert and lifeless substances, either solid or liquid, either simple or formed, comprise the second branch of natural objects, constitute the bulk of our globe, and are for the most part known as minerals.

* This division clearly presages the coining by Lamarck of the term "biology", which he first used in 1802. It was independently and simultaneously proposed by G. R. Treviranus.

They are governed by laws that are substantially known and which are very different from those to which living bodies are subject. One may say that there is an immense gap between inert substances and living bodies, which precludes the arrangement of these two sorts of matter in a single series and makes one feel that they have quite different origins.

Among the living beings, that is among those which comprise the first branch of natural objects, *plants,* lacking sensitivity, voluntary movement, and organs of digestion, are clearly distinguished from animals all of which have these faculties and organs. Plants, as you know, are the subject of that fine and important branch of Natural History known as *Botany.*

Similarly among living things *animals,* endowed with feeling, the faculty of moving their bodies or parts of them voluntarily, and provided with organs of digestion fall to the lot of that great and interesting branch of Natural History called *Zoology.* Now, as the many creatures, which I shall discuss and which I propose to examine with you in this course, form part of the field of Zoology, it would be well for us to pause for a moment to consider *animals* in general, to view the ensemble of these admirable creatures and, finally, not only to notice the excellence of their faculties, their pre-eminence over all other living things, but also to appreciate the strange and astounding gradation which they show in the aggregate in the construction and complexity of their organization, in the extent and number of their faculties, in brief in the facility, the rapidity, and the number of means, of their multiplication.

For several years I have pointed out in my lectures at the Museum that the presence or absence of a vertebral column in the animal body divides the whole animal kingdom into two great sections, which are very distinct and which one might regard as two great families of the first order.

I believe that I was the first to establish this important distinction, which had not occurred to any other Naturalist. It is today adopted by a number, and together with other of my observations, introduced into their written works without acknowledgement*.

All known animals can thus be divided in a remarkable manner.

1. Into *vertebrate animals.*
2. Into *invertebrate animals.*

* Possibly a reference to Cuvier.

The vertebrate animals all have an internal vertebral column, almost always bony, which renders the body firm, forms the base of the skeleton with which they are provided, and makes it difficult for them to contract. This vertebral column bears the animal's head at the anterior end, the pectoral ribs at the sides, and provides along its length a canal in which the pulpy strand known as the spinal marrow, which may be considered as a mass of united nerves, lies enclosed.

The animals possessing this vertebral column may be distinguished also by the red colour of their blood, or rather by the presence in the main vessels of their bodies of a red fluid that we call *blood,* and which is composed of three distinct elements intimately mixed. They have never more than four limbs; many of them have none at all.

One can make out in the vertebrate animals, as in the others, a gradual diminution both in organization and in the number of their faculties.

These animals are less numerous than others in nature, and all are contained in the first four classes of the animal kingdom, namely:

1. Mammals.	Viviparous and with mammae. With lungs.	Heart with 2 ventricles and warm blood.	Vertebrate animals.
2. Birds.	Oviparous and without mammae. With lungs.		
3. Reptiles.	Oviparous without hair or feathers. With lungs.	Heart with 1 ventricle and cold blood.	
4. Fish.	Oviparous with fins. With gills.		

These vertebrate animals are the most perfect, have a more complicated organization, enjoy more faculties, and are in general better known than the invertebrate ones.

The animals comprising the second branch of the animal kingdom, the second of the great families which make up this kingdom, are those which I have called *invertebrate animals,* and are those which we shall examine in greater detail. They are strongly marked off from the others in that they lack a vertebral column bearing the head and forming a base for an articulated skeleton.

Hence their bodies are flabby, highly contractile; and among them those which have somewhat rigid bodies owe this almost entirely to the consistency of their skins or external covering. If in some of them are found internal hard parts, these never form the basis of a true skeleton or provide a sheath for a spinal marrow. It is therefore wrong to compare these hard parts to a vertebral column, although the attempt has been made.

Among the *invertebrate animals* those possessing legs have at least six, and some have many more.

The *invertebrate animals* have no true blood, that is to say they have not actually that mixed and invariably red fluid composed of three distinct elements, which is made and exists above all in the principal vessels of the vertebrate animals. But in its place the invertebrates have a whitish sanies, rarely coloured red, which seems to be nothing more than an alimentary fluid more or less altered by the action of the organs.

It is then of this second branch of the animal kingdom, of this great family of *invertebrate animals,* that I propose to talk to you during this course. I shall attempt to list them for you, to tell you of their history and of their main distinctive characters ; and you will see that they form a special series, certainly the largest in numbers that the animal kingdom can offer us.

This great series, which alone comprises more species than all the other groups of the animal kingdom together, is at the same time the most fertile of all kinds of marvels, of the most peculiar and curious types of structure, of intriguing and even admirable details associated with the manner of life, or with the need to conserve, or reproduce the remarkable animals of which it consists. It is still, however, generally speaking the least known.

Certainly the study of this great part of the animal kingdom is full of attractions and varied interests. It provides useful knowledge from which the greatest advantage can be derived in many circumstances. Unhappily prejudice has caused this interesting branch of Natural History to be neglected. Apparently the general smallness of these animals and above all the prodigious numbers in which they occur have given rise to the contempt, or at least the indifference, which they have commonly enjoyed. It is, for all that, undeniable that these animals deserve from all points of view to secure the attention of Naturalists, and to become no less than other natural products the objects of their researches.

I would go further ; if we set aside the interest we have in knowing both those that may be of use to us either in themselves or for their products, and those against whose harmful or irksome activities we wish to guard ourselves, and of this I shall speak later, science may yet learn enormously from the study of these remarkable animals. For they show us still better than the others that astounding degradation in organization, and that progressive diminution in animal faculties which must greatly interest the philosophical Naturalist. Finally they take us gradually to the ultimate stage of animalization, that is to say to the most imperfect animals, the most simply organized, those indeed which are hardly to be suspected of animality. These are, perhaps, the ones with which nature began, while it formed all the others with the help of much time and of favourable circumstances.

If one considers the range of forms, of size, and of characters, which nature has given to its products, the variety of organs and of faculties with which it has endowed living beings, one cannot refrain from admiring the infinite resources which it has brought to the accomplishment of its task. For it almost appears as if everything imaginable has in fact occurred ; that all possible forms, faculties, and functions, have been exhausted in the creation and in the composition of this immense number of existing natural products. But if one examines carefully the means that nature seems to employ one feels that their power and abundance have sufficed to produce all the observed effects.

It seems, as I have already said, that *time* and *favourable circumstances* are the two principal means which nature uses in creating all its products. We know that time has no end for it, and is consequently always at its disposal.

As to the circumstances of which it has need, and which it still uses all the time to change its productions, one can say that they are in effect inexhaustible.

The main circumstances are derived from the influence of climate ; changes in the temperature of the atmosphere and of all the environment ; from the variety of places, of habits, movements, actions ; and lastly from the variety of modes of life, conservation, defence, multiplication, etc. etc. Now consequent upon these different influences the faculties are widened and strengthened by use, they are changed by new habits maintained for a long time ; and gradually the conformation, consistence, the nature and state of the parts of organs, affected by the results of all these influences, are conserved and are spread by reproduction*.

* 'Génération.' (tr.)

The bird, attracted by need to water in search of prey on which to live, spreads the digits of its feet when it wants to strike the water and to move across the surface. The skin, which joins these digits at their base thus gets the habit of stretching. So in time the large membranes joining the digits of ducks, geese, etc. were formed as we now see them.

But the bird whose manner of life accustoms it to perching on trees has necessarily, in the end, toes that are stretched and constructed in another way. Their claws are lengthened, sharpened, and curved into hooks for gripping the twigs on which they rest so often.

In the same way one feels that the bird of the shore that dislikes swimming, and which none the less needs to approach the water to find its prey, is continually exposed to sinking in the mud ; but, wishing to avoid the immersion of its body, its feet will get into the habit of stretching and lengthening. The effect of this, for those birds which continue to live in this manner over generations, will be that the individuals will be raised as if on stilts, on long naked legs, that is to say legs bare of feathers up to the thigh and often beyond.

I could here pass in review all the classes, orders, genera, and species of existing animals and show that the form of the individuals, and of their parts, their organs, faculties, etc. etc. are entirely the result of the circumstances to which the ancestry of each species has been subjected by nature.

I could prove that it is not the form of the body or its parts which determines the habits, the manner of life of animals ; but that on the contrary it is their habits, their manner of life, and all the effective circumstances which have, in time, established the form of the body and of its parts. With new forms, new faculties have been acquired, and little by little nature has come to the state which we see today.

It is proper to pay the greatest attention to this important consideration, the more so since the order which I briefly indicated in the animal kingdom, showing clearly a gradual degradation in organization as well as in the number of animal faculties, suggests the course which nature has taken in the formation of all living things.

Thus the vertebrate animals, and among them the mammals, show a maximum in the number and in the assembly of the major animal faculties ; while the invertebrate animals, and above all those of the last class (polyps) show, as you will later see, the minimum.

Indeed, if we consider first the simplest animal organization and rise afterwards by degrees to that which is most complex, as for instance to pass from the monad, which is, so to speak, nothing but an *animated point*, to the mammaliferous animals and amongst them to man, there is clearly an even gradation in the complexity of the organization of all animals, and in the nature of its results, that cannot be too much admired, and which we feel bound to study and to understand thoroughly.

Similarly among plants, from the byssus pulverulens* or from the simple mould[1] to the plant that has the most complicated structure and is the most fertile in organs of every kind, there is clearly an even gradation in some ways analogous to that seen in animals.

In speaking of this even gradation in the complexity of organization I do not mean at all to suggest the existence of a linear series with regular intervals between the species and genera: such a series does not exist; but I speak of a series in which the principal masses, such as the great families, are almost regularly spaced; a series which quite certainly exists among both animals and plants, but which when considered in terms of genera or more particularly of species, forms in many places lateral ramifications of which the terminal ends provide truly isolated points[2].

If there exists a graded series among living things, at least of the principal groups, in respect of the complexity or simplicity of their organization, it is clear that in a truly natural classification either of animals or plants, one must necessarily place at the two ends of the scale the forms which are most dissimilar, the furthest apart in affinities, and which consequently form the extreme terms presented by animal or plant organization.

Any classification which departs from this principle seems to me to be faulty because it cannot conform to the order of nature.

This important consideration will enable us to understand better the nature of the living beings with which we shall be

* A genus long since abandoned.

[1] Such perhaps as the *mucor viridescens* which appears to represent the *minimum* of vegetability.

[2] Several Naturalists in noting the more or less marked isolation of many species, of certain genera, and even of some small families, have thought that living things of either kingdom are related to one another in terms of their natural affinities as are different points on a geographical chart or map of the world. They regard the small, well-pronounced series which have been called *natural families* as being arranged among themselves as a network, in conformity with the order that they attribute to nature. This idea, precious to some moderns who have studied nature badly, is an error which will doubtless be dissipated in face of wider and more profound knowledge of the organization of living bodies.

concerned in this Course ; to judge more accurately their affinities with other living things ; and lastly to determine more easily the rank which each should occupy in the general series of living beings, and particularly in that of the animals already known.

You will see that the polyps, which form the last class of invertebrate animals and hence of the whole animal kingdom, and especially those comprising the last order of this class, show only the rudiments of animality, and you will become convinced that the polyps are in relation to other animals what the *cryptogamic* plants are to plants of other classes.

This gradation shown in the simplicity or complexity of the organization of living things is an undeniable fact which I stress because its recognition at the present time throws the greatest light on the natural order of living things, and at the same time sustains and guides the conception which embraces them all in the imagination, or which places each of them in true perspective when they are considered severally.

To this extremely interesting point must be added that which shows us that as animal structure becomes more complex, or composite, so animal faculties increase in number ; a result which is both simple and natural. But further, these animal faculties, in multiplying, lose something of their scope. Thus in those animals with most faculties, those of the faculties which are common to all animals are much less diffuse and are less potent than they are in animals with a simpler organization. This is something which observation teaches us and which it was important to notice. An example is the capacity to regenerate, met with in all animals however simple or complex their organization ; so also their means of multiplication are proportionately more numerous and easier as animals have a simpler organization, and *vice versa.*

In insects, even more in true worms, and above all in polyps, animal faculties actually are less numerous than in the animals of the first classes which are the most perfect, but they are in the former of wider extent since irritability is greater and more lasting, the capacity to regenerate the organs is more pronounced, and the reproductive capacity of individuals is much greater. Moreover, the place occupied in nature by the *invertebrate animals* is immense and by far exceeds that of all other animals together.

The extreme term of the animal scale at the end comprising the most simply organized forms is not known. We are necessarily ignorant also of the very smallest of these animals; but we may be sure that the further one goes towards this end of the animal scale the more vast becomes the number of individuals in each species, for their reproduction is proportionately the more rapid and the easier. Consequently the number of these animals is uncountable and has no limit other than those which nature sets in time, space, and circumstances[3].

This ease, this abundance, and this speed with which nature creates, multiplies, and propagates the most simply organized animals can be especially evident at such times and in all places as are favourable in this respect.

The earth, in fact, especially towards its surface, the waters, and even at certain times and in certain places the atmosphere, are populated to some extent by animated molecules whose organization, simple though it is, suffices for their existence. These animalcules reproduce and multiply, above all during hot periods and in hot climates, with frightful fecundity which is markedly greater than that of large animals whose structure is more complicated. It seems, so to speak, that matter is animalized throughout, so rapid are the results of this astonishing fecundity. Thus without the immense wastage of animals of the last orders of the animal kingdom that occurs in nature they would soon overwhelm, and perhaps annihilate, in consequence of their enormous numbers, the more organized and perfect animals of the first classes and orders, so great is the difference in the means and facility of reproduction between the former and the latter.

But nature has anticipated the dangerous consequences of this widespread capacity to reproduce and multiply. On the one hand, it has done this by limiting considerably the length of life of these simpler organisms of the last classes, and especially of the last orders. On the other hand, it has made these animals the prey of one another which constantly tends to reduce their numbers. Finally it has determined, by differences of climate, the places in which they can exist, and by the changing seasons, that is to say by the effects of different atmospheric conditions, even the periods during which they can do so.*

[3] What a point of view from which to assess nature! It has surely not proceeded from the complex to the simple in creating its productions. Let us rather consider what it has been able to achieve with time and circumstance.

* Later in the *Philosophie Zoologique* Lamarck was again to show that he appreciated clearly the importance of natural checks to the increase of animal population, It is interesting to consider the reasons for so convinced an evolutionist failing to be led from this to the conception of natural

By virtue of these wise precautions of nature, order prevails. Individuals multiply and propagate themselves, destroy themselves in various ways ; no species predominates to the extent of causing the ruin of another, except perhaps in the higher classes where the multiplication of individuals is slow and difficult ; and as a result of this state of affairs we can understand that species are in general conserved.

It follows nevertheless from this fecundity of nature, which is greater as the simplicity of living things increases, that the invertebrate animals should show, and actually do show, the greatest numbers of those living in nature, although they have at the same time the least vitality.

What is more remarkable is that among the changes which animals and plants continuously cause in the state and nature of the earth's surface by their products and debris, it is not the largest and most perfectly organized animals which have the most effect.

I have attempted to prove in my *Mémoires de Physique et d'Histoire Naturelle* (p. 342, no. 490) that the calcareous matter so abundant on the earth's surface is actually the product of animals which formed it.

But what should be our astonishment in realizing that most of the calcareous matter in existence, that, in brief, which in every country on earth forms great chains of calcareous mountains and enormous banks of chalk, is due only in small part to the shelled animals and is principally made of chalk produced by the polyps that live in polyparies, that is to say the madrepores, millipores, etc. which are almost the least perfect and smallest of animals?

Although these animals are so small, so simple in organization, so delicate, and of such low vitality, their reproductive capacity is so great that their huge numbers greatly surpass in their effects those which a greater volume and a longer life in other animals are capable of producing.

So that we can say that here what nature fails to obtain in quantity through the individual it obtains amply by the number of animals involved, by their enormous fecundity, by their

selection. I believe the most serious reason to be that selection, as it was presented for example by Buffon, was capable of an anti-evolutionary interpretation. Buffon had suggested that the differences between fossil and living faunas could be explained by the elimination of the less "fit" species in the course of time. Such a view clearly runs counter to Lamarck's belief that these differences are to be explained by the evolutionary transformation of species. It would indeed permit the belief that all known living species *and* all extinct ones had been originally created. Not until Charles Darwin were the two views to be reconciled.

admirable capacity for prompt reproduction by which they are enabled to multiply in a short time their rapidly accumulated generations; and finally by the aggregation of the products of these numerous animalcules.

It is a fact now well established that the coral-forming polyps (that is to say the great family of animals including the madrepores, millepores, astroites, meandrites, etc.) by the continuous excretion of their bodies, and by virtue of their astonishing numbers and accumulated generations, form on a large scale in the bosom of the ocean the greatest part of the calcareous matter that exists. The many polyparies which these animals produce, and to which they add continuously in volume and quantity, form in certain places islands of considerable size, fill in bays, gulfs, and the largest roadsteads, in a word they block up ports and completely alter the state of the coasts. These huge banks of madrepores, millepores, etc. piled on top of one another, covered and finally mixed up with serpulids, oysters, barnacles, and other shell-fish, form irregular submarine mountains of almost limitless extent.

The impressive conception of which I have spoken leads us to examine in living things the remarkable faculties of those that nature has endowed with animal life, and already we have learned from it, as I have just remarked, that as the organization of animals is simplified, so animal faculties become actually fewer, but also generally of greater scope.

The extraordinary metamorphoses of insects; the regeneration of the head in slugs, of the limbs in crustaceans, of the gills or rays in starfish, or of all the tentacles in actinians after these parts have been amputated; the reproduction of certain worms after section of a single individual; the multiplication of hydras or fresh-water polyps which is effected as in corals, the faculty of coral-forming polyps or zoophytes to multiply by a continual budding which ramifies their polypary, forming stems resembling those of plants in appearance and bearing; and finally the various methods of propagation and multiplication of all these animals, but above all of the amorphous or microscopic polyps, are phenomena not seen everywhere in the animal kingdom. The invertebrate animals, however, more simply organized than others, provide some examples.

As we approach the lower end of the animal series where animal life seems to come into existence, where, in brief, the first and most simple rudiments of organization are found, we feel that in so great a simplicity of structure reproduction by special organs can no longer find a place. Thus observation teaches us that in the animals with a very simple organization, such as the *polyps,* no organ special to reproduction is known.

These animals appear to be entirely devoid of sexuality ; the more organized of them multiply by a budding which usually ramifies their body or the polypary which they form and in which they live. But the least perfect of them, those with the simplest and in some respects the most problematic organization, multiply by a special fission which occurs slowly across or along the gelatinous body of these minute animals.

Thus reproduction in the least organized animals is reduced to a separation of one part of the body of the animal which becomes detached by a natural fission. In somewhat higher animals the portion of the body which breaks off is smaller, isolated, and shows in advance, in miniature, a body resembling that from which is has been derived. This method leads gradually to the setting aside of a special place in the animal's body, where the separations must take place of a kind of internal bud, which nature transforms little by little into eggs, as in the end it transforms these into organized *placentae.* This same method gives rise to organs special to reproduction, and soon afterwards the distinction between the sexes begins to establish itself. This at least is what observation suggests. I will not now prove these interesting points ; I will only say that the marvels which most invertebrate animals show us in the remarkable details of their structure, in their productions, in their ways, and in their habits and various methods of propagation, are not the only considerations which should bring us to study these singular animals. I can show that man has as well the greatest material interest in understanding them.

Indeed, it is known that many of the molluscs, insects, worms, etc. provide countless articles of use and often of the greatest importance for medicine, the arts, commerce, and domestic economy. Thus the silk worm, the Mexican and Polish cochineals, the kermes, the bee, the gallfly which produces gall nuts, the cochineals, producers of gum-lac, the leeches, the oysters, the crayfish, etc. etc. prove already that the invertebrate animals provide as much for our arts and our needs as do the objects of the other branches of Natural History and that they deserve to be studied and known.

But one can further show that apart from the considerable uses to which man can put a great many of these animals, or their products, he has the greatest interest in trying to know them well in order to shield himself from the ills which most of them inflict, and from the damage that they can cause. Plants, animals, and man himself, are not in any way spared. Many different insects comsume all parts of living plants ; bite, suck, and devour other living animals either by attaching themselves to the body or by introducing themselves into the interior ; destroy animal and plant products prepared and stored for our use, such as pelts, Natural History collections, etc. Finally most of the true worms inhabit the bodies of living animals, and that of man himself, and there multiply considerably and consume the substance of their hosts to such effect that one can say that the evils, the injuries, and the devastations, that all these animals inflict are often incalculable.

We must thus recognize that, some molluscs, many insects, most worms, and many other invertebrate animals being in general very harmful, man has the greatest interest in studying them in seeking to know them that he may find means to destroy them, to free himself of them, or at least to preserve himself from their ravages and the evils they can inflict.

Man is indeed able, by his industry, to reduce greatly the total of the ills that these animals can do him. But it is clear that it is in studying these kinds of animals well, in seeking knowledge of the places they inhabit, the times of their developments, their mode of life, etc. that he can hope to succeed in preventing both the excess of their multiplication, at least in his immediate neighbourhood, and that of the damage that they can cause. *V. Oliv. Journal d'Hist. Nat.* nos. 1 and 2.

Thus one feels that several powerful considerations should bring us to study the *invertebrate animals* and to know them as well as the others ; and they show that this study, while both amusing and curious, is not of less interest for us than other branches of Natural History.

The great interest which these important considerations arouse being now doubtless sufficiently known to you, I pass to the systematic distribution, that is to say the classification, of the animals of which I have to speak.

The celebrated Linné, and nearly all Naturalists up to the present have, as I have already said, divided the whole series of *invertebrate animals* into two classes only ; namely

Into *insects* and *worms*.

It follows that everything not regarded as an insect was without exception referred to the class of worms.

They placed the class of insects after that of fish, and that of worms after that of insects. The worms thus formed in this distribution the last class of the animal kingdom.

But known anatomical observations on the structure of these animals, and above all those made in the last few years, do not permit of the retention of this division of the invertebrate animals into *insects* and *worms*. It is now recognized that many of these animals, such as the molluscs which Linné placed with the worms are better, or less simply, organized than the *insects* and that in consequence they should be placed before them, that is to say immediately after the fish. While other invertebrate animals with an organization more simple still than that of the insects and even of the worms should be placed after them ; so that those with the most simple organization should really form the end of the animal kingdom.

It was thus necessary to take no further account of the division established by Linné, and it was necessary either to re-unite all these animals into a single class or to divide them into a certain number of well determined and distinct sections.

I have been continuously employed in this useful reform since I joined this establishment and, although the progress of my researches has meant successive different changes in the results of my work in this respect, I now believe that I can establish definitively the classification of the invertebrate animals, and that I am bound to characterize them in the following way.

Definition

Thus the *invertebrate* animals are those which lack a vertebral column, and consequently an articulated skeleton ; which have no true blood, having only in its place a sanies ordinarily white which seems to be nothing but a kind of lymph ; and which, finally, possess a flabby body which is in a high degree contractile. They are also those, as I have already said, in which the faculties of regenerating their parts and of multiplying by reproduction are most highly developed. They form the branch of the animal kingdom which not only has the greatest number of known species, but also is such that its extreme limit will doubtless never be determined because of the infinite smallness of the species which are near this limit and the coarseness of our senses, which stands in the way of our efforts to perceive them.

The division of the invertebrate animals

I divide the invertebrate animals, as you will see from the accompanying table, into seven classes and twenty orders, which I will describe in succession. The characters of these classes are derived from consideration of their organization and in particular of the three kinds of organ most essential to animal life, namely, 1st respiratory organs, 2nd those which serve the circulation or movement of fluids, 3rd and lastly those which provide for sensation.

These truly essential considerations bring together the animals with real affinities, and necessarily separate those which lack them. They establish moreover most precisely the course of the reduction in complexity of organization, a reduction which clearly grows from one extreme of the *invertebrate animals* to the other, as it does also in the *vertebrate* ones ; so that in the animals of the seventh and last class the organs of respiration, of circulation, and of feeling, are not at all distinguishable and even seem to be non-existent.

CLASSIFICATION

The seven classes which I have established among the *invertebrate animals* are :

1. The molluscs.
2. The crustaceans.
3. The arachnids.
4. The insects.
5. The worms.
6. The radiata.
7. The polyps.

These seven classes added to the four comprising the *vertebrate animals* give us eleven distinct classes for dividing the whole animal kingdom, each well isolated and all arranged in an order depending upon the progressive simplification of the organization of the animals comprising them.

The classification which I have just indicated seems to me such as must be established among the *invertebrate animals.* It is impossible usefully to add a single class to the seven which I have proposed* or to remove one from them and, above all, it is impossible to upset the order of affinities established by nature

* Lamarck later erected separate classes for the annelids (1802), and the infusarians (1807), and cirripedes (1809).

itself which is clearly shown by observation of structure and which I believe to be perfectly preserved in the order of the seven classes in question.

The *Molluscs,* although one stage lower than the fish in that they lack a vertebral column and consequently an articulated skeleton, together with true blood, are none the less the best organized of the *invertebrates.* They breathe through gills as do fish ; they all possess a brain and nerves, one or more muscular hearts, and a complete circulatory system.

The class of *Crustaceans,* the second class of the invertebrates, and that which includes animals hitherto confused with the insects because they possess, like insects, limbs and articulated antennae, follows that of the molluscs and it is no longer permissible to confound its members with those animals which really deserve the name of *insect.*

Indeed, great though the affinities of the crustaceans with the *insects* are, they are even greater with the *arachnids,* and they are essentially distinguished from both these groups in that they all breathe by gills as do the molluscs, and never have either stigmata or aerial tracheae, and are provided with a muscular heart for the circulation of their fluids.

The *Arachnids,* although nearer to the insects than are the crustaceans are no less to be separated from them to form a distinct class, because the animals of this class undergo no metamorphosis and from the earliest stages in their development possess articulated limbs and cephalic eyes. However as the *arachnids* have fairly numerous affinities with the crustaceans it is necessary to place them between the crustaceans and the insects. In this there is nothing arbitrary.

After the *arachnids* comes at once, and of necessity, the class of insects, that is to say that immense series of animals which undergo metamorphosis, which all possess six legs in the adult stage, and have antennae and eyes in the head, and stigmata and aerial tracheae for respiration.

These animals, of unlimited interest in respect of the peculiarities of their organization, their metamorphoses, and their singular habits, have a less complex structure than that of the *molluscs* or *crustaceans.* Thus there is to be found in them no muscular heart, but only a dorsal vessel, which has slight constrictions that contract alternately, and which does not appear to end in ramifications.

Respiration, which in *mammals, birds,* and *reptiles,* is effected by lungs and which is carried on by gills in *fish, mol-*

luscs, and *crustaceans*, is effected in *arachnids* and *insects* by tracheae only, that is by air-filled vessels which ramify throughout the whole extent of the body. It is only in the aquatic larvae of insects that gills are still found since tracheae would be of no use to them.

The *Worms* form the fifth class of the invertebrates. They have without doubt to follow immediately upon the insects, and not to precede them, on the basis of the nature of their organization. Still less should they be placed after the molluscs and before the crustaceans as a learned Naturalist has recently held.

Many worms, like insects, breathe only through tracheae of which the external apertures are the stigmata. Many others breathe by gills like aquatic insect larvae. In this respect, and in respect of their nervous system, since it is in them a knotted spinal marrow, they resemble insects. But the worms differ essentially in that they never have articulated limbs, and in that none of them suffer a true metamorphosis.

Worms, being without a muscular heart, cannot properly be placed after the molluscs and before the crustaceans. This is already so clear that the proofs that I might give in dealing with these animals are now unnecessary.

Finally the body form of the worms, much simpler than that of insects, places them of necessity after the latter, for their body seems to consist entirely of an elongated abdomen without a differentiated thorax. Usually they appear to have no head, no organ of vision, etc. etc.

After the *worms* must come the *Radiata*, which form the sixth class of the invertebrate animals.

Although these animals are extremely peculiar and even, in general, still little known, what we do know of their organization indicates clearly the place to which I assign them in the invertebrate series. Thus the essential organ of sensation with which the animals of all the preceding classes are endowed—there being still traces of it in the worms—is no longer distinguishable in them. It seems that they actually lack both longitudinal marrow and nerves and are only irritable in a simple way. Neither heart nor vessels for circulation have been found in them. Further their organ of respiration is so little obvious that one is forced to look for it in the multitude of absorbing contractile tubes to be seen in most of these animals; tubes which carry water into ramifying canals and there circulate it, or at least pass it through nearly all the parts of the animal's interior.

However the *radiata* do not comprise the last stage that can be established in the series of the animal kingdom. It is necessary to distinguish them from the *polyps*, which form the last link in this interesting chain.

In the *radiata*, which I have so named because their organs are usually radially arranged, we still see not only organs which appear to be respiratory, but also viscera other than the intestines such as ovaries of various forms, etc. The mouth which seems to be invariably ventral usually still possesses organs of mastication.

The *polyps* form the seventh class of the invertebrate animals, and the last in the animal kingdom. They represent the final grade which has been seen in this interesting kingdom, and it is among them that is to be found the undiscovered end of the animal ladder, in a word the rudiments of animal life which nature creates and multiplies with such facility in favourable circumstances; but which she also destroys with equal ease and speed by a simple departure from the right conditions for their origin.

Although the polyps are of all animals the least known, they are without doubt the most simple, and hence the ones with fewest faculties. No organs of sensation or respiration and no circulatory system are to be found in them. All their viscera are reduced to a simple alimentary canal which, as a sac of greater or lesser length, has only one opening—at once both mouth and anus. This alimentary canal is apparently enveloped by absorbing gobules containing fluids kept in some sort of motion by suction and transpiration.

The animalcules which are found at the end of the last order of the polyps are no more than animated particles, gelatinous corpuscles, simple in form, and contractile in nearly all directions.

Such, in brief, are the characters of the seven classes which can usefully be established among the invertebrates. I shall deal in order with these classes and with the genera comprising them, only limiting myself in the treatment of each genus by the time available.

Although the invertebrate animals seem at first to be of less interest than others, you have seen that they are no less worthy of your attention and curiosity, and that all sorts of reasons should lead you to study them and to understand them thoroughly. Their study is, moreover, so much the more fruitful of useful discoveries since our knowledge is here still so little advanced.

In the classification of invertebrate animals the respiratory organs have been most used as characters, and it seems to me proper to present succinctly here a definition of the various kinds of organ that appear to subserve the respiration of animals.

Animal respiration is effected by four different kinds of respiratory organs. That is to say that each animal in which such organs are apparent breathes by means of one of the following kinds of organ :—namely,

By lungs.
By gills.
By aerial tracheæ.
By aquatic tracheæ.

Of Lungs

Lungs are goupings of cells contained in a special cavity of the body of the animal possessing them, and to which are attached the ends of more or less ramified tubes known as *bronchi.* These tubes all end in a common tube to which is given the name of *trachea,* and which opens in the mouth at the base of the tongue. The cells and the bronchi alternately fill and empty with air, by the alternate inflation and collapse of the body cavity containing them.

To the walls of the cells and the bronchi cling the final ramifications of the pulmonary vessels, which are infinitely subdivided and are coiled in every direction. Doubtless the walls of the cells and the bronchi are full of pores, some absorbent some exhalent, which establish communication between the air introduced into the pulmonary cells and the blood which circulates in the vessels of the lung. (*See* my *Mémoires de Physique et d'Histoire Naturelle,* p. 311.) This is the respiratory organ of the animals of the first three classes.

Of Gills

Gills are naked respiratory organs which show no cells, no bronchi, and no tracheæ.

The vessels, which, in the lungs, creep over the walls of the cells and the bronchi to receive the influence of the air which is introduced through the trachea, creep nakedly in the *gills* on leaves or fringes that ramify and twist infinitely to present a large surface to the ambient fluid and receive its influence.

Animals with gills are usually aquatic, so that it is water itself that they breathe ; in other words liquid water is for them the ambient fluid.

Their respiration is thus entirely effected by the contact maintained between their gills and the continually replaced water. But it seems that these organs have the faculty of separating from the water the air held in solution in it, or which is constantly mingling with it, and that it absorbs this and introduces it into the fluids of the animal. There are doubtless aerial gills, that is to say those which do not function at all in water, but only in atmospheric air. Those of the slugs and snails are examples. Gills are the essential respiratory organs of fish, molluscs, and crustaceans.

Of Aerial Tracheæ

Aerial tracheæ are, in a sense, lungs without cells or bronchi, and equally without definite boundaries.

This respiratory organ consists of a multitude of aerial vessels, which ramify almost infinitely and extend throughout the interior of the animal and if its parts ; and which open to the exterior by holes or short slits called *stigmata*.

In animals with true lungs, the air enters a discrete organ ; it must there affect the blood which has itself to seek out the air in this organ.

In animals with aerial tracheæ the air, by contrast, introduces itself through an organ that is distributed everywhere ; it consequently itself seeks out the essential fluids of the animal to communicate its influence to them.

Aerial tracheæ are the respiratory organ of arachnids, insects, and many worms.

Of Aqueous Tracheæ

Aqueous tracheæ are to gills as aerial tracheæ are to lungs.

This organ, which seems to be respiratory, is not found except in aquatic animals of such simple organization that neither longitudinal spinal cord nor nerves are known in them. It consists of a number of aquatic vessels, which ramify and extend throughout the interior of the animal, and which open to the exterior by a host of small tubes which are extensible and contractile and take in and expel water. In this way the water perpetually circulates, so to speak, in the body of the animal and carries everywhere the effects of the air which the organ doubtless knows how to separate. This is the respiratory organ of the *radiata*, or at least of most of them.

The animals in which no respiratory organ is perceptible probably breathe by the absorption of air which they separate from the water by means of absorbing pores either on the surface of their bodies, or on that of their alimentary canal ; but they certainly have no special organ for effecting this separation. Such is the case in all the polyps.

FIN DU DISCOURS D'OUVERTURE

ON FOSSILS

I give the name of *fossil* to the remains of living bodies, changed by their long sojourn in the earth or beneath the waters, but whose form and organization are still recognizable.

From this point of view the bones of vertebrate animals and the remains of the shelled molluscs, of some crustaceans, of many of the echinoderms, of coral-forming polyps, and of the woody parts of plants, may be called fossils when, after a long period beneath earth or water, they have suffered changes which in debasing their substance have nevertheless failed to destroy their form, shape, and the characteristics of their organization.

It follows from this that, when a shell, as a consequence of a lengthy stay in the earth, has undergone changes partly denaturing its substance without destroying its form, it becomes a true *fossil*.

Of the different states of transformation in which fossil shells are found, the most usual is that in which the only change is the destruction of the animal portion, that is the gelatinous or membranous part which was mingled with the chalky part, in such a way that the shell after this destruction consists almost entirely of calcareous matter. Such a shell has thus lost its lustre, its colours, and often even its nacre, if it ever possessed it ; for we know that it owes these to the animal part mingled with the chalky part which it possessed in its freshwater or marine habitat. In this altered state such a shell is normally completely white. Sometimes, however, it has a colour, not its own, derived from the mud in which it has for long been buried.

In France the fossil shells of Courtagnon near Reims, of

Grignon near Versailles, of *ci-devant* Touraine, etc. are nearly all in this calcareous state and lack more or less entirely their animal part, that is their lustre, their original colours, and their nacre.

Some other fossils have suffered changes, which have not only robbed them of their animal parts but which have also transformed their own substance into silicious matter. To this second kind I give the name *silicious fossil.* In this state, as is known, are found various oysters (ostracites), many terebratulids (terebratulites), some trigonites, some ammonites, some echinites, some encrinites, etc.

When a calcareous fossil shell continues to change in nature, and is transformed into a silicious fossil, it suffers a contraction as the remaining parts of which it is composed come together. Thus there arises between it and the stony mass which contains it an empty space, which is, however, most often crossed by lateral adhesions joining the shell to the stone.

The fossils of which I have spoken are in some cases buried in the earth, in others lying here and there at its surface. They are found in all bare parts of our globe, even in the middle of the greatest continents, and, what is really remarkable, on mountains and at very considerable heights. In many places terrestrial fossils form banks several miles in length[4].

In earlier times little attention was paid to the discovery and study of the remains of living bodies found in the fossil state. They were only treated on their own merits and for this reason had little interest. A fossil shell, being necessarily without lustre, colours, or beauty, and being very often defaced, was thrown out of some collections as altered, dead as the conchologists put it, and without interest. But since it has been pointed out that these fossils are extremely precious *monuments* aiding the study of the revolutions undergone by the different parts of the earth's surface and the changes which living beings have themselves successively suffered (points which I have always stressed in my lectures), the study of fossils has become especially favoured and is now regarded by Naturalists as one of the most interesting of their pursuits.

The first fruits of the study of fossils have led a number of Naturalists to believe that the following proposition was well founded :

[4] On this subject see my work entitled *"De l'influence du mouvement des eaux sur la surface du globe terrestre, et des indices du deplacement continuel du bassin des mers, ainsi que de son transport successif sur less differens points de la surface du globe."*

That *all fossils appertain to remains of animals or plants of which the living counterparts no longer exist in nature.*

They have concluded from this, with regard to the earth's crust which has fossils in all its land parts and in all its various climates, that the earth has suffered a universal upheaval, a general catastrophe, as a result of which many diverse species of animals and plants are now absolutely lost or destroyed.

A universal upheaval, which would necessarily regularize nothing and would confuse and disperse everything, is a great convenience for those Naturalists who wish to explain everything without the trouble of observing and studying the order which nature follows in all its products and in everything pertaining to its domain. I have already dealt elsewhere with this supposed universal upheaval of the earth and I will now return to fossils.

It is quite true that, of the great numbers of fossil shells found in the various regions of the world, there are but very few species of which the living or marine counterparts are known. However, although the number is very small, it suffices, once accepted, to overthrow the universal applicability of the proposition referred to above.

It is well to notice that, among the fossil shells of which the living or marine counterparts are not known, there are many which have forms very similar to shells of the same genus which are known from a marine habitat. However, they differ more or less from the known living ones and cannot be regarded as of the same species because they do not resemble them perfectly ; these we are told are the lost species.

I agree that it is possible that fresh-water or marine shells perfectly equivalent to the fossil shells of which I have spoken will never be found. I believe the reason to be as I shall now give it briefly ; and I hope that it will then be felt that, although many fossil shells are different from all known marine shells, that does not prove in any way that the species of these shells are extinct, but only that they have changed in the course of time, and that contemporary ones have forms that differ from those of the individuals whose fossil remains we find.

Every observant and well-informed person knows that nothing on the earth's surface remains permanently in the same state. Everything in the course of time suffers various changes, sooner or later according to the nature of the object and of its circumstances. High regions are perpetually being reduced by the alternate actions of sun and rain ; all that comes away from them is carried to lower regions ; the beds of streams, of rivers, of the seas themselves, are gradually displaced[5] ; in brief everything at the earth's surface changes in place, in form, in nature, and in appearance.

Now if, as I will try to show elsewhere, differing conditions bring about for living things a diversity of habit, a different mode of existence, and consequently changes or developments in their organs and in the form of their parts, one must believe that every living thing whatsoever must change insensibly in its organization and in its form. One must also believe that all such change in organization and forms resulting from the influence of conditions is propagated in reproduction, and that, after many centuries, not only will new species, genera, and even orders, be created, but also each species will necessarily vary in its organization and in its forms.

It is then no longer surprising if, among the many fossils in all the land areas of the world, formed from the debris of so many animals that existed in other times, there are found so few the living counterparts of which we know. On the contrary, if anything is surprising, it is that we find, among these many fossil remains of bodies once alive, some whose living counterparts we do know. This fact, which is borne out by our collections, makes us suppose that *fossils* of animals with known living counterparts are the least ancient ones. The species to which they belong have doubtless not yet had sufficient time to change in any of their forms.

One must therefore never expect to find among living species all those which are found in the fossil state, and yet one may not assume that any species has really been lost or rendered extinct. It is certainly possible that among the largest animals some species have been extinguished as a consequence of the multiplication of man in the places which they inhabited. But this conjecture cannot be established from the consideration of fossils alone : we shall only be sure on this point when all the habitable globe is perfectly known.

[5] See my work on *l'influence du mouvement des eaux sur la surface du globe terrestre.*

BIOGRAPHICAL MEMOIR OF M. DE LAMARCK

GEORGES CUVIER

Among the men devoted to the noble employment of enlightening their fellows, a small number are to be found (and you have just witnessed an illustrious example[1]), who, gifted at the same time with a lofty imagination and a sound judgment, embracing in their vast conceptions the entire field of the sciences, and seizing with a steady eye whatever afforded the hope of discovery, have laid before the world nothing but certain truths, establishing them by evident demonstrations, and deducing from them no consequences but such as were irresistible, never allowing themselves to be led away by what is conjectural or doubtful ; men of unequalled genius, whose immortal writings will shed a light, like so many phari, on the paths of science, as long as the world is governed by the same laws.

Others, with minds not less ardent, nor less adapted to seize new relations, have been less severe in scrutinizing the evidence ; with real discoveries with which they have enriched science, they have mingled many fanciful conceptions ; and, believing themselves able to outstrip both experience and calculation, they have laboriously constructed vast edifices on imaginary foundations, resembling the enchanted palaces of our old romances, which vanished into air on the destruction of the talisman to which they owed their birth. But the history of these less favoured philosophers is not perhaps the least useful. While the former should be unreservedly held up to our admiration, it is equally important that the latter should form the subject of our study. Nature alone produces genius of the first order ; but it is competent to every laborious man to aspire to

[1] This Memoir, not yet published, was designed to follow that of Volta, read by Arago at the meeting on the 27th of June 1831, for which see the 16th vol. of this Journal. It was read at the meeting of the French Academy of Science, 26th November 1832.

a rank among those who have done service to science, and that rank will be the more elevated in proportion as he has learned to distinguish by marked examples the objects accessible to his exertions, and the difficulties which may oppose his progress. It is with this view, that, in sketching the life of one of our most celebrated naturalists, we have conceived it to be our duty, while bestowing the commendation they deserve on the great and useful works which science owes to him, likewise to give prominence to such of his productions in which too great indulgence of a lively imagination has led to results of a more questionable kind, and to indicate, as far as we can, the cause, or, if it may be so expressed, the genealogy of his deviations. This is the principle by which we have been guided in all our historical eloges, and, far from thinking that we have been thereby wanting in the respect due to the memory of our associates, we conceive that our homage is rendered purer, just because it is carefully freed from all that was unworthy of them.

JEAN BAPTISTE PIERRE ANTOINE DE MONET, otherwise named the Chevalier de Lamarck, was born at Bazantin, a village in Picardy between Albert and Bapaume, on the 1st August 1744. He was the eleventh child of Pierre de Monet, superior of the place, of an ancient house of Bearn, but whose patrimony was quite inadequate to the support of such a numerous offspring. The church, at that period, offered a ready resource, and sometimes a large fortune, to the cadets of noble families, and M. de Monet made an early choice of that destination for his son. As a preliminary step, he was sent to study under the Jesuits at Amiens ; but the boy's inclination by no means responded to his father's wishes. All that surrounded him spoke another language : for ages his relations had carried arms ; his eldest brother fell in the breach at the siege of Bergen-op-Zoom ; two others were still in the service ; and the moment when France was so actively engaged in the dismal struggle begun in 1756, was not one fitted to discourage a young man of spirit from following such examples. His father, however, opposed this desire ; but the good old man having died in 1760, no consideration could prevail on the youthful abbé to adhere to his profession. He set out for the army of Germany on a wretched horse, followed by a poor youth from his village, provided with no other passport than a letter from one of his neighbours, a Madame Lameth, directed to M. de Lastic, colonel of the regiment of Beaujolois. It is easy to conceive the annoyance of this officer at finding himself embarrassed with a boy, whose puny appearance caused him to be thought younger than he really was; he ordered him, however, to his quarters, and continued his duties. The moment in fact was a critical one. It was the 14th of July 1761, when the Marshal de Broglie, having united his army to that of the Prince of Soubise, designed next day to attack the allied army, commanded by Prince Ferdinand of Brunswick. At the dawn of day, M. de Lastic inspected his troops, and the first person whom he saw was the young stranger, who, without saying a word, had placed himself in the first rank of a company of grenadiers, and nothing could induce him to quit his station.

It is well known that this battle, which bears the name of the little village of Fissingshausen, between Ham and Lippstadt, was lost by the French, and that their two generals, mutually accusing each other of the defeat, immediately separated, and undertook no important measure during the rest of the campaign. In the vicissitudes of the contest, the company to which M. de Lamarck had attached himself was placed in a situation which exposed it to the whole fire of the enemy's artillery: in the confusion of the retreat it was forgotten and left there. Already all the officers were killed, and only fourteen men remaining, when the oldest grenadier perceiving that there were no longer any French within sight, proposed to the young volunteer so speedily become commander, to withdraw this little troop. "This post has been assigned us," replied the boy; "we must not quit it unless we are relieved"; and, in fact, he caused them to remain till the colonel, seeing that this company was wanting, sent an order, which could now reach its destination with the utmost difficulty. This instance of firmness having been reported to the Marechal, he instantly gave M. de Lamarck a commission, although his instructions required him to be very sparing in promotions of that nature. Soon after, M. de Lamarck was nominated to a lieutenancy; but such a successful commencement of his military career was not attended with the consequences that might have been expected, for a most unfortunate accident removed him altogether from the service, and entirely altered his destination. When his regiment was in garrison at Toulon and Monaco, one of his companions, in play, lifted him by the head, and occasioned a serious derangement in the glands of the neck. He was obliged to repair to Paris for more skilful treatment than these places afforded, but the efforts of the most celebrated surgeons had no effect; and the danger was become very imminent, when our late associate M. Tenon, with his usual penetration, perceived the nature of the disorder, and effected a cure by a complicated operation, the marks of which always continued visible. This confined him for a year, and during that time, the extreme slenderness of his resources kept him in solitude, which afforded ample leisure for reflection.

The profession of arms had not caused him to lose sight of the notions of physics he had acquired at college.

During his stay at Monaco, the singular vegetation of that rocky country had attracted his attention, and the Traité des Plantes Usuelles of Chomel having accidentally fallen into his hands, inspired him with some taste for botany. From his lodging in Paris, which, by his own account, was much higher than accorded with his wishes, the clouds formed almost his only spectacle, and their varied aspects suggested his earliest ideas of meteorology—a subject which could not fail to interest a mind always distinguished for activity and originality. He now began to perceive, as Voltaire has said of Condorcet, that lasting discoveries might confer on him a different kind of celebrity from a company of infantry.

The resolution which he formed in consequence, was not less firmly adhered to than the first. Reduced to an alimentary pension of 400 francs, he determined to become a doctor, and until he could obtain time for the requisite studies, he laboured assiduously for his daily bread in the office of a banker. His reflections, however, and the contemplations which he delighted to indulge, afforded him consolation, and when he found an opportunity of communicating his ideas to some friend, of discussing them, and defending them against objections, the real world was nothing to him; his warmth made him forget all his difficulties. It is in this way that many men have passed their youth, who have become the lights of the world. Too often is genius born to poverty; but there is in it a principle of resistance against misfortune, and adversity is perhaps the surest test by which it can be tried. Never ought the most unfortunate of young men to forget, that Linnaeus was preparing himself to become the reformer of Natural History, at the time when he was patching up for his own use the cast-off shoes of his companions.

At last, after having occupied ten years in preparing himself, M. de Lamarck made himself suddenly known, both to the world and men of science, by a work on a new plan, and executed in a manner full of interest.

He had been for a long time accustomed, when collecting plants, or visiting the Jardin du Roi, to engage in warm discussions with his fellow students on the imperfections of all the systems of arrangement then in vogue, and to maintain how easy it would be to form one which would lead with greater ease and certainty to the determination of plants. His friends in some measure defied him to the task ; he immediately set about proving his assertion, and after six months of unremitting labour, finished his "Flore Française."[2] This work has no pretensions to add to the number of species previously known as indigenous to France, nor even to give a more complete history of them. It is merely a guide which, by setting out from the most general forms, dividing and subdividing always by two, and only allowing the choice between two opposite characters, conducts the reader, however little he may understand descriptive language, as it were by the hand, with certainty, and even amusement, to the determination of the plant of which he desires the name. This kind of dichotomy or continual bifurcation, is implied in all methods of arrangement, and even forms the necessary foundation of them, but modern authors, for the sake of brevity, have attempted to present many ramifications together. M. de Lamarck, in imitation of some of the old botanists, developed and expressed them all, representing them by *accolades,* in such a manner that the most uninstructed reader, without any initiatory labour, by taking him for a guide, may suppose himself to be a botanist. His book appeared at a time when botany had become a popular science, the example of J. J. Rousseau, and the enthusiasm which he inspired, having even caused it to be studied by ladies and people of fashion ; the success of the work was therefore rapid. M. de Buffon, not perhaps unwilling to shew by this example how easily systems, on which he set so little value, could be framed, and at the same time their indifferent consequence, used his interest to have the Flore Française printed by the royal press. A place having become open in the botanical department of the Academy of Sciences, and M. de Lamarck being presented in the second rank, the Minister caused it to be given to him by the King in 1775, in preference (a thing almost unexampled,) to M. Descemet, who was presented first, and who has never been

[2] Flore Française, ou description succincte des toutes les Plantes qui croissent naturellement en France. 3 vols. in 8vo, Paris, 1778.

able, during a long life, to recover the station of which the preference deprived him. In short, the poor officer, so little regarded since the commencement of the peace, all of a sudden attained to the good fortune, always of rare occurrence, and particularly so then, of being at the same time an object of favour with the court and with the public. The partiality of M. de Buffon obtained for him another advantage. When his son was about to set out on his travels, after finishing his studies, M. de Buffon proposed to M. de Lamarck to accompany him ; and not wishing that the latter should appear merely in the character of a preceptor, he procured for him the commission of botanist to the King, for the purpose of visiting foreign gardens and cabinets, and opening a correspondence between them and similar establishments in Paris. In consequence of this he travelled in company with the younger Buffon during part of the years 1781 and 1782, through Holland, Germany, and Hungary ; visited Gleditsch at Berlin, Jacquin at Vienna, Murray at Goettingen, and obtained an idea of the magnificent establishments devoted to botany in many foreign countries, to which our own do not yet approach, notwithstanding all that has been done for them for the last thirty years.

Shortly after his return, he commenced more important works than his Flora, although less widely known, and which have procured for him a more eminent rank among botanists,—I mean his *Dictionary of Botany*,[3] and his *Illustrations of Genera*,[4] both of which form a part of the Encyclopedie Methodique.

These generic illustrations are perhaps better calculated than any other work for conveying a speedy and accurate knowledge of this beautiful science. The precision of the descriptions and definitions of Linnæus is accompanied, as in the institutions of Tournefort, with figures fitted to embody their abstractions, and to present them to the eye as well as to the mind. Nor will the student have the means of becoming acquainted with the fruits and flowers only ; the whole appear-

[3] Encyclopedie Methodique (Botany). The first vol., 1783, and the second, 1786, are by M. de Lamarck ; the third, 1789, is by him and M. Desrousseaux, who likewise assisted with the fourth, 1795, along with MM. Poiret and Savigny ; the fifth, 1804, is by MM. Poiret and De Candolle ; the sixth, seventh, and eighth, from 1804 to 1808, are by M. Poiret, as well as the five supplementary vols. from 1810 to 1817.

[4] Illustrations of Genera, or an exposition of the characters of all the genera of plants established by botanists, arranged according to the sexual system of Linnæus, with figures displaying the characters of these genera, and a table of all the known species referable thereto, the description of which is found in the Botanical Dictionary of the Encyclopedie. The first vol., 1791, second, 1793, third, 1800, containing 900 plates, are by M. de Lamarck, and a Supplement by Poiret, in 1823, contains the last hundred plates.

ance and habits of one or two of the principal species are often represented, the whole consisting of two thousand genera on a thousand quarto plates, and accompanied at the same time with abridged characters of an infinity of species. The Dictionary contains a more detailed history of them, with careful descriptions, critical investigations of their synonymy, and many interesting observations on their uses, and the peculiarities of their organization. All is not original, it is true, in these two works; but the selection of figures is judicious, the descriptions are derived from the best authors, and a very considerable number of both these are to be found, which refer to species and even genera previously unknown.

It may excite surprise that M. de Lamarck, who had hitherto occupied himself with botany merely as an amateur, should so soon have been in a condition to produce such considerable works, containing representations and descriptions of the very rarest plants. The reason is, that the moment he undertook the task, he entered upon it with all the ardour of his character, occupying himself exclusively with plants, seeking them in all the gardens and in every herbarium. He spent his time among such botanists as could supply him with information, and was often in the company of M. de Jussieu, whose enlightened hospitality rendered his residence for a very long period the favourite resort of all who devoted their attention to the amiable science of plants. Whoever arrived in Paris with specimens, might be certain that M. de Lamarck would be the first to pay him a visit. His eagerness was the means of procuring him one of the finest presents he could have desired. When the celebrated traveller Sonnerat returned the second time from India in 1781, with valuable collections of objects in natural history, he imagined that all who cultivated that science would eagerly assemble round him; he could not learn at Pondicherry, or among the Moluccas, that the philosophers of this capital are too often as much engrossed as men of the world. No one appeared but M. de Lamarck, and Sonnerat in his disappointment presented him with the magnificent herbarium which he had brought with him. He likewise availed himself of that of Commerson, and the collections accumulated in the house of M. de Jussieu were generously laid open to his inspection.

It may likewise appear surprising, although in a different way, that M. de Lamarck has not adopted in these his great works, the more perfect modes of arrangement, the rules for which he has so accurately laid down in the preface to his

Flora ; and that he confined himself, in the one case, to the sexual system, and in the other, to mere alphabetical order. Such, however, were the conditions which the manager of the Encyclopædia had imposed on him, for it must be acknowledged that M. de Lamarck was still obliged to labour for booksellers, and according to their direction. This kind of labour, indeed, constituted his only resource.

The attachment of M. de Buffon, and even that of the minister, had not procured him any settled occupation ; nor was any thing done for him till M. de la Billardiere, Buffon's successor, and related to M. de Lamarck's family, created for him the paltry place of keeper of the herbaria in the king's cabinet ; a place of which he was continually on the point of being deprived, for strong opposition was made to its establishment, and the National Assembly was even required to suppress it, as I learn from two pamphlets which he was obliged to publish in its defence. If he obtained some years afterwards a less precarious means of support, it was only to be attained by again changing his vocation.

In 1793, the King's Garden and Cabinet were re-established, under the title of Museum of Natural History. All the superior functionaries were appointed professors, and charged with the superintendence of those departments most in unison with their preceding employments or personal studies. M. de Lamarck, being the last appointed, had to content himself with the branch not selected by the others, and was nominated to the chair relating to the two last classes of the animal kingdom, according to the Linnean division,—those, namely, which were then called Insects and Worms. He was at that time nearly fifty years of age, and the only preparatory knowledge which he possessed of this vast department of zoology, consisted of some acquaintance with shells, which he had often studied with Bruguiere, and of which he had made a small collection. But his former courage did not desert him ; he began the study of these new objects with unremitting ardour. Availing himself of the aid of some of his friends, and applying, at least to all that related to shells and corals, that sagacity which a long exercise had given him in reference to plants, he laboured so successfully in this new field of inquiry, that his works on those animals will confer on his name perhaps a more lasting reputation than all that he has published on botany. Before we give an analysis of these, however, we have first to speak of other writings, which will not probably enjoy the same advantage.

During the thirty years which had elapsed since the peace of 1763, all his time had not been occupied with botany. In the long solitudes to which his restricted circumstances confined him, all the great questions which for ages had fixed the attention of men, passed through his mind. He had meditated on the general laws of physics and chemistry, on the phenomena of the atmosphere, on those of living bodies, and on the origin of the globe and its revolutions. Psychology, and the higher branches of metaphysics, were not beyond the range of his contemplations; and on all these subjects he had formed a number of definite ideas, original in respect to himself, because conceived by the unaided power of his own mind, but which he believed to be equally new to others, and not less certain in themselves, than calculated to place every branch of knowledge on a new foundation. In this respect, he resembled so many others who spend their lives in solitude, who never entertain a doubt of the accuracy of their opinions, because they never happen to be contradicted. These views he began to lay before the public as soon as he had obtained a fixed occupation; and for twenty years he continued to reproduce them in every variety of form, introducing them even into such of his works as appear most foreign to them. It is the more necessary that we should point them out, as without them some of his best writings would be unintelligible. Even the character of the man himself could not otherwise be understood; for so intimately did he identify himself with his systems, and such was his desire that they should be propagated, that all other objects seemed to him subordinate, and even his greatest and most useful works appeared in his own eyes merely as the slight accessories of his lofty speculations.

Thus, while Lavoisier was creating in his laboratory a new chemistry, founded on a beautiful and methodical series of experiments, M. de Lamarck, without attempting experiment, and destitute of the means of doing so, imagined that he had discovered another, which he did not hesitate to set in opposition to the former, although nearly the whole of Europe had received it with the warmest approbation.

As early as 1780, he had ventured to present this theory in manuscript to the Academy of Sciences; but it was not till 1792 that he published it, under the title of Recherches sur les Causes des Principaux Faits Physiques.[5] It reappeared in an improved

[5] Researches on the causes of the most important physical facts, and particularly on those of combustion; of the raising of water in the state of vapour; of the heat produced by the friction

order in the Memoires de Physique et d'Histoire Naturelle,[6] which he hastened to read to the Institute shortly after its establishment, and which he collected into a volume in 1797. According to him, "matter is not homogeneous ; it consists of simple principles, essentially different among themselves. The connection of these principles in compounds varies in intensity ; they mutually conceal each other, more or less, according as each of them is more or less predominant. The principle of no compound is ever in a natural state, but always more or less modified. As, however, it is not agreeable to reason that a substance should have a tendency to pass from its natural condition, it must be concluded, that combinations are not produced by nature ; but that, on the contrary, she tends unceasingly to destroy the combinations which exist, and each principle of a compound body tries to disengage itself according to the degree of its energy. From this tendency, favoured by the presence of water, dissolutions result : affinities have no influence ; and all experiments by which it is attempted to be proved that water decomposes, that there are many kinds of air, are mere illusions, and it is fire which produces them. The element of fire[7] is subject, like the others, to modification, when combined. In its natural state, every where diffused, and penetrating every substance, it is absolutely imperceptible ; only, when it is put into vibration, it becomes the essence of sound ; for air is not the vehicle of sound, as natural philosophers believe.[8] But fire is fixed in a great number of bodies, where it accumulates, and becomes, in its highest degree of condensation, *carbonic fire,* the basis of all combustible substances, and the cause of all colours. When less condensed, and more liable to escape, it is acidific fire (*feu acidifique*), the cause of causticity when in great abundance, and of tastes and smells when less so. At the moment when it disengages itself, and in its transitory state of expansive motion, it is *caloric fire.* It is in this

---

of solid bodies against each other ; of the heat which becomes sensible in sudden decompositions, in effervescences, and in the bodies of many living animals ; of causticity, and of the taste and smell of certain compounds ; of the colour of bodies, and of the origin of compounds and of all minerals ; finally, remarks on the life of organic beings, their growth, strength, decay, and death. Paris, 1794, 2 vols. 8vo.

[6] Memoirs on physics and natural history, founded on reason, independently of all theory, with the exposition of new considerations on the general cause of dissolutions, on the substance of fire, on the colour of bodies, on the formation of compounds, on the origin of minerals, and on the organization of living bodies. Paris, 1797, 1 vol. 8vo.

[7] Memoir on the substance of Fire, considered as a chemical agent in analysis. Journal de Physique, Floreal an VII.

[8] Memoir on the substance of Sound. Journal de Physique, 16 & 26 Brumaire an. VII.

form that it dilates, warms, liquefies, and volatilizes bodies, by surrounding their molecules ; that it burns them, by destroying their aggregation ; and that it calcines or acidifies them, by again becoming fixed in them. In the greatest force of its expansion, it possesses the power of emitting light, which is of a white, red, or violet-blue colour, according to the force with which it acts ; and it is, therefore, the origin of the prismatic colours ; as also of the tints seen in the flame of candles. Light, in its turn, has likewise the power of acting upon fire, and it is thus that the sun continually produces new sources of heat. Besides, all the compound substances observed on the globe, are owing to the organic powers of beings endowed with life, of which, consequently, it may be said, that they are not conformable to nature, and are even opposed to it, because they unceasingly reproduce what nature continually tends to destroy. Vegetables form direct combinations of the elements ; animals produce more complicated compounds, by combining those formed by vegetables ; but there is in every living body a power which tends to destroy it ; all, therefore, die, each in his appointed season, and all mineral substances, and all inorganic bodies whatsoever, are nothing but the remains of bodies which once had life, and from which the more volatile principles have been successively disengaged. The products of the most complex animals are calcareous substances, those of vegetables, soils, or clays. Both of these pass into a siliceous state, by freeing themselves more and more from their less fixed principles, and at last are reduced to rock-crystal, which is earth in its greatest purity. Salts, pyrites, metals, differ from other minerals only because certain circumstances have had the effect of accumulating in them, in different proportions, a greater quantity of carbonic or acidific fire.

With respect to life, the only cause of all compositions,—the mother, not only of animals and vegetables, but all bodies which now occupy the surface of the earth,—M. de Lamarck yet admitted, in these his two earliest works, that all we know of it is, that living beings all come from individuals similar to themselves, but that it is impossible for us to ascertain the physical cause which has given birth to the first individual of each species.

To these two writings he added a third of a polemical description, viz. a refutation of the pneumatic theory,[9] in which

[9] Refutation of the Pneumatic Theory, or of the new doctrine of modern chemists, presented article after article, in a series of answers to the principles published by C. Fourcroy in his

he, in some measure, challenged the new chemists to the combat : conceiving, like so many other authors of system, that to keep silence would be to cause his system to be forgotten, and not doubting that if he could only enter it in the lists, it would obtain an easy triumph, and the public, attracted by the eclat of the dispute, would not hesitate to adopt a system of which they could scarcely otherwise be aware of the existence.

To his great regret, neither this refutation, nor his exposition, met with any reply ; no one considered it necessary. He was himself, in fact, too well aware, that the whole of this edifice rested on two assertions equally conjectural ; the one, that substances do not enter into combinations, unless modified in their nature ; the other, that it is not reasonable to believe, that nature impresses on them a tendency to such a change. Deprived of one of these foundations, the whole falls to the ground.

We have mentioned that M. de Lamarck at this period still conceived it impossible to remount to the first origin of living beings : this was a great step yet remaining for him, and he was not long in making it. In 1802 he published his Researches on Living Bodies,[10] containing a physiology peculiar to himself, in the same way that his researches on the principal facts of physics contained a chemistry of that character. In his opinion, the egg contains nothing prepared for life before being fecundated, and the embryo of the chick becomes susceptible of vital motion only by the action of the seminal vapour : but, if we admit that there exists in the universe a fluid analogous to this vapour, and capable of acting upon matter placed in favourable circumstances, as in the case of the embryon, which it organizes and fits for the enjoyment of life, we will then be able to form an idea of spontaneous generations. Heat alone is perhaps the agent employed by Nature to produce these incipient organizations ; or it may act in concert with electricity. M. de Lamarck did not believe that a bird, a horse, nor even an insect, could directly form themselves in this manner ; but, in regard to the most simple living bodies, such as occupy the extremity of the scale in the different kingdoms, he perceived no

---

Chemical Philosophy ; preceded by a Supplement to the theory explained in the work, entitled, Researches on the Causes of the Principal Facts in Physics, to which this forms a necessary appendage. Paris, 1830, 1 vol. 8vo.

[10] Researches on the organization of living bodies, and particularly on its origin, on the cause of its developments and the progress of its composition, and on that which, by continually tending to destroying it in every individual, necessarily brings on death. Preceded by a discourse delivered at the opening of the Zoological Course in the Museum of Natural History. Paris, 1802. 1 vol. 8vo.

difficulty; for a monad or a polypus are, in his opinion, a thousand times more easily formed than the embryo of a chick. But how do beings of more complicated structure, such as spontaneous generation could never produce, derive their existence? Nothing, according to him, is more easy to be conceived. If the orgasm, excited by this organizing fluid, be prolonged, it will augment the consistency of the containing parts, and render them susceptible of re-acting on the moving fluids which they contain, and an irritability will be produced, which will consequently be possessed of feeling. The first efforts of a being thus beginning to develop itself must tend to procure it the means of subsistence, and to form for itself a nutritive organ. Hence the existence of an alimentary canal! Other wants and desires, produced by circumstances, will lead to other efforts, which will produce other organs: for, according to a hypothesis inseparable from the rest, it is not the organs, that is to say, the nature and the form of the parts, which give rise to habits and faculties; but it is the latter which in process of time give birth to the organs. It is the desire and the attempt to swim that produces membranes in the feet of aquatic birds; wading in the water, and at the same time the desire to avoid wet, has lengthened the legs of such as frequent the sides of rivers; and it is the desire of flying that has converted the arms of all birds into wings, and their hairs and scales into feathers. In advancing these illustrations, we have used the words of the author, that we may not be suspected either of adding to his sentiments or detracting anything from them.

These principles once admitted, it will easily be perceived that nothing is wanting but time and circumstances to enable a monad or a polypus gradually and indifferently to transform themselves into a frog, a stork, or an elephant. But it will also be perceived that M. de Lamarck could not fail to come to the conclusion that species do not exist in nature; and he likewise affirms, that if mankind think otherwise, they have been led to do so only from the length of time which has been necessary to bring about those innumerable varieties of form in which living nature now appears. This result ought to have been a very painful one to a naturalist, nearly the whole of whose long life had been devoted to the determination of what had hitherto been believed to be species, whether in reference to plants or animals, and whose most acknowledged merit, it must be confessed, consisted in this very determination.

However this may be, M. de Lamarck reproduced this theory of Life in all the zoological works which he afterwards published ; and whatever interest these works may have excited by their positive merits, no one conceived their systematic part sufficiently dangerous to be made the subject of attack. It was left undisturbed like his theory of Chemistry, and for the same reason, because every one could perceive that, independently of many errors in the details, it likewise rested on two arbitrary suppositions ; the one, that it is the seminal vapour which organizes the embryo ; the other, that efforts and desires may engender organs. A system established on such foundations may amuse the imagination of a poet ; a metaphysician may derive from it an entirely new series of systems ; but it cannot for a moment bear the examination of any one who has dissected a hand, a viscus, or even a feather.

But his theory of chemistry and of living bodies is by no means the whole that M. de Lamarck accomplished in this way. In his Hydrogeology,[11] published in 1802, he advanced a corresponding theory of the formation of the globe and its changes, founded on the supposition that all composite minerals are the remains of living beings. The seas, unceasingly agitated by the tides, which the action of the moon produces, are continually hollowing out their bed ; and in proportion as the latter deepens in the crust of the earth, it necessarily follows that their level lowers, and their surface diminishes ; and thus the dry land, formed, as has been already said, by the debris of living creatures, is more and more disclosed. As the lands emerge from the sea, the water from the clouds forms currents upon their surface, by which they are rent and excavated, and divided into valleys and mountains. With the exception of volcanoes, our steepest and most elevated ridges have formerly belonged to plains, even their substance once made a part of the bodies of animals and plants ; and it is in consequence of being so long purified from foreign principles that they are reduced to a siliceous nature. But running waters furrow them in all directions, and carry their materials into the bed of the sea, and the latter, from continual efforts to deepen its bottom, necessarily throws them out on some side or other. Hence there results a

[11] Hydrogeology, or researches on the influence exerted by water on the surface of the terrestrial globe ; on the causes of the existence of the basin of seas, and its successive shifting to different points of the globe ; finally, on the changes which living bodies produce on the nature and condition of the surface. 1 vol. 8vo. 1802.

general movement, and a constant transposition of the ocean, which has perhaps already made several circuits of the globe. This shifting cannot occur without displacing the centre of gravity in the globe ; a circumstance which, according to Lamarck, would have the effect of displacing the axis itself, and changing the temperatures of the different climates. If none of these things have fallen under our observation, it is on account of the excessive slowness with which these operations are carried on. Time is always necessary to account for them ; unlimited time, which plays such an important part in the religion of the magi, is no less necessary to Lamarck's physics, and it was to it that he had recourse to silence his own doubts, and to answer all the objections of his readers.

The case was no longer the same, when he ventured to make an application of his systems to phenomena capable of being appreciated by near intervals. He had soon an opportunity of convincing himself how far nature sometimes rebels against doctrines conceived *a priori.* The atmosphere, according to him, may be compared to the sea,—it has a surface, waves, storms ; it ought likewise to have a flux and reflux, for the moon ought to heave it upwards as it does the ocean. In the temperate and frigid zones, therefore, the wind, which is only the tide of the atmosphere, must depend greatly on the declination of the moon ; it ought to blow towards the pole which is nearest to it, and advancing in that direction only, in order to reach every place, traversing dry countries or extended seas, it ought then to render the sky serene or stormy. If the influence of the moon on the weather is denied, it is only that it may be referred to its phases ; but its position in the ecliptic will afford probabilities much nearer the truth.[12]

In order to demonstrate this theory in some measure by facts, and to attract the attention of the public to it, M. de Lamarck thought it would be useful to present it under the form of predictions. He had even the perseverance to print almanacs

[12] Of the influence of the moon on the earth's atmosphere ; Journal de Physique, prairial, an VI.—On the variations in the state of the sky in the mean latitudes between the equator and the pole, and on the principal causes which produce them ; Journal de Physique, frimaire, an XI.—On the mode of drawing up and notifying meteorological observations, in order to obtain from them useful results, and on the considerations which ought to be kept in view for this purpose ; (ibid.)—On tempests, storms, hurricanes, and on the character of destructive winds ; Journal de Physique, 18 brumaire, an IX.—Researches on the presumed periodicity of the principal variations in the atmosphere, and on the means of determining its existence, (ibid) ; read to the Institute, 26 ventose, an IX. In a note to his memoir on Sound, he promised to advance a theory of the earth's atmosphere, at which, he says, he had laboured for more than thirty years ; but this was never published.

for eleven years successively,[13] announcing the probable state of the temperature for each day; but it may be said that the weather took pleasure in exposing his fallacies. In vain did he attempt every year to introduce some new consideration, such as the phases, the apogee and perigee of the moon, and the relative position of the sun; in vain did he seek thereby to explain his false reckonings, and to rectify his calculations. The very succeeding season taught him, to his disappointment, that our atmosphere is subjected to influences far too complicated for mankind to calculate upon its phenomena. At last he renounced this fruitless labour, and, returning to that which he ought never to have neglected, occupied himself with the direct object of his professorship,—the history of invertebrate animals,—in which he at last found an indisputable source of reputation, and a lasting title to the gratitude of posterity. It is to him that we are indebted for the above name, *invertebrate animals,* which expresses perhaps the only circumstance in their organization which is common to them all. He was the first to use it in preference to that of *white-blooded animals,* hitherto employed; and the accuracy of his views was not long in being confirmed by observations, which prove that an entire class of these animals possess red blood. A new classification, founded on their anatomy, had been published in 1795; this he in a great measure adopted in 1797,[14] and substituted it in the room of those of Linnæus and Bruguiere, which at first formed the base of his course. After that period, he modified it in various ways, but without entirely changing it.[15] His ana-

[13] Annual of Meteorology for the year VIII. (1800) of the Republic, containing an exposition of the probabilities acquired by a long series of observations on the state of the weather, and variations of the atmosphere, in different seasons of the year; an indication of the times when it may be expected to be fine weather or rain, storms and tempests, frost, &c.; finally, an enumeration, according to probabilities, of the times favourable for fêtes, journeys, voyages, harvest, and other undertakings in which it is of importance not to be interrupted by the weather; with simple and concise directions regarding those new measures. Paris, 1800, continued till 1810, forming altogether 11 vols.

[14] See the table inserted at the 314th page of his Memoires de Physique et d'Histoire Naturelle, and the subjoined note, the only testimony he has left of the source whence he derived it. This table differs from the arrangement in question, only in establishing a class of radiarii which cannot be maintained, and in leaving the crustacea with insects, a union which he afterwards regarded as improper.

[15] In his system of Animaux sans Vertebres, in 1810,[a] he adopted the class of Crustacea, and created that of Arachnides, in consequence of some observations which had been communicated to him on the heart and pulmonary sacs of spiders. In 1812, in his Researches on the Organi-

[a] System of Invertebrate Animals, or general table of the classes, orders, and genera, of these animals; presenting their essential characters, and their distribution according to their natural relations and organization, after the mode of arrangement adopted with preserved specimens in the galleries of the Museum of Natural History; preceded by a Discourse delivered at the opening of the Zoological Course in the Museum, year VIII. of the Republic. 1 vol. 8vo. Paris, year IX.

tomical knowledge was not of such a kind as to enable him to advance many new views ; it may even be said that the general distribution of these animals into *apathetic, sensible*, and *intelligent*, which he at last introduced into his method, was neither founded on their organization, nor exact observation of their faculties. But what was peculiarly his own, and will continue to be of fundamental importance in all ulterior researches on these subjects, are his observations on shells and polypi, whether of a stony or flexible nature. The sagacity with which he circumscribed and characterized the genera, according to the circumstances of form, proportion, surface, and structure, judiciously selected and easily recognised ; the perseverance he displayed in comparing and distinguishing the species, fixing the synonyms, and furnishing clear and detailed descriptions, have rendered each of his successive works the regulator of this department of natural history. It was chiefly according to his views that such as have written on the same subject, have named and arranged their species ; and even at present, we should in vain seek for a more complete account of sponges (for example), of alcyons, and many other kinds of corals, than what is afforded by his *Histoire des Animaux sans Vertebres*. There is one branch of knowledge in particular to which he has given a remarkable impulse, the history, namely, of shells found in the

---

zation of Living Bodies,[b] he admits the class of Annelides, established, as he acknowledges in the 24th page, on my observations respecting their circulating organs, and the colour of their blood. In 1809, in his Philosophical Zoology,[c] he creates two classes in addition, viz. the infusoria disjoined from the polypi, and the centripedes separated from the molluscae. In this work, also, he for the first time presented animals in the inverse ratio of their organization, beginning with the most simple.—He preserves this order and arrangement in the *Extract from his Course*, published in 1812 ;[d] and besides, he separates in that work the classes of animals into the grand divisions *Apathiques, Sensibles*, and *Intelligents*.—It is on this plan that he drew up his grand history of invertebrate animals, begun in 1815.[e]

[b] See Supp. p. 36.

[c] Zoological Philosophy, or exposition of considerations relating to the natural history of animals ; of the diversity of their organization, and faculties resulting therefrom ; of the physical causes which support life in them, and give rise to the movements which they execute ; and those which produce, sometimes feeling and at other times intelligence, on such as are so endowed. 2 vols. 8vo. Paris, 1809.

[d] Extract from the Zoological Course in the Museum of Natural History, on the invertebrate animals ; presenting the arrangement of animals, the characters and principal divisions, together with a simple list of genera. 1 vol. 8vo. Paris, 1812.

[e] Natural History of Invertebrate Animals, presenting the generic and particular characters of these animals, their distribution, their classes, their families, their genera, and the principal species ; preceded by an introduction, determining the essential characters of an animal, and its distinction from vegetables and other natural bodies ; and finally, an exposition of the fundamental principles of zoology. 7 vols. 8vo. Paris, 1815 to 1822. This is M. de Lamarck's capital work. A part of the 6th, and the whole of the 7th, volumes were drawn up by his daughter from his papers. In the 6th, the Mytilacés, the Malleacés, the Pectinides, and the Ostracés, are by M. Valenciennes. The first five are written by M. de Lamarck himself, assisted in the part relating to insects by the advice of M. Latreille.

bowels of the earth. These had attracted the attention of geologists from the time that the chimerical notion was exploded, which attributed their origin to the plastic force of a mineral nature. It was perceived that a comparison of such as belong to the different beds, and their approximation to those now living in different seas, could alone throw light on this anomalous phenomenon,—the deepest, perhaps, of all the mysteries which inanimate nature presents to our view. This comparison, however, had scarcely been attempted, or, if it were, it was made in the most superficial manner. The study had been regarded as a trifling object of curiosity. Whence do they come ? Have they lived in our climate, or, have they been transported hither ? Are they still in a living state elsewhere ? All these important questions could not be answered but by carefully examining them one by one. The prosecution of this inquiry was the more tempting to M. de Lamarck, on account of the basin of Paris being, perhaps, the only spot in the world where such a vast number of these productions are accumulated in so small a space. At Grignon, which does not exceed a few square toises in extent, no fewer than 600 different species of shells have been collected.

M. de Lamarck entered upon this examination with that profound knowledge which he had acquired of living shells, and the excellent figures and careful descriptions which he produced, caused those beings, deprived of life for so many ages, again, as it were, to reappear in the world.[16]

It was thus that M. de Lamarck, by resuming occupations analogous to those which first procured him reputation, at last raised for himself a monument which will endure as long as the objects on which it rests. Fortunate had it been for him if he had been able to render it more perfect. But we have already seen that he was late in devoting himself to zoology ; and from the first, the weakness of his eyes obliged him to have recourse for the investigation of insects to our celebrated associate M. de Latreille, whom Europe recognises as his master in this immense department of Natural History. The clouds thickened upon him by degrees, and allowed but an imperfect glimpse of all those delicate organizations, the observation of

[16] Memoir on the fossils of the neighbourhood of Paris, comprising the determination of species which belong to marine invertebrate animals, and of which the greater part are figured in the collection of drawings in the museum.—This memoir, begun in the Annals of the Museum, vol. i, and continued in the subsequent volumes, was never brought to a conclusion. It was accompanied with a collection of plates of fossil shells found near Paris, with their explanation. Vol. i. 4to. Paris, 1823.

which constituted his only enjoyment. No art could stop the inroads of this calamity, nor administer a remedy ; that light, which had been so much the subject of his study, at last entirely failed him, and he passed many of his last years in absolute blindness. This misfortune was the more distressing, because it overtook him in such circumstances that he could obtain none of those means of distraction or alleviation which might have otherwise been procured. He had been married four times, and was the father of seven children. The whole of his little patrimony, and even the fruits of his early economy, were lost in one of those hazardous investments, which are so often held out as baits to credulity by shameless speculators.

His retired life, the consequence of his youthful habits, and attachments to systems so little in accordance with the ideas which prevailed in science, were not calculated to recommend him to those who had the power of dispensing favours. When numberless infirmities, brought on by old age, had increased his wants, nearly his whole means of support consisted of a small income derived from his chair. The friends of science, attracted by the high reputation which his botanical and zoological works had obtained for him, witnessed this with surprise. It appeared to them that a government which protects the sciences, ought to have been more careful to become better acquainted with the situation of a celebrated individual ; but their esteem for him was doubled, when they saw the courage with which the illustrious old man bore up against the assaults both of fortune and of nature. They particularly admired the devotedness which he inspired in such of his children as remained with him. His eldest daughter, entirely devoted to the duties of filial affection for many years, never left him for an instant, readily engaged in every study which might supply his want of sight, wrote to his dictation a portion of his last works, and accompanied and supported him as long as he was able to take some exercise. Her sacrifices, indeed, were carried to a degree which it is impossible to express ; when the father could no longer leave his room, the daughter never once left the house. When she afterwards did so for the first time, she was incommoded by the free air, the use of which had been so long unfamiliar to her. It is rare to see virtue carried to such a degree, and it is not less so to inspire it to that degree ; and it is adding to the praise of M. de Lamarck to recount what his children did for him.

M. de Lamarck died on the 18th December 1829, at the age of eighty-five years, leaving only two sons and two daughters. The eldest of these sons occupies an important place in the Corps des Ponts et Chaussées. His place in the Institute has been given to M. Auguste de Saint Hilaire, whose travels in America have procured so many interesting plants, and which he has studied so profoundly. His chair in the Museum of Natural History, the object of which was too extensive for the exertions of one individual, has been, at the request of his colleagues, divided into two by the government; M. Latreille taking the charge of Insects and Crustacea; and M. de Blainville of all the other divisions which constituted the Linnean Class of Vermes.

INDEX

INDEX